A Complete Solution Guide to Complex Analysis

by Kit-Wing Yu, PhD

kitwing@hotmail.com

ISBN: 978-988-74155-0-3 (eBook)
ISBN: 978-988-74155-1-0 (Paperback)

About the author

Dr. Kit-Wing Yu received his B.Sc. (1st Hons), M.Phil. and Ph.D. degrees in Math. at the HKUST, PGDE (Mathematics) at the CUHK. After his graduation, he has joined United Christian College to serve as a mathematics teacher for at least seventeen years. He has also taken the responsibility of the mathematics panel since 2002. Furthermore, he was appointed as a part-time tutor (2002 – 2005) and then a part-time course coordinator (2006 – 2010) of the Department of Mathematics at the OUHK.

Between 2012 and 2014, Dr. Yu was invited to be a Judge Member by the World Olympic Mathematics Competition (China). In the research aspect, he has published over twelve research papers in international mathematical journals, including some well-known journals such as J. Reine Angew. Math., Proc. Roy. Soc. Edinburgh Sect. A and Kodai Math. J.. His research interests are inequalities, special functions and Nevanlinna's value distribution theory. In the area of academic publication, he is the author of five books:

- *A Complete Solution Guide to Real and Complex Analysis I.*

- *Problems and Solutions for Undergraduate Real Analysis I.*

- *Problems and Solutions for Undergraduate Real Analysis II.*

- *Mock Tests for the ACT Mathematics.*

- *A Complete Solution Guide to Principles of Mathematical Analysis.*

Preface

There are many books entitled *Complex Analysis*. For examples, Ahlfors [1], Bak and Newman [4], Freitag and Rolf [9], Gamelin [10], Lang [14], and Stein & Shakarchi [24]. (I believe that you can find more if you search your library website.) As an introductory textbook for complex analysis, I would like to recommend the book by Bak and Newman [4]. The reasons are that this book gives readers some important, insightful and interesting background to build up the theory of complex analysis. Furthermore, their presentation is quite clear and concise, so I believe that you can grasp the main concepts and skills in an easier way. Finally, the book contains a lot of helpful examples and problems.

There are a total of 300 exercises in the third edition [4], but only 225 of them are provided by solutions. (The exercises without solutions are marked an asterisk.) In my opinion, some solutions are a bit "brief" and I guess that some students still have difficulties when reading them. To provide assistance, I decide to write a solution manual for this book and I hope that students / instructors can benefit from this solution book.

Before you read this book, I have a gentle reminder for you. As a mathematics instructor at a college, I understand that the growth of a mathematics student depends largely on how hard he/she does exercises. When your instructor asks you to do some exercises of Bak and Newman's book, you are not suggested to read my solutions unless you have tried your best to prove them yourselves.

The features of this book are as follows:

- It covers all the 300 exercises with *detailed* and *complete* solutions. As a matter of fact, my solutions show every detail, every step and every theorem that I applied.

- There are 34 illustrations for explaining the mathematical concepts or ideas used behind the questions or theorems.

- Different colors are used in order to highlight or explain problems, lemmas, remarks, main points/formulas involved, or show the steps of manipulation in some complicated proofs. (ebook only)

- Necessary lemmas with proofs are provided.

- Useful or relevant references are provided to some questions for interested readers.

Finally, if you find such any typos or mistakes, please send your valuable comments or opinions to

kitwing@hotmail.com

so that I will post the updated errata on my website

$$\texttt{https://sites.google.com/view/yukitwing/}$$

from time to time.

Kit-Wing Yu
January 2020

List of Figures

Contents

CHAPTER **1**

The Complex Numbers

Bak and Newman Chapter 1 Exercise 1.

Proof.

(a) It is clear that
$$
\frac{1}{6+2i} = \frac{1}{6+2i} \times \frac{6-2i}{6-2i} = \frac{6-2i}{36+4} = \frac{3}{20} - \frac{1}{20}i.
$$

(b) It is easy to see that
$$
\frac{(2+i)(3+2i)}{1-i} = \frac{4+7i}{1-i} = \frac{4+7i}{1-i} \times \frac{1+i}{1+i} = \frac{-3+11i}{2} = -\frac{3}{2} + \frac{11}{2}i.
$$

(c) Since $-\frac{1}{2} + \frac{\sqrt{3}}{2}i = \operatorname{cis}\frac{2\pi}{3}$, we have
$$
\left(-\frac{1}{2} + \frac{\sqrt{3}}{2}i\right)^4 = \operatorname{cis}\left(4 \times \frac{2\pi}{3}\right) = \operatorname{cis}\frac{6\pi + 2\pi}{3} = \operatorname{cis}\left(2\pi + \frac{2\pi}{3}\right) = \operatorname{cis}\frac{2\pi}{3} = -\frac{1}{2} + \frac{\sqrt{3}}{2}i.
$$

(d) We have
$$
i^n = \begin{cases}
i, & \text{if } n = 4m+1 \text{ for some } m \in \mathbb{Z}; \\
-1, & \text{if } n = 4m+2 \text{ for some } m \in \mathbb{Z}; \\
-i, & n = 4m+3 \text{ for some } m \in \mathbb{Z}; \\
1, & \text{if } n = 4m \text{ for some } m \in \mathbb{Z}.
\end{cases}
$$

We have completed the proof of the problem. ∎

Bak and Newman Chapter 1 Exercise 2.

Proof. Suppose that $(x + iy)^2 = -8 + 6i$. By the discussion on [4, p. 3], we see that
$$
x = \pm\sqrt{\frac{-8 + \sqrt{64 + 36}}{2}} = \pm 1 \quad \text{and} \quad y = \pm\sqrt{\frac{8 + \sqrt{64 + 36}}{2}} \cdot \operatorname{sgn}(6) = \pm 3.
$$

Hence the two values of $\sqrt{-8 + 6i}$ are $\pm(1 + 3i)$, completing the proof of the problem. ∎

1

Problem 1.3

Bak and Newman Chapter 1 Exercise 3.

Proof. We have

$$z = \frac{-\sqrt{32}i \pm \sqrt{(\sqrt{32}i)^2 - 4 \times (1) \times (-6i)}}{2} = \frac{-\sqrt{32}i \pm \sqrt{-32 + 24i}}{2} = -2\sqrt{2}i \pm \sqrt{-8 + 6i}.$$

By Problem 1.2, we see immediately that

$$z = -2\sqrt{2}i \pm (1 + 3i) = 1 + (3 - 2\sqrt{2})i \quad \text{or} \quad -1 - (3 + 2\sqrt{2})i.$$

This completes the proof of the problem. $\blacksquare$

Problem 1.4

Bak and Newman Chapter 1 Exercise 4.

Proof. Suppose that $z_1 = a + bi$ and $z_2 = c + di$, where a, b, c and d are real. Furthermore, we suppose that $P(z) = a_0 z^n + a_1 z^{n-1} + \cdots + a_{n-1}z + a_n$, where $a_0, a_1, \ldots, a_n \in \mathbb{R}$ and $n \in \mathbb{N} \cup \{0\}$.

(a) Then $\overline{z_1 + z_2} = \overline{(a + c) + (b + d)i} = (a + c) - (b + d)i = a - bi + c - di = \overline{z_1} + \overline{z_2}$.

(b) We have $\overline{z_1 z_2} = \overline{(a + c)(b + d)i} = \overline{ac - bd + (ad + bc)i} = ac - bd - (ad + bc)i = \overline{z_1} \cdot \overline{z_2}$.

(c) Since $a_0, a_1, \ldots, a_n \in \mathbb{R}$, we have $\overline{a_k} = a_k$ for all $0 \le k \le n$. By repeated applications of parts (a) and (b), we have

$$\overline{P(z)} = \overline{\sum_{k=0}^{n} a_k z^{n-k}} = \sum_{k=0}^{n} \overline{a_k z^{n-k}} = \sum_{k=0}^{n} \overline{a_k} \cdot \overline{z^{n-k}} = \sum_{k=0}^{n} a_k \cdot \overline{z}^{n-k} = P(\overline{z}).$$

(d) If $z = A + Bi$, then we have $\overline{\overline{z}} = \overline{A - Bi} = A + Bi = z$.

We complete the proof of the problem. $\blacksquare$

Problem 1.5

Bak and Newman Chapter 1 Exercise 5.

Proof. Since $P(z) = 0$ if and only if $\overline{P(z)} = 0$, it follows from Problem 1.4(c) that $P(z) = 0$ if and only if $P(\overline{z}) = 0$. This completes the proof of the problem. $\blacksquare$

Problem 1.6

Bak and Newman Chapter 1 Exercise 6.

Proof. Using rectangular coordinates, let $z = a + bi$. Then $z^2 = (a^2 - b^2) + 2abi$ so that

$$|z^2| = \sqrt{(a^2 - b^2)^2 + 4a^2b^2} = a^2 + b^2 = |z|^2.$$

Using polar coordinates, if $z = r\operatorname{cis}\theta$, then we have $z^2 = r^2 \operatorname{cis} 2\theta$ so that $|z^2| = r^2 = |z|^2$. We end the analysis of the problem. $\blacksquare$

Problem 1.7

Bak and Newman Chapter 1 Exercise 7.

Proof. Suppose that $z = r\operatorname{cis}\theta$.

(a) For every $n \in \mathbb{Z}$, we have $z^n = r^n \operatorname{cis} n\theta$ so that $|z^n| = r^n = |z|^n$.

(b) Since $\overline{z} = r\operatorname{cis}(-\theta)$, we have $z \cdot \overline{z} = r^2 \operatorname{cis}(\theta - \theta) = r^2 = |z|^2$.

(c) Let $z = a + bi$. Then we have $\operatorname{Re} z = a$, $\operatorname{Im} z = b$ and $|z| = \sqrt{a^2 + b^2}$. Since $a^2 \le a^2 + b^2$ and $b^2 \le a^2 + b^2$, we get $|\operatorname{Re} z| \le |z|$ and $|\operatorname{Im} z| \le |z|$. The fact $a^2 + 2|a| \cdot |b| + b^2 \ge a^2 + b^2$ implies that $(|a| + |b|)^2 \ge a^2 + b^2$ which reduces to exactly

$$|\operatorname{Re} z| + |\operatorname{Im} z| \ge |z|.$$

Now we study when an equality occurs. For example, $|\operatorname{Re} z| = |z|$ if and only if $|a|^2 = a^2 + b^2$ if and only if $b = 0$. The case for $|\operatorname{Im} z| = |z|$ is similar. Next, $|z| = |\operatorname{Re} z| + |\operatorname{Im} z|$ if and only if $a^2 + b^2 = |a|^2 + 2|a| \cdot |b| + |b|^2$ if and only if $|a| \cdot |b| = 0$ if and only if

$$|a| = 0 \quad \text{or} \quad |b| = 0.$$

In conclusion, an equality occurs if and only if $\operatorname{Re} z = 0$ or $\operatorname{Im} z = 0$.

This completes the proof of the problem. $\blacksquare$

Problem 1.8

Bak and Newman Chapter 1 Exercise 8.

Proof.

(a) By Problem 1.7(b) and then Problem 1.4(a), we have

$$|z_1 + z_2|^2 = (z_1 + z_2)\overline{(z_1 + z_2)} = (z_1 + z_2)(\overline{z_1} + \overline{z_2}). \tag{1.1}$$

By expanding the right-hand side of the equation (1.1) and then use Problems 1.7(a) and (c), we get

$$\begin{aligned}
|z_1 + z_2|^2 &= |z_1|^2 + |z_2|^2 + 2\operatorname{Re}(z_1\overline{z_2}) \tag{1.2}\\
&\le |z_1|^2 + |z_2|^2 + 2|z_1\overline{z_2}| \tag{1.3}\\
&= |z_1|^2 + |z_2|^2 + 2|z_1||z_2|\\
&= (|z_1| + |z_2|)^2
\end{aligned}$$

which gives the triangle inequality.

(b) Note that we apply Problem 1.7(c) to get the inequality (1.3), so the equality (1.3) occurs if and only if $\operatorname{Im}(z_1\overline{z_2}) = 0$ if and only if $z_1\overline{z_2} \in \mathbb{R}$.

(c) Replace z_1 by $z_1 - z_2$ in the triangle inequality, we know that $|z_1 - z_2 + z_2| \le |z_1 - z_2| + |z_2|$ which is equivalent to

$$|z_1| - |z_2| \le |z_1 - z_2|.$$

This completes the proof of the problem. $\blacksquare$

Problem 1.9

Bak and Newman Chapter 1 Exercise 9.[*]

Proof.

(a) Let $z_1 = a + bi$ and $z_2 = c + di$, where $a, b, c, d \in \mathbb{Z}$. Thus $z_1 z_2 = ac - bd + (ad + bc)i$, where $ac - bd, ad + bc \in \mathbb{Z}$. Since $|z_1 z_2| = |z_1| \cdot |z_2|$, we have $|z_1 z_2|^2 = |z_1|^2 \cdot |z_2|^2$ so that

$$(a^2 + b^2)(c^2 + d^2) = (ac - bd)^2 + (ad + bc)^2. \tag{1.4}$$

If we define $u = ac - bd$ and $v = ad + bc$, then we get the desired result.

(b) Since all a, b, c and d are nonzero, without loss of generality, we may assume they are all positive. Besides the formula (1.4), we also note that[a]

$$(a^2 + b^2)(c^2 + d^2) = (ac + bd)^2 + (ad - bc)^2. \tag{1.5}$$

Now $ad + bc > 0$ and $ac + bd > 0$, we follow from the two formulas (1.4) and (1.5) that it suffices to check that *at least* one of $ad - bc$ and $ac - bd$ is nonzero. Assume that

$$ad = bc \quad \text{and} \quad ac = bd. \tag{1.6}$$

Then the product of the same sides of the equations (1.6) gives $a^2 cd = b^2 cd$ which implies $a^2 = b^2$. Similarly, the product of the opposite sides of the equations (1.6) gives $abd^2 = abc^2$ so that $c^2 = d^2$. This is a contradiction to the hypotheses and hence we have $ad - bc \neq 0$ or $ac - bd \neq 0$.

(c) We label

$$u = ac - bd, \quad v = ad + bc, \quad s = ac + bd \quad \text{and} \quad t = ad - bc.$$

Direct computation gives

$$u^2 - s^2 = (u - s)(u + s) = (-2bd)(2ca) = -4abcd.$$

Since a, b, c and d are nonzero, we have $u^2 \neq s^2$. Similarly, we can show that

$$u^2 - t^2 = (u - t)(u + t) = (a^2 - b^2)(c^2 - d^2).$$

By the hypotheses, we know that $u^2 \neq t^2$. Furthermore, we also have $v^2 \neq s^2$ and $v^2 \neq t^2$. Otherwise, $u^2 + v^2 = s^2 + t^2$ will imply either $u^2 = s^2$ or $u^2 = t^2$, a contradiction. Consequently, the sets $\{(ac - bd)^2, (ad + bc)^2\}$ and $\{(ac + bd)^2, (ad - bc)^2\}$ are distinct.

(d) The assumption $a, b, c, d \neq 0$ means that the vectors from 0 to the complex numbers z_1 and z_2 are *not* parallel to the real axis or the imaginary axes. If we write

$$z_1 = r_1 \operatorname{cis} \theta_1 \quad \text{and} \quad z_2 = r_2 \operatorname{cis} \theta_2, \tag{1.7}$$

then both θ_1 and θ_2 are *not* a multiple of $\frac{\pi}{2}$. Next, notice that $a^2 \neq b^2$ if and only if $a \neq \pm b$. Therefore, the vector from 0 to z_1 is *not* parallel to the lines $y = \pm x$. In other words, θ_1 is *not* a multiple of $\frac{\pi}{4}$. Similarly, $c^2 \neq d^2$ if and only if θ_2 is *not* a multiple of $\frac{\pi}{4}$.

---

[a]This comes from the fact that $|z_1 \overline{z_2}| = |z_1| \cdot |z_2|$.

– **Geometric interpretation of part (b).** We obtain directly from the representations (1.7) that

$$z_1 z_2 = r_1 r_2 \operatorname{cis}\left(\theta_1 + \theta_2\right) \quad \text{and} \quad z_1 \overline{z_2} = r_1 r_2 \operatorname{cis}\left(\theta_1 - \theta_2\right).$$

Assume that both the vectors from 0 to $z_1 z_2$ and $z_1 \overline{z_2}$ were parallel to the real axis or the imaginary axis. Then the previous analysis shows that both $\theta_1 \pm \theta_2$ are multiples of $\frac{\pi}{2}$, i.e.,

$$\theta_1 + \theta_2 = \frac{m\pi}{2} \quad \text{and} \quad \theta_1 - \theta_2 = \frac{n\pi}{2}$$

but they imply that

$$\theta_1 = \frac{(m+n)\pi}{4}$$

which is a contradiction. Thus *at least* one of the vectors from 0 to $z_1 z_2$ and $z_1 \overline{z_2}$ is *not* parallel to the axes. Consequently, we have

$$\operatorname{cis}\left(\theta_1 + \theta_2\right) \neq 0 \quad \text{or} \quad \operatorname{cis}\left(\theta_1 - \theta_2\right) \neq 0. \tag{1.8}$$

Finally, we recall from the equations (1.4) and (1.5) that $ac - bd$ and $ad - bc$ are exactly the real and imaginary parts of $z_1 z_2$ or $z_1 \overline{z_2}$ respectively, so the results (1.8) ensure that both u and v are nonzero.

– **Geometric interpretation of part (c).** To show that $\{u^2, v^2\}$ and $\{s^2, t^2\}$ are distinct, it is equivalent to showing that

$$(|u|, |v|) \neq (|s|, |t|) \quad \text{and} \quad (|u|, |v|) \neq (|t|, |s|). \tag{1.9}$$

In Figure 1.1, we see easily that $(|u|, |v|) = (|s|, |t|)$ if and only if the difference between the arguments of $z_1 z_2$ and $z_1 \overline{z_2}$ is a multiple of π.

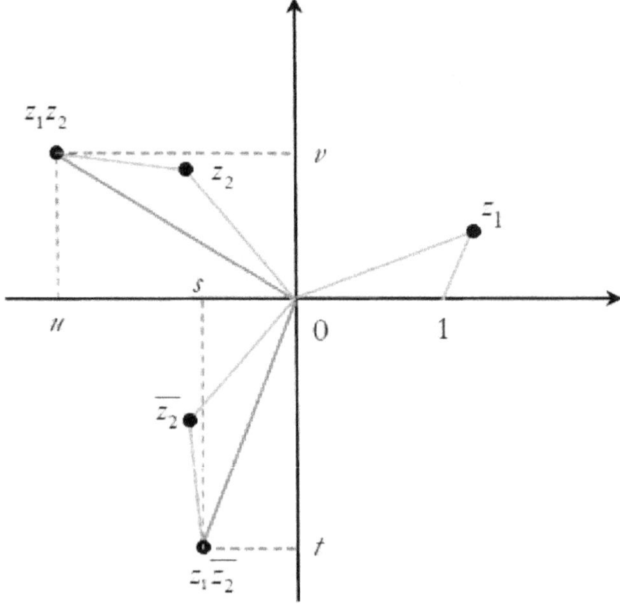

Figure 1.1: The graph of $z_1 z_2$ and $z_1 \overline{z_2}$.

Assume that $\mathrm{Arg}\,(z_1 z_2) = \mathrm{Arg}\,(z_1 \overline{z_2}) + m\pi$ for some $m \in \mathbb{N}$. Then we have

$$2\theta_2 = (\theta_1 + \theta_2) - (\theta_1 - \theta_2) = \mathrm{Arg}\,(z_1 z_2) - \mathrm{Arg}\,(z_1 \overline{z_2}) = m\pi$$

so that $\theta_2 = \frac{m\pi}{2}$ which is a contradiction by the description at the beginning of part (d). Next, it is also true that $(|u|, |v|) = (|t|, |s|)$ if and only if the difference between the arguments of $z_1 z_2$ and $\overline{z_1} z_2$ is a multiple of $\frac{\pi}{2}$. In this case, we have $\mathrm{Arg}\,(z_1 z_2) = \mathrm{Arg}\,(\overline{z_1} z_2) + \frac{n\pi}{2}$ for some $n \in \mathbb{Z}$ and then similar argument can show that

$$2\theta_1 = (\theta_1 + \theta_2) - (\theta_2 - \theta_1) = \mathrm{Arg}\,(z_1 z_2) - \mathrm{Arg}\,(\overline{z_1} z_2) = \frac{n\pi}{2},$$

a contradiction to the fact that $\theta_1 \neq \frac{n\pi}{4}$. Hence we obtain the results (1.9) which mean that $\{u^2, v^2\}$ and $\{s^2, t^2\}$ are different.

This completes the proof of the problem.

> **Problem 1.10**
>
> *Bak and Newman Chapter 1 Exercise 10.*[*]

Proof. We replace z_2 by $-z_2$ in the equation (1.2) to obtain

$$|z_1 - z_2|^2 = |z_1|^2 + |-z_2|^2 + 2\mathrm{Re}\,(z_1 (\overline{-z_2})) = |z_1|^2 + |z_2|^2 - 2\mathrm{Re}\,(z_1 \overline{z_2}). \qquad (1.10)$$

Then the sum of the equations (1.2) and (1.10) imply that

$$|z_1 + z_2|^2 + |z_1 - z_2|^2 = 2(|z_1|^2 + |z_2|^2).$$

Geometrically, the sum of the squares of the lengths of the diagonals (left-hand side) is double to the sum of the squares of the lengths of the sides (right-hand side).[b] This completes the proof of the problem.

> **Problem 1.11**
>
> *Bak and Newman Chapter 1 Exercise 11.*

Proof. Suppose that $z = x + iy \neq 0$. By the definition, we notice that

$$-\frac{\pi}{2} \leq \tan^{-1} \frac{y}{x} \leq \frac{\pi}{2}.$$

However, recall that $\mathrm{Arg}\,z$ is the angle which the vector (originating from 0) to z makes with the positive x-axis (see the figure on [4, p. 6]), so it takes values $[-\pi, \pi]$ (modulo 2π). Thus their connection can be expressed as

$$\mathrm{Arg}\,z = \mathrm{Arg}\,(x + iy) = \begin{cases} \tan^{-1} \dfrac{y}{x}, & \text{if } x > 0; \\[2ex] \tan^{-1} \dfrac{y}{x} + \pi & \text{if } x < 0 \text{ and } y \geq 0; \\[2ex] \tan^{-1} \dfrac{y}{x} - \pi, & \text{if } x < 0 \text{ and } y < 0; \\[2ex] \dfrac{\pi}{2}, & \text{if } x = 0 \text{ and } y > 0; \\[2ex] -\dfrac{\pi}{2}, & \text{if } x = 0 \text{ and } y < 0. \end{cases}$$

---

[b]See also [26, Exercise 1.17, p. 9].

This completes the proof of the problem.

Problem 1.12

Bak and Newman Chapter 1 Exercise 12.

Proof. Let $z = r\operatorname{cis}\theta$. Here we mainly follow the steps in the example on [4, pp. 8, 9].[c]

(a) Since $r^6\operatorname{cis}6\theta = 1\operatorname{cis}0$, we have $r = 1$ and $6\theta = 0$ (modulo 2π). Then we have

$$6\theta = 0, 2\pi, 4\pi, 6\pi, 8\pi, 10\pi.$$

Hence the six solutions are given by

$$z_1 = 1, \quad z_2 = \operatorname{cis}\frac{\pi}{3}, \quad z_3 = \operatorname{cis}\frac{2\pi}{3}, \quad z_4 = \operatorname{cis}\pi = -1, \quad z_5 = \operatorname{cis}\frac{4\pi}{3} \quad \text{and} \quad z_6 = \operatorname{cis}\frac{5\pi}{3}.$$

(b) Since $r^4\operatorname{cis}4\theta = 1\operatorname{cis}\pi$, we have $r = 1$ and $4\theta = \pi$ (modulo 2π). Then we obtain

$$4\theta = \pi, 3\pi, 5\pi, 7\pi.$$

Hence the four solutions are given by

$$z_1 = \operatorname{cis}\frac{\pi}{4}, \quad z_2 = \operatorname{cis}\frac{3\pi}{4}, \quad z_3 = \operatorname{cis}\frac{5\pi}{4} \quad \text{and} \quad z_4 = \operatorname{cis}\frac{7\pi}{4}.$$

(c) Since $r^4\operatorname{cis}4\theta = 2\operatorname{cis}\frac{2\pi}{3}$, we have $r = \sqrt[4]{2}$ and $4\theta = \frac{2\pi}{3}$ (modulo 2π). Then we have

$$4\theta = \frac{2\pi}{3}, \frac{8\pi}{3}, \frac{14\pi}{3}, \frac{20\pi}{3}.$$

Hence the four solutions are given by

$$z_1 = \sqrt[4]{2}\operatorname{cis}\frac{\pi}{6}, \quad z_2 = \sqrt[4]{2}\operatorname{cis}\frac{2\pi}{3}, \quad z_3 = \sqrt[4]{2}\operatorname{cis}\frac{7\pi}{6} \quad \text{and} \quad z_4 = \sqrt[4]{2}\operatorname{cis}\frac{5\pi}{3}.$$

The locations of the roots in parts (a) to (c) have been shown in Figure 1.2 below. This ends the proof of the problem.

Problem 1.13

Bak and Newman Chapter 1 Exercise 13.

Proof. Let ω be an nth root of unity other than 1. Since

$$0 = \omega^n - 1 = (\omega - 1)(\omega^{n-1} + \omega^{n-2} + \cdots + 1),$$

we conclude that ω satisfies the equation

$$z^n + z^{n-1} + \cdots + 1 = 0,$$

completing the proof of the problem.

---

[c]In fact, there is a general formula of finding the nth roots of the equation $z^n = r$, see [1, p. 16].

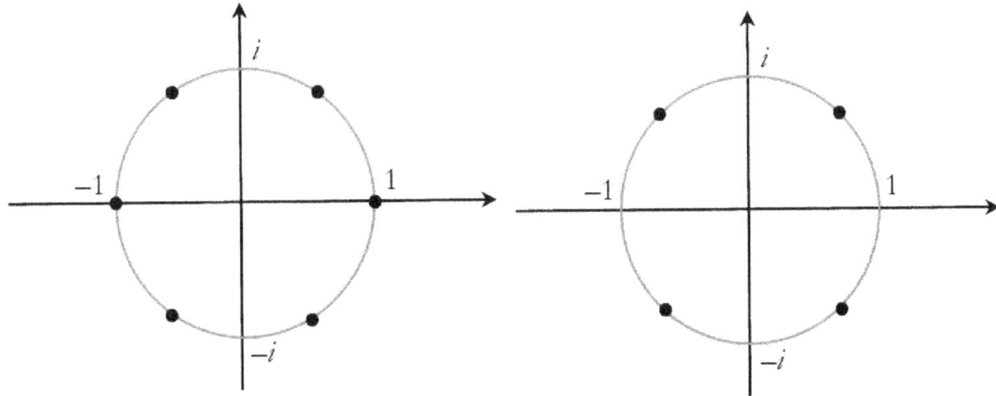

(a) The locations of the roots in part (a).

(b) The locations of the roots in part (b).

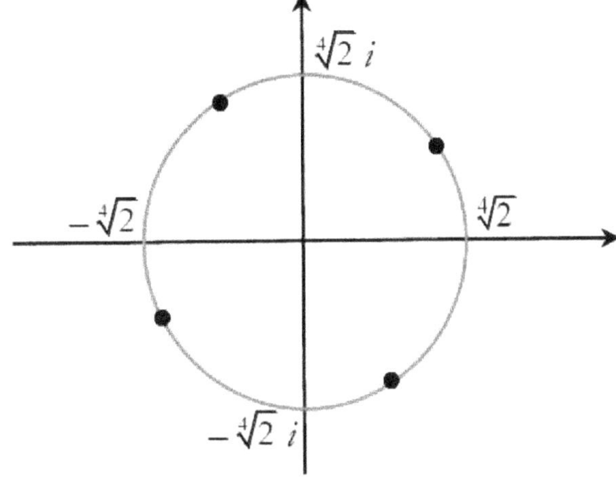

(c) The locations of the roots in part (c).

Figure 1.2: The locations of the roots.

Problem 1.14

Bak and Newman Chapter 1 Exercise 14.

Proof. By the comments on [4, p. 9] or [14, pp. 13 - 16], the n-th roots of 1 are located at the vertices of the regular n-gon inscribed in the unit circle. Let $z_1, z_2, \ldots, z_n$ be the n-th roots of

$$z^n = 1$$

with $z_1 = 1$. Without loss of generality, we may assume that the vertices all be connected to z_1. (For example, see Figure 1.3 for the regular 5-gon.) Then the lengths of the diagonals are exactly $|z_1 - z_2|, |z_1 - z_3|, \ldots, |z_1 - z_n|$. By Problem 1.13, we know that

$$z^{n-1} + z^{n-2} + \cdots + 1 = (z - z_2)(z - z_3) \cdots (z - z_n)$$

which certainly implies that

$$|1 - z_2| \times |1 - z_3| \times \cdots \times |1 - z_n| = 1^{n-1} + 1^{n-2} + \cdots + 1 = n.$$

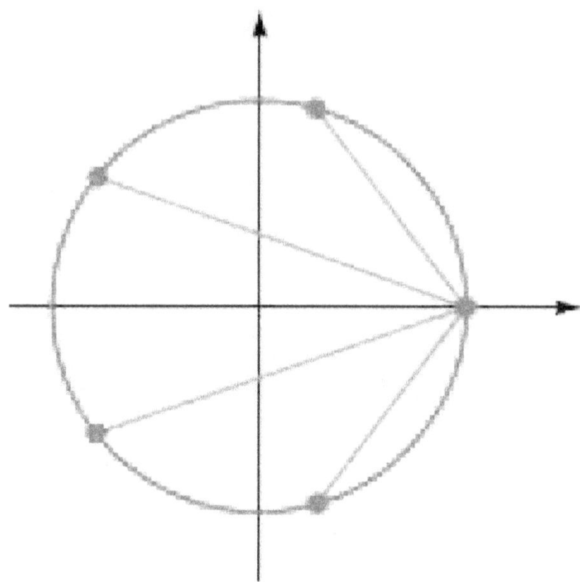

Figure 1.3: The product of the 4 diagonals of a regular 5-gon.

This ends the proof of the problem.

Problem 1.15

Bak and Newman Chapter 1 Exercise 15.

Proof. In the following, we suppose that $z = x + iy$.

(a) This is a closed disc centred at i with radius 1. By Definition 1.6, it is not a region.

(b) The equation becomes $|(x - 1) + iy| = |(x + 1) + iy|$. By the definition of the modulus and after squaring, we get

$$(x - 1)^2 + y^2 = (x + 1)^2 + y^2. \tag{1.11}$$

Thus we must have $x = 0$, i.e., the set of solutions of the equation is exactly the imaginary axis. Since squaring an equation *may* create new solutions of the original equation, we have to check that *every* solution of the equation (1.11) is also a solution of the original equation.[d] In fact, if $z = iy$, then it is easy to see that

$$\left| \frac{z - 1}{z + 1} \right| = \frac{|-1 + iy|}{|1 + iy|} = 1.$$

---

[d]For example, if $x = \sqrt{2 - x}$, then $x^2 + x - 2 = 0$ whose solutions are $x = 1$ or $x = -2$. However, note that $-2 = \sqrt{2 + 2} = 2$ which is impossible, so $x = -2$ is *not* a solution of the original solution.

In other words, all solutions of the equation (1.11) are also solutions of the original equation. By Definition 1.6 again, it is not a region.

(c) We have $(x-2)^2 + y^2 > (x-3)^2 + y^2$ which gives $x > \frac{5}{2}$. Similar to part (b), if $x > \frac{5}{2}$, then $z = x + iy$ satisfies the original inequality. Hence the solution set of the inequality is $\operatorname{Re} z > \frac{5}{2}$. By Definition 1.6, it is a region.

(d) Note that $|z| < 1$ is the open unit disc centred at 0 and $\operatorname{Im} z > 0$ is the upper half plane, so the solution set is the upper half of the open unit disc centred at 0. By Definition 1.6, it is a region.

(e) The equation is equivalent to $z \cdot \overline{z} = 1$ which is $x^2 + y^2 = 1$. This is a unit circle centred at 0, so it is not a region by Definition 1.6.

(f) We have $x^2 + y^2 = y$ which is $x^2 + (y - \frac{1}{2})^2 = \frac{1}{2^2}$. Hence it is a circle centred at $\frac{i}{2}$ and the radius is $\frac{1}{2}$. Similar to part (e), it is not a region.

(g) By the definition, we have $|(x^2 - y^2 - 1) + 2xyi| < 1$ so that

$$\sqrt{(x^2 - y^2 - 1)^2 + 4x^2y^2} < 1$$

which is equivalent to

$$(x^2 - y^2)^2 - 2(x^2 - y^2) + 1 + 4x^2y^2 < 1.$$

After simplification, we obtain

$$(x^2 + y^2)^2 - 2(x^2 - y^2) < 0$$

which is known as a **lemniscate of Bernoulli** [15, pp. 121 - 123], see Figure 1.4, which is generated by "Desmos" (`https://www.desmos.com/`), below:

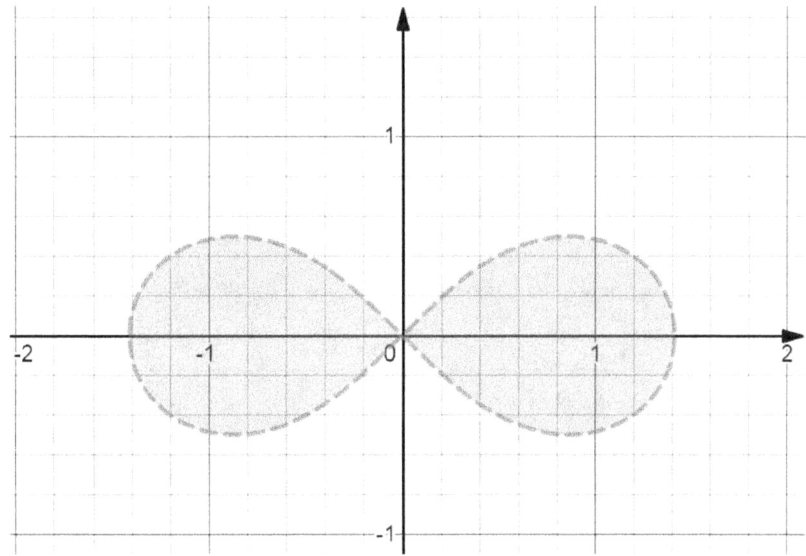

Figure 1.4: The graph of lemniscate of Bernoulli.

We notice that this set is open but not connected. By Definition 1.6, it is not a region.

We have completed the proof of the problem.

Problem 1.16

*Bak and Newman Chapter 1 Exercise 16.**

Proof.

(a) Let $z = x + iy$. Then the equation becomes

$$\sqrt{x^2 + y^2} = x + 1 \tag{1.12}$$

which, by squaring both sides, reduces to

$$y^2 = 2x + 1. \tag{1.13}$$

Clearly, this is a parabola which opens rightward. Similar to Problem 1.15(b), we have to check that *every* solution of the equation (1.13) is also a solution of the equation (1.12). To see this, let (x, y) satisfy the equation (1.13). Since $y^2 \geq 0$, we have $x \geq -\frac{1}{2}$ so that $x + 1 > 0$. Next, we substitute the equation (1.13) into the left-hand side of the equation (1.12) to get

$$\sqrt{x^2 + y^2} = \sqrt{x^2 + 2x + 1} = \sqrt{(x+1)^2} = |x + 1| = x + 1.$$

This shows that the solutions of the equations (1.12) and (1.13) are identical.

(b) Geometrically, it says that the sum of the distances from z to 1 and from z to -1 is always 4, so the locus of the points is the ellipse with foci $(\pm 1, 0)$ and the semi-major axis 2, i.e.,

$$\frac{x^2}{2^2} + \frac{y^2}{b^2} = 1,$$

where b is the semi-minor axis. By the definition, we have $b^2 = 2^2 - 1^2 = 3$ so that the set of points satisfying the given equation is

$$\frac{x^2}{4} + \frac{y^2}{3} = 1.$$

(c) Notice that $n \in \mathbb{Z}$. There are several cases for consideration.

- **Case (i):** $n = 0$. The equation becomes $\frac{1}{z} = \bar{z}$ which is Problem 1.15(e).

- **Case (ii):** $n = 1$. In this case, $\bar{z} = 1$. If $z = x + iy$, then it implies that $x - iy = 1$ so that $x = 1$ and $y = 0$.

- **Case (iii):** $n = 2$. Now we have $z = \bar{z}$. If $z = x + iy$, then $x + iy = x - iy$ so that $y = 0$. In other words, the set of solutions in this case is the real axis.

- **Case (iv):** $n \geq 3$. Obviously, $z = 0$ is a solution of the equation. Suppose that $z \neq 0$. Then $|z^{n-1}| = |z|$ implies that $|z| = 1$. Next, $z^{n-1} = \bar{z}$ also gives

$$z^n = z \cdot \bar{z} = |z|^2 = 1,$$

i.e., z is an n-th root of unity. Hence the set of solutions in this case consists of the n-th roots of unity and $z = 0$.

– **Case (v):** $n \le -1$. In this case, $z \ne 0$. Similar to **Case (iv)**, $|z^{n-1}| = |z|$ implies that $|z| = 1$ and $z^{n-1} = \bar{z}$ gives

$$z^n = 1, \tag{1.14}$$

where $n \le -1$. If we rewrite the equation (1.14) as $z^{-|n|} = 1$, then we get

$$z^{|n|} = 1$$

so that z is an $|n|$-th root of unity.

This completes the proof of the problem.

Bak and Newman Chapter 1 Exercise 17.

Proof. Let $z = x + iy$, where $x^2 + y^2 = 1$. Then it is easy to check that

$$\begin{aligned}
\frac{z-1}{z+1} &= \frac{x-1+iy}{x+1+iy} \\
&= \frac{x-1+iy}{x+1+iy} \cdot \frac{x+1-iy}{x+1-iy} \\
&= \frac{x^2+y^2-1}{(x+1)^2+y^2} + i\frac{2y}{(x+1)^2+y^2} \\
&= i\frac{2y}{(x+1)^2+y^2}.
\end{aligned}$$

Note that if $y > 0$, then $\frac{z-1}{z+1}$ lies on the positive imaginary axis; if $y < 0$, then $\frac{z-1}{z+1}$ lies on the negative imaginary axis. By the definition of $\operatorname{Arg} w$ given in the problem, we obtain immediately that

$$\operatorname{Arg}\left(\frac{z-1}{z+1}\right) = \begin{cases} \dfrac{\pi}{2}, & \text{if } \operatorname{Im} z > 0; \\[2mm] -\dfrac{\pi}{2}, & \text{if } \operatorname{Im} z < 0. \end{cases}$$

We have completed the proof of the problem.

*Bak and Newman Chapter 1 Exercise 18.**

Proof. By [4, Eqn. (4), p. 10], the solutions of $x^3 - 6x = 4$ are the three real-valued *possibilities* for $\sqrt[3]{2+2i} + \sqrt[3]{2-2i}$. Suppose that $a = \sqrt[3]{2+2i}$ and $b = \sqrt[3]{2-2i}$. Then we have $a^2 = \sqrt[3]{(2+2i)^2} = 2\sqrt[3]{i}$, $b^2 = \sqrt[3]{(2-2i)^2} = 2\sqrt[3]{-i}$ and $ab = 2$. We note that

$$\sqrt[3]{i} = \frac{\sqrt{3}}{2} + \frac{i}{2}, \quad -\frac{\sqrt{3}}{2} + \frac{i}{2}, \quad -i$$

and

$$\sqrt[3]{-i} = \frac{\sqrt{3}}{2} - \frac{i}{2}, \quad -\frac{\sqrt{3}}{2} - \frac{i}{2}, \quad i.$$

Thus we have

$$a^2 - ab + b^2 = 2\sqrt{3} - 2, \quad -2\sqrt{3} - 2 \quad \text{or} \quad -2.$$

By the hint, we conclude that the three real-valued possibilities for $\sqrt[3]{2+2i}+\sqrt[3]{2-2i}$ are given by

$$x = a + b = \frac{a^3 + b^3}{a^2 - ab + b^2} = \frac{2}{\sqrt{3}-1}, \quad \frac{2}{-\sqrt{3}-1} \quad \text{or} \quad -2 = 1 + \sqrt{3}, \quad 1 - \sqrt{3} \quad \text{or} \quad -2,$$

completing the proof of the problem. ■

> **Remark 1.1**
>
> By direct substitution, we see that $x = -2$ is a root of the equation $x^3 - 6x - 4 = 0$. Therefore, we have
> $$x^3 - 6x - 4 = (x+2)(x^2 - 2x - 2)$$
> so that the other real roots of the equation $x^3 - 6x - 4 = 0$ are exactly $1 \pm \sqrt{3}$.

> **Problem 1.19**
>
> *Bak and Newman Chapter 1 Exercise 19.*

Proof. Suppose that α, β, γ are the roots of the equation

$$x^3 + px - q = 0. \tag{1.15}$$

Then we have

$$\begin{cases} \alpha + \beta + \gamma = 0, \\ \alpha\beta + \beta\gamma + \gamma\alpha = p, \\ \alpha\beta\gamma = q. \end{cases} \tag{1.16}$$

If $\gamma = 0$, then $q = 0$ and thus the equation $x^3 + px = 0$ has three roots $0, \pm\sqrt{-p}$. Now the roots are real if and only if $p < 0$ if and only if $4p^3 + 27q^2 = 4p^3 < 0$. Therefore, we may suppose that all the roots of the equation (1.15) are nonzero. Thus it follows from the equations (1.16) that

$$\begin{aligned}
(\alpha - \beta)^2 &= \alpha^2 + \beta^2 + \gamma^2 - \gamma^2 - \frac{2\alpha\beta\gamma}{\gamma} \\
&= (\alpha + \beta + \gamma)^2 - 2(\alpha\beta + \beta\gamma + \gamma\alpha) - \gamma^2 - \frac{2\alpha\beta\gamma}{\gamma} \\
&= -2p - \gamma^2 - \frac{2q}{\gamma}.
\end{aligned} \tag{1.17}$$

Similarly, we also have

$$(\beta - \gamma)^2 = -2p - \alpha^2 - \frac{2q}{\alpha} \quad \text{and} \quad (\gamma - \alpha)^2 = -2p - \beta^2 - \frac{2q}{\beta}. \tag{1.18}$$

Suppose that $(\alpha - \beta)^2, (\beta - \gamma)^2$ and $(\gamma - \alpha)^2$ are roots of the new equation

$$y^3 + Ay^2 + By + C = 0,$$

where A, B and C will be determined very soon. By the formulas (1.17) and (1.18), it yields that

$$y = -2p - x^2 - \frac{2q}{x}$$

which is equivalent to

$$xy = -2px - x^3 - 2q$$

$$(x^3 + px - q) + xy = -px - 3q$$
$$xy + px = -3q$$
$$x = -\frac{3q}{y+p}. \tag{1.19}$$

Substituting the expression (1.19) into the original equation (1.15), we arrive at

$$\left(-\frac{3q}{y+p}\right)^3 + p\left(-\frac{3q}{y+p}\right) - q = 0$$
$$-27q^3 - 3pq(y+p)^2 - q(y+p)^3 = 0$$
$$-qy^3 - 6qpy^2 - 9qp^2 y - 27q^3 - 4qp^3 = 0. \tag{1.20}$$

Recall that $q \neq 0$, so the equation (1.20) reduces to

$$y^3 + 6py^2 + 9p^2 y + (27q^2 + 4p^3) = 0$$

so that the product of its roots is given by

$$(\alpha - \beta)^2 (\beta - \gamma)^2 (\gamma - \alpha)^2 = -(4p^3 + 27q^2). \tag{1.21}$$

Hence the formula (1.21) implies the following results:

- **Case (i):** The equation (1.15) has three distinct real roots if and only if $4p^3 + 27q^2 < 0$.

- **Case (ii):** The equation (1.15) has a repeated (real) root if and only if $4p^3 + 27q^2 = 0$.

- **Case (iii):** The equation (1.15) has two complex roots if and only if $4p^3 + 27q^2 > 0$.

Hence we complete the proof of the problem.

Remark 1.2

(a) By **Case (ii)** in the proof of Problem 1.19, it is believed that "three real roots" should be replaced by "three distinct real roots".

(b) If we assume that the equation (1.15) has a real root, then the algebra will become much simpler. In fact, let α be a real zero of the function $f(x) = x^3 + px - q$. Then we can write

$$f(x) = (x - \alpha)(x^2 + ax + b),$$

where $a, b \in \mathbb{R}$. Direct expansion gives $f(x) = x^3 + (a - \alpha)x^2 + (b - a\alpha)x - b\alpha$, so we have $\alpha = a$ and then

$$p = b - a^2 \quad \text{and} \quad q = ba.$$

Now $4p^3 + 27q^2 < 0$ if and only if

$$4(b - a^2)^3 + 27b^2 a^2 < 0$$
$$4b^3 - 12b^2 a^2 + 12ba^4 - 4a^6 + 27b^2 a^2 < 0$$
$$4b^3 + 15b^2 a^2 + 12ba^4 - 4a^6 < 0$$
$$(-a^2 b^2 - 4ba^4 - 4a^6) + (4b^3 + 16a^2 b^2 + 16ba^4) < 0$$
$$-a^2(2a^2 + b)^2 + 4b(b + 2a^2)^2 < 0$$
$$(b + 2a^2)^2(4b - a^2) < 0$$

if and only if $a^2 - 4b > 0$ which is equivalent to saying that the equation $x^2 + ax + b = 0$ has two (distinct) real roots. Hence, we have completed the proof of the problem.

> ### Problem 1.20
>
> *Bak and Newman Chapter 1 Exercise 20.**

Proof.

(a) It is clear that

$$(1-z)P(z) = 1 + 2z + 3z^2 + \cdots + nz^{n-1} - z - 2z^2 - \cdots - (n-1)z^{n-1} - nz^n$$
$$= 1 + z + z^2 + \cdots + z^{n-1} - nz^n. \tag{1.22}$$

Assume that z_0 was a zero of $P(z)$ with $|z_0| = r > 1$. Then the expression (1.22) implies that

$$1 + z_0 + z_0^2 + \cdots + z_0^{n-1} = nz_0^n$$

and then

$$nr^n \leq 1 + r + \cdots + r^{n-1}. \tag{1.23}$$

However, since $r > 1$, it is easy to see that $r^k - 1 > 0$ for every $k \in \mathbb{N}$ so that

$$nr^n - (1 + r + \cdots + r^{n-1}) = (r^n - 1) + (r^n - r) + \cdots + (r^n - r^{n-1})$$
$$= (r^n - 1) + r(r^{n-1} - 1) + \cdots + r^{n-1}(r - 1)$$
$$> 0$$

which contradicts the inequality (1.23). Hence no such z_0 exists and all the zeros of $P(z)$ lie in $|z| \leq 1$.

(b) Let $P(z) = a_0 + a_1 z + \cdots + a_n z^n$. If $a_0 = a_1 = \cdots = a_n = 0$, then $P(z) \equiv 0$ which is meaningless. Therefore, we may assume that $a_k > 0$ for some $k \in \{0, 1, 2, \ldots, n\}$. We consider

$$(1-z)P(z) = a_0 + a_1 z + \cdots + a_n z^n - a_0 z - a_1 z^2 - \cdots - a_{n-1} z^n - a_n z^{n+1}$$
$$= a_0 + (a_1 - a_0)z + (a_2 - a_1)z^2 + \cdots + (a_n - a_{n-1})z^n - a_n z^{n+1}. \tag{1.24}$$

Assume that w was a zero of $P(z)$ with $|w| = r > 1$. On the one hand, the expression (1.24) implies that

$$a_0 + (a_1 - a_0)w + (a_2 - a_1)w^2 + \cdots + (a_n - a_{n-1})w^n = a_n w^{n+1}$$

so that

$$a_n r^{n+1} \leq a_0 + (a_1 - a_0)r + (a_2 - a_1)r^2 + \cdots + (a_n - a_{n-1})r^n. \tag{1.25}$$

On the other hand, we know that $r^k - 1 > 0$ for every $k \in \mathbb{N}$ and since $a_k > 0$ for some k, we see that

$$a_n r^{n+1} - [a_0 + (a_1 - a_0)r + (a_2 - a_1)r^2 + \cdots + (a_n - a_{n-1})r^n]$$
$$= a_n r^n(r - 1) + a_{n-1} r^{n-1}(r - 1) + \cdots + a_1 r(r - 1) + a_0(r - 1)$$
$$= (r - 1)(a_n r^n + a_{n-1} r^{n-1} + \cdots + a_1 r + a_0)$$
$$> 0,$$

but this contradicts the inequality (1.25). Hence no such w exists and all the zeros of $P(z)$ lie in $|z| \leq 1$.

Consequently, we complete the proof of the problem. $\blacksquare$

Remark 1.3

Problem 1.20(b) is the well-known Eneström and Kakeya Theorem. See, for examples, [17, pp. 136, 137] and [18].

Problem 1.21

Bak and Newman Chapter 1 Exercise 21.

Proof.

(a) Let z_0 be a fixed point such that $|z_0| < 1$. Then there exists a $r > 0$ such that $|z_0| < r < 1$. It is clear that $|kz^k| \le kr^k$ for every $|z| \le r$ and $k = 0, 1, 2, \ldots$. Let $M_k = kr^k$. Since

$$\alpha = \limsup_{k \to \infty} \sqrt[k]{kr^k} = \lim_{k \to \infty} \sqrt[k]{k}\, r = r < 1,$$

the Root Test [27, Theorem 6.7, p. 76] ensures that the series $\displaystyle\sum_{k=0}^{\infty} kr^k$ converges. By Theorem 1.9 (The Weierstrass M-test), the series $\displaystyle\sum_{k=0}^{\infty} kz^k$ converges uniformly to a continuous function $f(z)$ on $D(0; r)$. Since z_0 is arbitrary, we conclude that $f(z)$ is continuous on $D(0; 1)$.

(b) Let $z = x + iy$ with $x > 0$. Obviously, we know that

$$\left| \frac{1}{k^2 + z} \right| = \frac{1}{\sqrt{(k^2 + x)^2 + y^2}} \le \frac{1}{k^2}.$$

Since $\displaystyle\sum_{k=1}^{\infty} \frac{1}{k^2}$ converges, Theorem 1.9 (The Weierstrass M-test) guarantees that the series

$$\sum_{k=1}^{\infty} \frac{1}{k^2 + z}$$

converges uniformly to a continuous function $g(z)$ in the right half-plane $\operatorname{Re} z > 0$.

We complete the proof of the problem. $\blacksquare$

Problem 1.22

Bak and Newman Chapter 1 Exercise 22.

Proof. Suppose that S is a polygonally connected set. Assume that S was disconnected. Then there are disjoint open sets U and V of $\mathbb{C}$ such that

$$S \subseteq U \cup V, \quad S \cap U \neq \varnothing \quad \text{and} \quad S \cap V \neq \varnothing. \tag{1.26}$$

Select $a \in S \cap U$ and $b \in S \cap V$. Since U and V are disjoint, $a \neq b$. Since S is polygonally connected, there exists a polygonally line $L : [0, 1] \to S$ connecting a and b, i.e., $L(0) = a$ and $L(1) = b$. By the definition, we have

$$L = [a, z_1] \cup [z_1, z_2] \cup \cdots \cup [z_n, b]$$

for some $n \in \mathbb{N}$. Since each line segment is continuous, L is continuous and then $L([0,1])$ is connected by [27, Theorem 7.12, p. 100]. However, the set relations (1.26) imply that

$$L([0,1]) \subseteq U \cap V, \quad L([0,1]) \cap U \neq \varnothing \quad \text{and} \quad L([0,1]) \cap V \neq \varnothing.$$

In other words, $L([0,1])$ is disconnected, a contradiction. Hence we end the proof of the problem.

> **Problem 1.23**
>
> *Bak and Newman Chapter 1 Exercise 23.*

Proof. First of all, the graph of S is shown in Figure 1.5.

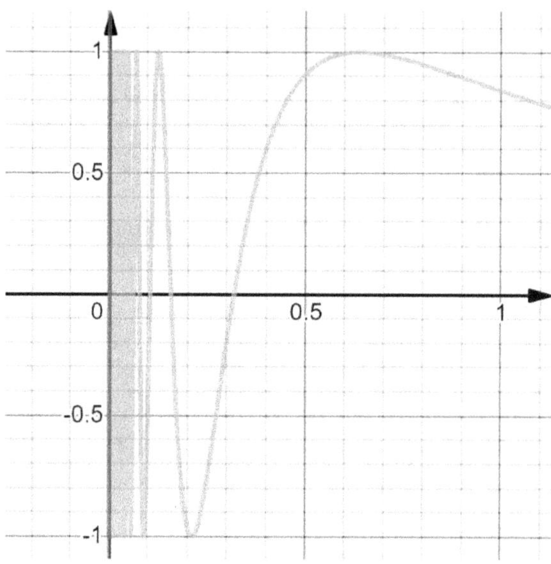

Figure 1.5: The graph of S in the plane $\mathbb{C}$.

Define
$$T = \{x + iy \mid x > 0 \text{ and } y = \sin \tfrac{1}{x}\} \quad \text{and} \quad I = \{iy \mid y \in \mathbb{R}\}.$$

Then we have $S = I \cup T$. We assume that S was disconnected, i.e., there are disjoint open sets U and V such that
$$S \subseteq U \cup V, \quad S \cap U \neq \varnothing \quad \text{and} \quad S \cap V \neq \varnothing. \tag{1.27}$$

Since $(0,0) \in S$, we may suppose that $(0,0) \in U$ so that $I \cap U \neq \varnothing$. Since I is obviously polygonally connected, it is connected by Problem 1.22. If $I \cap V \neq \varnothing$, then the definition shows that I is disconnected, a contradiction. Hence we know that

$$I \subseteq U. \tag{1.28}$$

Next, the openness of U implies that there exists a $\delta > 0$ such that
$$\{z = a + ib \mid |z| < \delta\} \subseteq U.$$

We note that the points $x_n = \frac{1}{2n\pi}$ with $n \in \mathbb{N}$ satisfy $\sin \frac{1}{x_n} = \sin 2n\pi = 0$, so we can select N large enough such that the points $z_n = x_n + i \sin \frac{1}{x_n}$ satisfy

$$|z_n| = \sqrt{(x_n - 0)^2 + \left(\sin \frac{1}{x_n} - 0\right)^2} = x_n < \delta$$

for all $n \geq N$.[e] In other words, it means that $z_n \in U$ for all $n \geq N$ or equivalently, $T \cap U \neq \varnothing$. Recall that $\sin \frac{1}{x}$ is continuous on $\mathbb{R} \setminus \{0\}$, so the function $f : (0, \infty) \to \mathbb{C}$ defined by

$$f(x) = x + i \sin \frac{1}{x}$$

is continuous on $(0, \infty)$ and its image $f((0, \infty))$ is connected by [27, Theorem 7.12, p. 100]. Since $T = f((0, \infty))$, T is also connected and the argument in the previous paragraph shows that

$$T \subseteq U. \tag{1.29}$$

Combining the set relations (1.28) and (1.29), we conclude that $S \cap V = \varnothing$ which contradicts the assumption (1.27). Hence S is connected which completes the proof of the problem. ∎

Remark 1.4

In Problem 1.23, the union of the set T with its limit point $(0, 0)$ is called the **Topologist's sine curve**, see [19, Example 7, pp. 156, 157]. Furthermore, it is obvious that the point $2i$ cannot be connected by any curve in S because $|\sin \frac{1}{x}| \leq 1$.

Problem 1.24

Bak and Newman Chapter 1 Exercise 24.

Proof. Recall from [4, Eqn. (3), p. 17] that

$$\xi = \frac{x}{x^2 + y^2 + 1}, \quad \eta = \frac{y}{x^2 + y^2 + 1} \quad \text{and} \quad \zeta = \frac{x^2 + y^2}{x^2 + y^2 + 1}$$

so that

$$x^2 + y^2 = \frac{\zeta}{1 - \zeta}. \tag{1.30}$$

Since $\zeta \geq \zeta_0$, we have $\frac{1}{1-\zeta} \geq \frac{1}{1-\zeta_0}$. Hence it follows from the equation (1.30) that

$$x^2 + y^2 \geq \frac{\zeta_0}{1 - \zeta_0},$$

i.e., T is the exterior of the circle centred at 0 with radius $\frac{\zeta_0}{1-\zeta_0}$. This completes the proof of the problem. ∎

Problem 1.25

Bak and Newman Chapter 1 Exercise 25.

Proof.

---

[e] In fact, this argument also shows that $(0, 0)$ is a limit point of S.

(a) Suppose that T is a circle on $\mathbb{C}$. Then $\zeta \neq 1$ and T has the form

$$x^2 + y^2 + Dx + Ey + F = 0. \tag{1.31}$$

Using the expression (1.30) and [4, Eqn. (2), p. 17], the equation (1.31) reduces to

$$\frac{\zeta}{1-\zeta} + \frac{D\xi}{1-\zeta} + \frac{E\eta}{1-\zeta} + F = 0$$

$$D\xi + E\eta + (1 - F)\zeta = -F. \tag{1.32}$$

Recall that a circle on $\sum$ is the intersection of $\sum$ with a plane $A\xi + B\eta + C\zeta = G$, where $A, B, C, G \in \mathbb{R}$. As the equation (1.32) is in this form, the corresponding set S is a circle on $\sum$ that doesn't contain $(0, 0, 1)$ as required. See Figure 1.6 for an illustration.

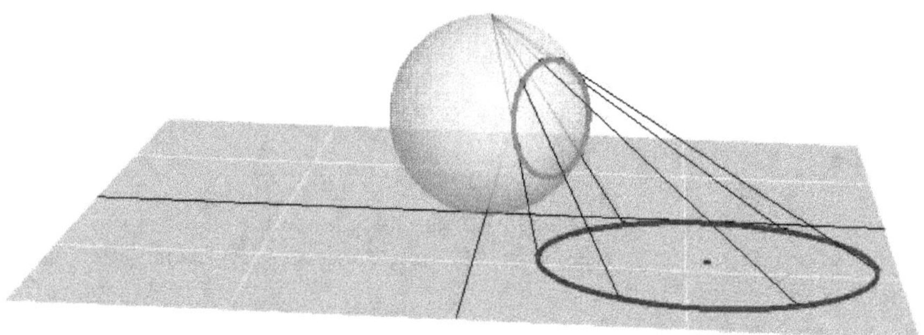

Figure 1.6: S is a circle when T is a circle.

(b) If T is a line in the form $ax + by + c = 0$ for some $a, b, c \in \mathbb{R}$, then we obtain

$$\frac{a\xi}{1-\zeta} + \frac{b\eta}{1-\zeta} + c = 0$$

$$a\xi + b\eta - c\zeta = -c.$$

Thus S is again a circle on $\sum$ that minus the pole $(0, 0, 1)$, see Figure 1.7.

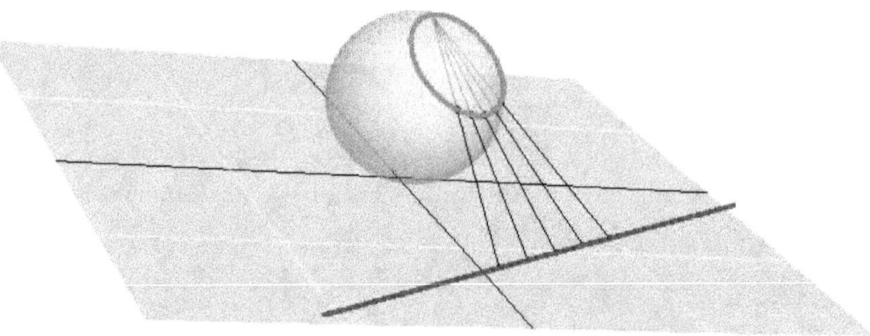

Figure 1.7: S is a circle minus $(0, 0, 1)$ when T is a line.

This completes the proof of the problem.

Problem 1.26

Bak and Newman Chapter 1 Exercise 26.

Proof. Let $P(z) = a_n z^n + a_{n-1} z^{n-1} + \cdots + a_1 z + a_0$, where $a_n \neq 0$. Then we have

$$|P(z)| = \left| z^n \left(a_n + \frac{a_{n-1}}{z} + \cdots + \frac{a_0}{z^n} \right) \right| = |z|^n \times \left| a_n + \frac{a_{n-1}}{z} + \cdots + \frac{a_0}{z^n} \right|. \qquad (1.33)$$

By [1, Eqn. (12), p.10], we know that $\big| |z_1| - |z_2| \big| \leq |z_1 - z_2|$ holds for every complex z_1 and z_2. Apply this result to the right-hand side of the expression (1.33), we deduce that

$$|P(z)| \geq |z|^n \times \left| |a_n| - \left| \frac{a_{n-1}}{z} + \cdots + \frac{a_0}{z^n} \right| \right|. \qquad (1.34)$$

Let $M = \max(|a_0|, |a_1|, \ldots, |a_{n-1}|)$. If $|z| > 1$, then it follows from the triangle inequality that

$$\left| \frac{a_{n-1}}{z} + \cdots + \frac{a_0}{z^n} \right| \leq \frac{|a_{n-1}|}{|z|} + \frac{|a_{n-2}|}{|z|^2} + \cdots + \frac{|a_0|}{|z|^n} \leq \frac{Mn}{|z|}.$$

Therefore, if $|z| \geq \max(1, \frac{2Mn}{|a_n|})$, then $\frac{Mn}{|z|} \leq \frac{|a_n|}{2}$ and the inequality (1.34) becomes

$$|P(z)| \geq |z|^n \times \left(|a_n| - \left| \frac{a_{n-1}}{z} + \cdots + \frac{a_0}{z^n} \right| \right) \geq |z|^n \times \frac{|a_n|}{2}$$

which tends to ∞ as $|z| \to \infty$. Consequently, we have $P(z) \to \infty$ as $z \to \infty$ which completes the proof of the problem. $\blacksquare$

Problem 1.27

Bak and Newman Chapter 1 Exercise 27.

Proof. Let $z = x + iy$. Then we have $\frac{1}{z} = \frac{1}{x+iy} = \frac{x}{x^2+y^2} - i\frac{y}{x^2+y^2}$.

(a) Using [4, Eqn. (3), p. 17], we have

$$\xi = \frac{x}{x^2 + y^2 + 1}, \quad \eta = \frac{y}{x^2 + y^2 + 1} \quad \text{and} \quad \zeta = \frac{x^2 + y^2}{x^2 + y^2 + 1}.$$

By [4, Eqn. (3), p. 17] again, we see that

$$\xi' = \frac{\frac{x}{x^2+y^2}}{(\frac{x}{x^2+y^2})^2 + (\frac{-y}{x^2+y^2})^2 + 1} = \frac{x}{x^2 + y^2 + 1} = \xi$$

and

$$\eta' = \frac{\frac{-y}{x^2+y^2}}{(\frac{x}{x^2+y^2})^2 + (\frac{-y}{x^2+y^2})^2 + 1} = \frac{-y}{x^2 + y^2 + 1} = -\eta.$$

Finally, we get

$$\zeta' = \frac{(\frac{x}{x^2+y^2})^2 + (\frac{-y}{x^2+y^2})^2}{(\frac{x}{x^2+y^2})^2 + (\frac{-y}{x^2+y^2})^2 + 1} = \frac{x^2 + y^2}{x^2 + y^2 + (x^2 + y^2)^2} = \frac{1}{x^2 + y^2 + 1} = 1 - \zeta.$$

(b) Obviously, $(-\frac{1}{2}, 0, \frac{1}{2})$ and $(\frac{1}{2}, 0, \frac{1}{2})$ are on $\sum$. By part (a), if the point z corresponds to (ξ_1, η_1, ζ_1) on $\sum$, then the point $\frac{1}{z}$ corresponds to $(\xi_1, -\eta_1, 1 - \zeta_1)$ on $\sum$. The form of the corresponding points show that they lie on the circle C which is the intersection of the plane $\xi = \xi_1$ and $\sum$. It is clear that the centre of circle C is $(\xi_1, 0, \frac{1}{2})$. See Figure 1.8 for an illustration.

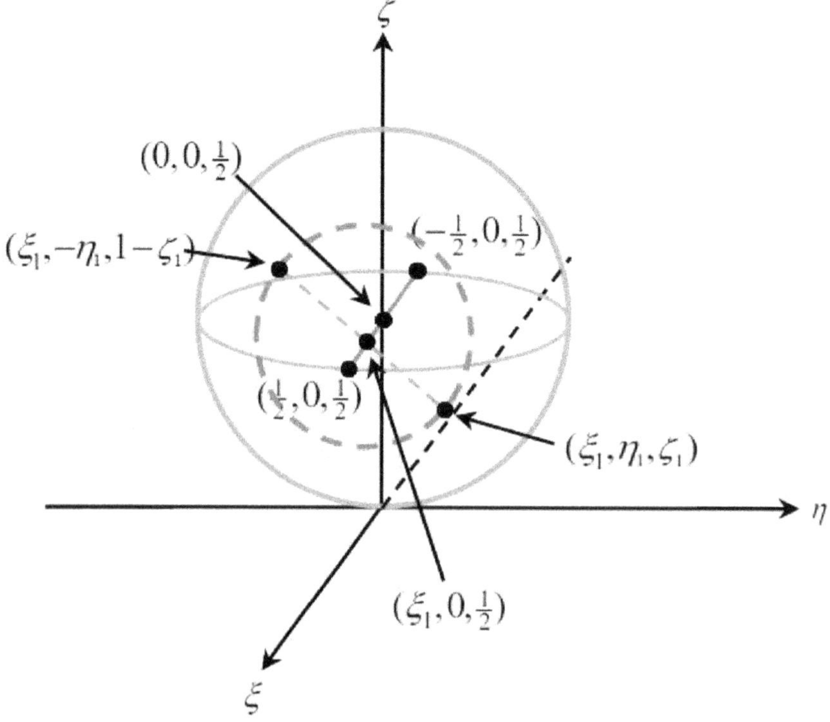

Figure 1.8: The effect of $f(z) = \frac{1}{z}$ on $\sum$.

By considering the projections of the points (ξ_1,η_1,ζ_1), $(\xi_1,-\eta_1,1-\zeta_1)$ and the circle $(\xi_1,0,\frac{1}{2})$ on the $\eta-\zeta$ plane, we observe that (η_1,ζ_1) and $(-\eta_1,1-\zeta_1)$ are points on a circle centred at $(0,\frac{1}{2})$. Therefore, the equation of the straight line L passing through $(0,\frac{1}{2})$ and (η_1,ζ_1) is

$$(2\zeta_1 - 1)\eta - 2\eta_1\zeta + \eta_1 = 0. \tag{1.35}$$

By substituting the point $(-\eta_1,1-\zeta_1)$ into the equation (1.35), it is easy to check that it lies on L. In other words, the point $(-\eta_1,1-\zeta_1)$ is the image of the $180°$ rotation of the point (η_1,ζ_1) about the centre $(0,\frac{1}{2})$. Going back to the three dimensional situation, it implies that the point $(\xi_1,-\eta_1,1-\zeta_1)$ is the image of the $180°$ rotation of the point (ξ_1,η_1,ζ_1) about the diameter with endpoints $(-\frac{1}{2},0,\frac{1}{2})$ and $(\frac{1}{2},0,\frac{1}{2})$.

We complete the proof of the problem. ■

Remark 1.5

Another way to intercept Problem 1.27(b) can be found in [20, p. 143].

Problem 1.28

Bak and Newman Chapter 1 Exercise 28.

Proof. Let T be a circle or a line in $\mathbb{C}$. By Problem 1.25, its corresponding $S \subset \sum$ is also a circle. By Definition 1.11, S has the form

$$A\xi + B\eta + C\zeta = D, \tag{1.36}$$

where $A, B, C, D \in \mathbb{R}$. Substituting the result of Problem 1.27(a) into the equation (1.36), it leads

$$A\xi' + B(-\eta') + C(1 - \zeta') = D$$
$$A\xi' - B\eta' - C\zeta' = D - C. \tag{1.37}$$

By Definition 1.11 again, the set S' with the form (1.37) is also a circle on $\sum$. By Proposition 1.12, its corresponding projection T' is a circle or a line in $\mathbb{C}$. In conclusion, we have shown that $f(z) = \frac{1}{z}$ maps circles and lines in $\mathbb{C}$ onto other circles and lines. This ends the proof of the problem.

Remark 1.6

The map in Problem 1.28 is sometimes called the **inversion through the unit circle**, see [14, Exercise I.3.1, p. 17].

CHAPTER 2

Functions of the Complex Variable z

Bak and Newman Chapter 2 Exercise 1.

Proof. We consider the monomial $P(x, y) = (x + iy)^n$ first, where $n \in \mathbb{N} \cup \{0\}$. It is clear that $P_x = n(x + iy)^{n-1}$ and $P_y = in(x + iy)^{n-1}$ which imply

$$P_y = iP_x.$$

Thus the necessity of Proposition 2.3 holds for a monomial. Since an analytic polynomial is a finite linear combination of monomials, the necessity of Proposition 2.3 also holds for an analytic polynomial. This completes the proof of the problem. $\blacksquare$

Bak and Newman Chapter 2 Exercise 2.[*]

Proof.

(a) For every *real* z, it follows from Definition 2.4 that

$$f'(z) = \lim_{\substack{h \to 0 \\ h \in \mathbb{C}}} \frac{f(z + h) - f(z)}{h} \tag{2.1}$$

exists. In particular, we may take h to be real in the limit (2.1) and the resulting limit still equals to $f'(z)$, i.e.,

$$f'(z) = \lim_{\substack{h \to 0 \\ h \in \mathbb{R}}} \frac{f(z + h) - f(z)}{h}. \tag{2.2}$$

Since both z and h are real, $f(z + h)$ and $f(z)$ are also real so that the quotient in the limit (2.2) is real-valued. Hence $f'(z)$ is real-valued for real z.

(b) The limit (2.1) exists for all imaginary points z. If $z = iy$, where $y \in \mathbb{R}$, then we have

$$f'(iy) = \lim_{\substack{h \to 0 \\ h \in \mathbb{C}}} \frac{f(iy + h) - f(iy)}{h} = \frac{1}{i} \cdot \lim_{\substack{r \to 0 \\ r \in \mathbb{R}}} \frac{f(iy + ir) - f(iy)}{r}. \tag{2.3}$$

By the hypothesis, both $f(iy + ir)$ and $f(iy)$ are real-valued so that the quotient in the limit (2.3) is also real-valued. Hence $f'(iy)$ is purely imaginary.

We have completed the proof of the problem.   ◼

> **Problem 2.3**
>
> Bak and Newman Chapter 2 Exercise 3.

Proof.

(a) Since $P_x = 3x^2 - 3y^2 - 1 + 6ixy$ and $P_y = -6xy + i(3x^2 - 3y^2 - 1)$, we have $P_y = iP_x$. By Proposition 2.3, P is analytic.

(b) Since $P_x = 2x$ and $P_y = 2iy$, we have $P_y \neq iP_x$. By Proposition 2.3, P is not analytic.

(c) Since $P_x = 2y - 2ix$ and $P_y = 2x + 2iy$, we have $P_y = iP_x$. By Proposition 2.3, P is analytic.

This completes the proof of the problem.   ◼

> **Problem 2.4**
>
> Bak and Newman Chapter 2 Exercise 4.

Proof. Assume that P was a nonconstant analytic polynomial which takes imaginary values only. Then $Q = iP$ takes only real values. Since $Q_y = iP_y = i(iP_x) = iQ_x$, Q is a nonconstant analytic polynomial. By the Example 1 on p. 24, no such Q exists. Hence no such P exists and we complete the proof of the problem.   ◼

> **Problem 2.5**
>
> Bak and Newman Chapter 2 Exercise 5.

Proof. For Problem 2.3(a), we have $P(z) = z^3 - z$ so that

$$P'(z) = 3z^2 - 1 = 3(x + iy)^2 - 1 = 3x^2 - 3y^2 - 1 + 6ixy = P_x.$$

Similarly, for Problem 2.3(c), we have $P(z) = -iz^2$ so that

$$P'(z) = -2iz = -2i(x + iy) = 2y - 2ix = P_x.$$

We end the proof of the problem.   ◼

> **Problem 2.6**
>
> Bak and Newman Chapter 2 Exercise 6.

Proof. By Definition 2.4, we have

$$h_1'(z) = \lim_{t \to 0} \frac{h_1(z+t) - h_1(z)}{t} = \lim_{t \to 0} \frac{f(z+t) - f(z)}{t} + \lim_{t \to 0} \frac{g(z+t) - g(z)}{t} = f'(z) + g'(z)$$

and

$$h_2'(z) = \lim_{t \to 0} \frac{h_2(z+t) - h_2(z)}{t}$$

$$= \lim_{t \to 0} \frac{f(z+t)g(z+t) - f(z)g(z)}{t}$$

$$= \lim_{t \to 0} \frac{f(z+t)g(z+t) - f(z)g(z+t) + f(z)g(z+t) - f(z)g(z)}{t}$$

$$= \lim_{t \to 0} \frac{f(z+t) - f(z)}{t} \cdot g(z+t) + \lim_{t \to 0} \frac{g(z+t) - g(z)}{t} \cdot f(z)$$

$$= f'(z)g(z) + f(z)g'(z). \tag{2.4}$$

If $g(z) \neq 0$, then we have

$$[g^{-1}(z)]' = \lim_{t \to 0} \frac{\frac{1}{g(z+t)} - \frac{1}{g(z)}}{t}$$

$$= \lim_{t \to 0} \frac{g(z) - g(z+t)}{tg(z)g(z+t)}$$

$$= -\lim_{t \to 0} \frac{1}{g(z)g(z+t)} \cdot \lim_{t \to 0} \frac{g(z+t) - g(z)}{t}$$

$$= -\frac{g'(z)}{g^2(z)}. \tag{2.5}$$

Using the results (2.4) and (2.5), we see that

$$h_3'(z) = [f(z)g^{-1}(z)]'$$

$$= f'(z)g^{-1}(z) + f(z)([^{-1}(z)]'$$

$$= \frac{f'(z)}{g(z)} - \frac{f(z)g'(z)}{g^2(z)}$$

$$= \frac{f'(z)g(z) - f(z)g'(z)}{g^2(z)},$$

completing the proof of the problem.

Problem 2.7

Bak and Newman Chapter 2 Exercise 7.

Proof. By Definition 2.4 and the identity $a^n - b^n = (a-b)(a^{n-1} + a^{n-2}b + \cdots + b^{n-1})$, we know that

$$(z^n)' = \lim_{h \to 0} \frac{(z+h)^n - z^n}{h} = \lim_{h \to 0} [(z+h)^{n-1} + (z+h)^{n-2}z + \cdots + z^{n-1}] = nz^{n-1}. \tag{2.6}$$

Thus Proposition 2.6 holds for a monomial. If $P(z) = \alpha_0 + \alpha_1 z + \cdots + \alpha_N z^N$, then the formula (2.6) and repeated applications of Proposition 2.5 imply that

$$P'(z) = \alpha_1 + 2\alpha_2 z + \cdots + N\alpha_N z^{N-1}.$$

This completes the proof of the problem.

Problem 2.8

Bak and Newman Chapter 2 Exercise 8.

Proof. Let $S_n = n^{\frac{1}{n}}$, where $n \geq 2$. Then $\log S_n = \frac{\log n}{n} > 0$ which tends to 0 as $n \to \infty$. Since $S_n = e^{\log S_n}$ and e^x is continuous on $\mathbb{R}$, it has led that

$$\lim_{n\to\infty} S_n = \lim_{n\to\infty} e^{\log S_n} = \exp\left(\lim_{n\to\infty} \log S_n\right) = e^0 = 1.$$

This ends the proof of the problem.

Problem 2.9

Bak and Newman Chapter 2 Exercise 9.

Proof.

(a) Now $C_{n!} = 1$ so that $|C_{n!}|^{\frac{1}{n!}} = 1$. Therefore, we get

$$L = \limsup_{k\to\infty} |C_k|^{\frac{1}{k}} = \lim_{n\to\infty} |C_{n!}|^{\frac{1}{n!}} = 1$$

and so $R = 1$ by Theorem 2.8.

(b) Now $C_n = (n + 2^n)$ so that $|C_n|^{\frac{1}{n}} = (n + 2^n)^{\frac{1}{n}} = 2 \cdot (1 + \frac{n}{2^n})^{\frac{1}{n}}$. Since $\frac{n}{2^n} \to 0$ as $n \to \infty$, we have

$$\limsup_{n\to\infty} \left(1 + \frac{n}{2^n}\right)^{\frac{1}{n}} = 1$$

and then

$$L = \limsup_{n\to\infty} |C_n|^{\frac{1}{n}} = \limsup_{n\to\infty} 2 \cdot \left(1 + \frac{n}{2^n}\right)^{\frac{1}{n}} = 2.$$

By Theorem 2.8, we conclude that $R = \frac{1}{2}$.

We have completed the proof of the problem.

Problem 2.10

Bak and Newman Chapter 2 Exercise 10.

Proof. By the hypothesis, we have

$$L = \limsup_{k\to\infty} |c_k|^{\frac{1}{k}} \quad \text{and} \quad R = \frac{1}{L},$$

where $0 < L < \infty$.

(a) By Problem 2.8, we have

$$\limsup_{k\to\infty} |k^p c_k|^{\frac{1}{k}} = \limsup_{k\to\infty} k^{\frac{p}{k}} \cdot |c_k|^{\frac{1}{k}} = \limsup_{k\to\infty} k^{\frac{p}{k}} \cdot \limsup_{k\to\infty} |c_k|^{\frac{1}{k}} = L.$$

Therefore, the radius of convergence of the power series is also R.

(b) Now it is clear that

$$\limsup_{k\to\infty} ||c_k||^{\frac{1}{k}} = \limsup_{k\to\infty} |c_k|^{\frac{1}{k}} = L,$$

so the radius of convergence of the power series is also R.

(c) Obviously, we have

$$\limsup_{k\to\infty} |c_k^2|^{\frac{1}{k}} = \limsup_{k\to\infty} (|c_k|^{\frac{1}{k}})^2 = L^2$$

so that the radius of convergence of the power series is R^2.

This has completed the proof of the problem. ◼

> **Problem 2.11**
>
> *Bak and Newman Chapter 2 Exercise 11.*

Proof. Let R be the radius of convergence of the sum of the two power series. Now we claim that

$$R \geq \min(R_1, R_2). \tag{2.7}$$

To this end, we suppose first that $R_1 < R_2$. Thus for every z with $|z| \leq R_1$, the hypotheses imply that both

$$\sum_{n=0}^{\infty} a_n z^n \quad \text{and} \quad \sum_{n=0}^{\infty} b_n z^n \tag{2.8}$$

converge so that their sum

$$\sum_{n=0}^{\infty} (a_n + b_n) z^n = \sum_{n=0}^{\infty} a_n z^n + \sum_{n=0}^{\infty} b_n z^n \tag{2.9}$$

converges for such z. In other words, $R \geq R_1$. However, $R_1 < R < R_2$ is impossible. Otherwise, if w satisfies $R_1 < |w| = R < R_2$, then

$$\sum_{n=1}^{\infty} (a_n + b_n) w^n - \sum_{n=0}^{\infty} b_n w^n = \sum_{n=0}^{\infty} a_n w^n$$

converges which means that the radius of convergence of the first power series (2.8) is greater than R_1, a contradiction. Consequently, we obtain $R = R_1$ which satisfies the inequality (2.7) in this case. Next, if $R_1 = R_2$, then the formula (2.9) is still valid for any z with $|z| \leq R_1 = R_2$. By the definition, it implies the inequality (2.7) in this case.

To show that the strict inequality can hold in the estimate (2.7), we consider the power series

$$\sum_{n=0}^{\infty} z^n \quad \text{and} \quad \sum_{n=0}^{\infty} (-z^n)$$

both have radius of convergence 1, but their sum

$$\sum_{n=0}^{\infty} (1 - 1) z^n = 0$$

has radius of convergence ∞. This has completed the proof of the problem. ◼

Problem 2.12

Bak and Newman Chapter 2 Exercise 12.

Proof. Let $|z| = 1$. If $z = 1$, then the series

$$\sum_{n=1}^{\infty} \frac{z}{n} = \sum_{n=1}^{\infty} \frac{1}{n}$$

is divergent. Without loss of generality, we may assume that $z \neq 1$. Now we want to apply Dirichlet's Test [27, Theorem 6.14, p. 78]:

Lemma 2.1 (Dirichlet's Test)

Suppose that $\sum a_n$ is a series of complex numbers whose partial sums $\{A_n\}$ is bounded. If $\{b_n\}$ is a monotonically decreasing sequence and $\lim_{n \to \infty} b_n = 0$, then the series $\sum a_n b_n$ converges.

Let $a_n = z^n$ and $b_n = \frac{1}{n}$. Then the sequence $\{b_n\}$ satisfies the conditions of Lemma 2.1 (Dirichlet's Test). Next, it follows from the triangle inequality that

$$|A_n| = \left| \sum_{k=1}^{n} a_k \right| = \left| \sum_{k=1}^{n} z^k \right| = \left| z \cdot \frac{z^n - 1}{z - 1} \right| \leq \frac{2|z|}{|z - 1|} = \frac{2}{|z - 1|},$$

so it is bounded for $z \neq 1$. Hence Lemma 2.1 (Dirichlet's Test) asserts that the power series

$$\sum_{n=1}^{\infty} \frac{z^n}{n}$$

converges at all points on the unit circle except $z = 1$. This completes the proof of the problem. $\blacksquare$

Problem 2.13

Bak and Newman Chapter 2 Exercise 13.

Proof.

(a) Suppose that L is finite. For every $n \geq 1$, we know that

$$a_{n+1} = a_1 \times \left(\frac{a_2}{a_1} \right) \times \left(\frac{a_3}{a_2} \right) \times \cdots \times \left(\frac{a_{n+1}}{a_n} \right). \tag{2.10}$$

By the definition[a], for every $\epsilon > 0$, there is an $N \in \mathbb{N}$ such that $k \geq N$ implies

$$\left| \frac{a_{k+1}}{a_k} - L \right| < \epsilon$$

or equivalently,

$$L - \epsilon < \frac{a_{k+1}}{a_k} < L + \epsilon \tag{2.11}$$

---

[a]For example, [27, p. 49].

for all $k \geq N$. Rewrite the expression (2.10) as

$$0 < a_{n+1} = a_N \times \left(\frac{a_{N+1}}{a_N}\right) \times \cdots \times \left(\frac{a_{n+1}}{a_n}\right)$$

and then using the inequalities (2.11) to obtain

$$a_N(L-\epsilon)^{n-N} < a_n < a_N(L+\epsilon)^{n-N}$$
$$a_N^{\frac{1}{n}}(L-\epsilon)^{1-\frac{N}{n}} < a_n^{\frac{1}{n}} < a_N^{\frac{1}{n}}(L+\epsilon)^{1-\frac{N}{n}}. \tag{2.12}$$

Taking $n \to \infty$ to each inequality (2.12) to deduce

$$L - \epsilon \leq \lim_{n\to\infty} a_n^{\frac{1}{n}} \leq L + \epsilon.$$

Since ϵ is arbitrary, we conclude that

$$\lim_{n\to\infty} a_n^{\frac{1}{n}} = L$$

as desired.

Next, if $L = \infty$, then for every $M > 0$, there exists an $N \in \mathbb{N}$ such that $k \geq N$ implies

$$\frac{a_{k+1}}{a_k} > M.$$

Using similar argument as above, it gives

$$a_{n+1} > a_N \cdot M^{n+1-N}$$
$$a_{n+1}^{\frac{1}{n+1}} > a_N^{\frac{1}{n+1}} \cdot M^{1-\frac{N}{n+1}}$$

which shows that

$$\lim_{n\to\infty} a_{n+1}^{\frac{1}{n+1}} > M.$$

Since M is arbitrary large, we conclude that

$$\lim_{n\to\infty} a_n^{\frac{1}{n}} = \infty.$$

(b) Let $a_n = \frac{1}{n!} > 0$. Then we have

$$\lim_{n\to\infty} \frac{a_{n+1}}{a_n} = \lim_{n\to\infty} \frac{1}{n+1} = 0.$$

Hence the required result follows immediately from part (a).

This ends the proof of the problem. $\blacksquare$

> ### Problem 2.14
>
> Bak and Newman Chapter 2 Exercise 14.

Proof.

(a) Take $a_n = \frac{(-1)^n}{n!}$. Then we have

$$\lim_{n\to\infty} \frac{a_{n+1}}{a_n} = \lim_{n\to\infty} \frac{-1}{n+1} = 0.$$

By Problem 2.13(a) and Theorem 2.8, it can be seen easily that $R = \frac{1}{L} = \infty$.

(b) Suppose that

$$f(\omega) = \sum_{n=0}^{\infty} \frac{\omega^{n+1}}{(2n+1)!} \quad \text{and} \quad g(z) = \sum_{n=0}^{\infty} \frac{z^{2n+1}}{(2n+1)!}.$$

Then it is obvious that $g(z) = f(z^2)$. Take $a_n = \frac{1}{(2n+1)!}$. Since

$$\lim_{n\to\infty} \frac{a_{n+1}}{a_n} = \lim_{n\to\infty} \frac{(2n+1)!}{(2n+3)!} = 0,$$

Problem 2.13(a) asserts that $\lim_{n\to\infty} a_n^{\frac{1}{n}} = 0$ and then Theorem 2.8 ensures that

$$\sum_{n=0}^{\infty} \frac{\omega^{n+1}}{(2n+1)!}$$

converges for $|\omega| < \infty$. Therefore, the relation $g(z) = f(z^2)$ implies that the radius of convergence of the power series of g is also ∞.

(c) Let $a_n = \frac{n!}{n^n}$. Using [12, §215, Eqn. (1)] or [27, Eqn. (5.24), p. 61], we find that

$$\lim_{n\to\infty} \frac{a_{n+1}}{a_n} = \lim_{n\to\infty} \frac{(n+1)!}{(n+1)^{n+1}} \times \frac{n^n}{n!} = \lim_{n\to\infty} \left(\frac{n}{n+1}\right)^n = e^{-1}.$$

By Problem 2.13(a) and Theorem 2.8, we conclude that $R = e$.

(d) Let $a_n = \frac{2^n}{n!}$. Then

$$\lim_{n\to\infty} \frac{a_{n+1}}{a_n} = \lim_{n\to\infty} \frac{2^{n+1}}{(n+1)!} \times \frac{n!}{2^n} = \lim_{n\to\infty} \frac{2}{n+1} = 0.$$

Thus Problem 2.13(a) and Theorem 2.8 yield that $R = \infty$.

This completes the proof of the problem.   ■

Problem 2.15

*Bak and Newman Chapter 2 Exercise 15.**

Proof.

(a) Let $x \in \mathbb{R}$ be such that $|\sin x| \le \frac{1}{2}$. We claim that $|\sin(x+2)| > \frac{1}{2}$. To see this, the condition $|\sin x| \le \frac{1}{2}$ implies that there is a $k \in \mathbb{Z}$ such that

$$-\frac{\pi}{6} \le x - k\pi \le \frac{\pi}{6}.$$

Since $3 < \pi < 4$, we have $\frac{\pi}{2} < 2 < \frac{2\pi}{3}$ so that

$$\frac{\pi}{3} = -\frac{\pi}{6} + \frac{\pi}{2} < x + 2 - k\pi < \frac{\pi}{6} + \frac{2\pi}{3} = \frac{5\pi}{6}.$$

Consequently, we have the claim that

$$|\sin(x+2)| > \frac{1}{2}. \tag{2.13}$$

Next, since $0 \leq |\sin n| \leq 1$ for every $n \in \mathbb{N}$, we certainly have

$$|\sin n|^{\frac{1}{n}} \leq 1 \tag{2.14}$$

for every $n \in \mathbb{N}$. Now we claim that given $\epsilon > 0$, one can find a positive integer N such that

$$|\sin N|^{\frac{1}{N}} > 1 - \epsilon. \tag{2.15}$$

To prove this claim, we first notice that $\lim\limits_{n \to \infty} 2^{-\frac{1}{n}} = 1$, so there exists an $N' \in \mathbb{N}$ such that $n \geq N'$ implies

$$2^{-\frac{1}{n}} > 1 - \epsilon. \tag{2.16}$$

On the one hand, if $|\sin N'| > \frac{1}{2}$, then we get from the inequality (2.16) that

$$|\sin N'|^{\frac{1}{N'}} > 2^{-\frac{1}{N'}} > 1 - \epsilon.$$

On the other hand, if $|\sin N'| \leq \frac{1}{2}$, then the previous claim (2.13) and the inequality (2.16) imply that

$$|\sin(N' + 2)|^{\frac{1}{N'+2}} > 2^{-\frac{1}{N'+2}} > 1 - \epsilon.$$

Hence they mean that our claim (2.15) is true or equivalently, there exists a sequence $\{n_k\}$ of positive integers such that

$$\lim_{k \to \infty} |\sin n_k|^{\frac{1}{n_k}} \geq 1. \tag{2.17}$$

Finally, the estimates (2.14) and (2.17) guarantee that

$$\limsup_{n \to \infty} |\sin n|^{\frac{1}{n}} = 1.$$

By Theorem 2.8, it concludes that $R = 1$.

(b) By Theorem 2.8, we have

$$L = \limsup_{n \to \infty} (e^{-n^2})^{\frac{1}{n}} = \lim_{n \to \infty} e^{-n} = 0$$

so that $R = \infty$.

We complete the proof of the problem. ◼

Problem 2.16

*Bak and Newman Chapter 2 Exercise 16.**

Proof. By the definition, we have

$$\lim_{k \to \infty} c_{2k}^{\frac{1}{2k}} = \lim_{k \to \infty} (2^k)^{\frac{1}{2k}} = \lim_{k \to \infty} \sqrt{2} = \sqrt{2}$$

and

$$\lim_{k \to \infty} c_{2k-1}^{\frac{1}{2k-1}} = \lim_{k \to \infty} \left[\left(1 + \frac{1}{k}\right)^{k^2} \right]^{\frac{1}{2k-1}} = \lim_{k \to \infty} \left[\left(1 + \frac{1}{k}\right)^{k} \right]^{\frac{k}{2k-1}} = \sqrt{e}.$$

Since $e > 2$, we establish from Theorem 2.8 that

$$R = \frac{1}{\sqrt{e}}.$$

This completes the analysis of the problem. ◼

Bak and Newman Chapter 2 Exercise 17.

Proof. The following proof is basically adopted from [22, Theorem 3.50, pp. 74, 75]. Suppose that $\beta_n = B_n - B$ for all $n = 0, 1, 2, \ldots$. It is clear that

$$
\begin{aligned}
C_n &= \sum_{k=0}^{n} c_k \\
&= a_0 b_0 + (a_0 b_1 + a_1 b_0) + \cdots + (a_0 b_n + a_1 b_{n-1} + \cdots + a_n b_0) \\
&= a_0 B_n + a_1 B_{n-1} + \cdots + a_n B_0 \\
&= a_0 (B + \beta_n) + a_1 (B + \beta_{n-1}) + \cdots + a_n (B + \beta_0) \\
&= (a_0 + a_1 + \cdots + a_n) B + a_0 \beta_n + a_1 \beta_{n-1} + \cdots + a_n \beta_0 \\
&= A_n B + a_0 \beta_n + a_1 \beta_{n-1} + \cdots + a_n \beta_0.
\end{aligned}
\tag{2.18}
$$

Since $A_n B \to AB$ as $n \to \infty$, it follows from the expression (2.18) that $C_n \to AB$ if and only if

$$
\lim_{n \to \infty} (a_0 \beta_n + a_1 \beta_{n-1} + \cdots + a_n \beta_0) = 0.
\tag{2.19}
$$

Since $\sum_{k=0}^{\infty} a_k$ converges absolutely, $\sum_{k=0}^{\infty} |a_k|$ is finite and we denote this number by α. Given $\epsilon > 0$. Since $\beta_n \to 0$ as $n \to \infty$, there exists an $N \in \mathbb{N}$ such that $n \geq N$ implies $|\beta_n| < \epsilon$. Therefore, the triangle inequality shows that

$$
\begin{aligned}
|a_0 \beta_n + a_1 \beta_{n-1} + \cdots + a_n \beta_0| &\leq |a_0 \beta_n + \cdots + a_{n-(N+1)} \beta_{N+1}| + |a_{n-N} \beta_N + \cdots + a_n \beta_0| \\
&\leq |a_{n-N} \beta_N + \cdots + a_n \beta_0| + (|a_0| \cdot |\beta_n| + \cdots + |a_{n-(N+1)}| \cdot |\beta_{N+1}|) \\
&< |a_{n-N} \beta_N + \cdots + a_n \beta_0| + \epsilon(|a_0| + |a_1| + \cdots + |a_{n-(N+1)}|) \\
&\leq |a_{n-N} \beta_N + \cdots + a_n \beta_0| + \epsilon \alpha.
\end{aligned}
\tag{2.20}
$$

Since $a_n \to 0$ as $n \to \infty$, we fix N in the inequality (2.20) and then take $n \to \infty$ to establish

$$
\limsup_{n \to \infty} |a_0 \beta_n + a_1 \beta_{n-1} + \cdots + a_n \beta_0| \leq \epsilon \alpha.
$$

Since ϵ is arbitrary, this proves the desired result (2.19) and thus

$$
\lim_{n \to \infty} C_n = AB.
$$

This completes the proof of the problem. $\blacksquare$

Bak and Newman Chapter 2 Exercise 18.

Proof. If the complex number z satisfies $|z| < R_1$, then the power series

$$
\sum_{n=0}^{\infty} |a_n| \cdot |z^n|
$$

converges. Similarly, if $|z| < R_2$, then the power series

$$\sum_{n=0}^{\infty} |b_n| \cdot |z^n|$$

converges. Therefore, both series converges within $|z| < \min(R_1, R_2)$.

Clearly, by replacing a_n and b_n by $a_n z^n$ and $b_n z^n$ respectively in Problem 2.17, we get

$$\sum_{j=0}^{k}(a_j z^j)(b_{k-j} z^{k-j}) = \sum_{j=0}^{k} a_j b_{k-j} \cdot z^k = c_k z^k$$

and then Problem 2.17 ensures that

$$\sum_{n=0}^{\infty} c_n z^n$$

converges within $|z| < \min(R_1, R_2)$. We have completed the proof of the problem.

> **Problem 2.19**
>
> Bak and Newman Chapter 2 Exercise 19.

Proof.

(a) Suppose that $|z| < 1$. Apply the given identity, we have

$$\lim_{N \to \infty} (1 - z)(1 + z + z^2 + \cdots + z^N) = \lim_{N \to \infty} (1 - z^{N+1})$$

$$(1 - z) \sum_{n=0}^{\infty} z^n = 1$$

$$\sum_{n=0}^{\infty} z^n = \frac{1}{1 - z}.$$

(b) Let $\displaystyle\sum_{n=0}^{\infty} c_n z^n$ be the Cauchy product of $\displaystyle\sum_{n=0}^{\infty} z^n$ with itself. By the definition[b], we have

$$c_n = \sum_{k=0}^{n}(1 \times 1) = n + 1.$$

Combining this fact and part (a), we see that

$$\frac{1}{(1 - z)^2} = \left(\sum_{n=0}^{\infty} z^n\right)\left(\sum_{n=0}^{\infty} z^n\right) = \sum_{n=0}^{\infty}(n + 1)z^n = \sum_{n=0}^{\infty} n z^n + \frac{1}{1 - z}$$

which asserts that

$$\sum_{n=0}^{\infty} n z^n = \frac{1}{(1 - z)^2} - \frac{1}{1 - z} = \frac{z}{(1 - z)^2}.$$

Hence we complete the proof of the problem.

---

[b] On p. 28.

Problem 2.20

Bak and Newman Chapter 2 Exercise 20.

Proof. Let the power series under consideration is

$$\sum_{n=0}^{\infty} C_n z^n. \tag{2.21}$$

Recall from [1, p. 53] that x is an accumulation point of the set S if and only if every neighborhood of x contains infinitely many points of S. Particularly, since S has an accumulation point at 0, each neighborhood $D(0; \frac{1}{n})$ contains a point $z_n \neq 0$ of S. Thus we have

$$|z_n| < \frac{1}{n}$$

which means that $\{z_n\}$ converges to 0. By Theorem 2.12 (The Uniqueness Theorem for Power Series), we conclude that the power series (2.21) is identically zero. This completes the proof of the problem. $\blacksquare$

Problem 2.21

Bak and Newman Chapter 2 Exercise 21.

Proof. Consider the power series $g(z) = \displaystyle\sum_{n=0}^{\infty} D_n z^n$, where

$$D_n = \begin{cases} 1, & \text{if } n = 0; \\ 0, & \text{if } n \in \mathbb{N}. \end{cases}$$

Since $f(\frac{1}{n}) = g(\frac{1}{n})$ for every $n = 2, 3, \ldots$ and $\frac{1}{n} \to 0$ as $n \to \infty$, we deduce from Corollary 2.14 that $C_n = D_n$ for all $n = 0, 1, 2, \ldots$. In other words, we have $f(z) = 1$ and then

$$f'(z) \equiv 0,$$

so $f'(0) > 0$ is impossible. Hence we have completed the proof of the problem. $\blacksquare$

Problem 2.22

Bak and Newman Chapter 2 Exercise 22.

Proof. We write

$$g(z) = f(z + \alpha) = \sum_{n=0}^{\infty} C_n z^n.$$

Since $\limsup_{n \to \infty} |C_n|^{\frac{1}{n}} < \infty$, Theorem 2.8 ensures that the radius of convergence of g is nonzero. Hence it follows from Corollary 2.11 that

$$C_n = \frac{g^{(n)}(0)}{n!} = \frac{f^{(n)}(\alpha)}{n!}$$

for all $n \in \mathbb{N} \cup \{0\}$, completing the proof of the problem. $\blacksquare$

Problem 2.23

Bak and Newman Chapter 2 Exercise 23.

Proof.

(a) By Problem 2.8, we have $\limsup\limits_{n\to\infty} n^{\frac{1}{n}} = 1$ so that the comment following Corollary 2.14 on p. 32 implies that the series converges throughout $|z - 1| < 1$.

(b) As we have shown in Problem 2.14(a) that the radius of convergence of the series there is ∞, so the comment following Corollary 2.14 on p. 32 implies that the radius of convergence of

$$\sum_{n=0}^{\infty} \frac{(-1)^n}{n!}(z+1)^n$$

is also ∞.

(c) Note that

$$\sum_{n=0}^{\infty} n^2(2z-1)^n = \sum_{n=0}^{\infty} 2^n n^2 \left(z - \frac{1}{2}\right)^n. \tag{2.22}$$

By Problem 2.8, we see that

$$\limsup_{n\to\infty}(2^n n^2)^{\frac{1}{n}} = \lim_{n\to\infty} 2n^{\frac{2}{n}} = 2.$$

Again the comment following Corollary 2.14 on p. 32 implies that the series (2.22) converges throughout $|z - \frac{1}{2}| < \frac{1}{2}$.

This completes the proof of the problem. $\blacksquare$

CHAPTER 3

Analytic Functions

> **Problem 3.1**
>
> *Bak and Newman Chapter 3 Exercise 1.*

Proof. Notice that $f(z) = f(x, y)$ and $z = x + iy$, so it is easy to see that

$$f_x = \lim_{\substack{h \to 0 \\ h \in \mathbb{R}}} \frac{f(x + h, y) - f(x, y)}{h} = \lim_{\substack{h \to 0 \\ h \in \mathbb{R}}} \frac{f(z + h) - f(z)}{h}$$

and

$$f_y = \lim_{\substack{h \to 0 \\ h \in \mathbb{R}}} \frac{f(x, y + h) - f(x, y)}{h} = \lim_{\substack{h \to 0 \\ h \in \mathbb{R}}} \frac{f(z + ih) - f(z)}{h}.$$

This completes the proof of the problem. ∎

> **Problem 3.2**
>
> *Bak and Newman Chapter 3 Exercise 2.*

Proof.

(a) Obviously, $f_x = 2x$ and $f_y = 2iy$ exist in a neighborhood of any point on the line $y = x$ and furthermore, they are continuous at all points on the line $y = x$. Since $f_y = if_x$ if and only if $y = x$, Proposition 3.2 asserts that f is differentiable at all points on the line $y = x$.

(b) By Definition 3.1, if f was analytic at $z = (x, y)$, then f is differentiable in a neighborhood of $z = (x, y)$ which is prohibited by part (a).

Hence we have completed the proof of the problem. ∎

> **Problem 3.3**
>
> *Bak and Newman Chapter 3 Exercise 3.*

Proof. If g is differentiable at w, then Definition 2.4 implies

$$g(s) - g(w) = (s - w)[g'(w) + \epsilon(s)],$$

where $\epsilon(s) \to 0$ as $s \to w$. Therefore, we get

$$g(f(z + h)) - g(f(z)) = [f(z + h) - f(z)] \cdot [g'(f(z)) + \epsilon(h)], \tag{3.1}$$

where $\epsilon(h) \to 0$ as $h \to 0$. By dividing both sides of the expression (3.1) by h, we see that

$$\frac{g(f(z + h)) - g(f(z))}{h} = [g'(f(z)) + \epsilon(h)] \times \frac{f(z + h) - f(z)}{h}$$

and thus

$$\begin{aligned}
(g \circ f)'(z) &= \lim_{h \to 0} \frac{g(f(z + h)) - g(f(z))}{h} \\
&= \lim_{h \to 0} [g'(f(z)) + \epsilon(h)] \times \lim_{h \to 0} \frac{f(z + h) - f(z)}{h} \\
&= g'(f(z)) \cdot f'(z).
\end{aligned}$$

This completes the analysis of the problem.

> **Remark 3.1**
>
> Problem 3.3 is the Chain Rule for the complex variable case.

> **Problem 3.4**
>
> *Bak and Newman Chapter 3 Exercise 4.*

Proof. Since $g^2(z) = z$, if $z_0 \in \mathbb{C}$, then we have

$$\frac{g^2(z) - g^2(z_0)}{z - z_0} = 1$$

which gives

$$\frac{g(z) - g(z_0)}{z - z_0} = \frac{1}{g(z) + g(z_0)}. \tag{3.2}$$

Since g is continuous at $\sqrt{z_0}$, it deduces from the expression (3.2) that

$$g'(z_0) = \lim_{z \to z_0} \frac{g(z) - g(z_0)}{z - z_0} = \lim_{z \to z_0} \frac{1}{g(z) + g(z_0)} = \lim_{z \to z_0} \frac{1}{\sqrt{[g(z)]^2} + g(z_0)} = \frac{1}{2g(z_0)} = \frac{1}{2\sqrt{z_0}}.$$

Since z_0 is arbitrary, we obtain our expected result and we have completed the proof of the problem.

> **Problem 3.5**
>
> *Bak and Newman Chapter 3 Exercise 5.*

Proof. Suppose that f is analytic in the region $D \subseteq \mathbb{C}$. By Definition 3.3, f is differentiable in D. Since $f' \equiv 0$ in D, Proposition 3.1 implies that $\frac{f_y}{i} = f_x = f' \equiv 0$ in D, i.e.,

$$f_x \equiv f_y \equiv 0 \qquad (3.3)$$

in D. If $f = u + iv$, then the identities (3.3) mean that

$$u_x \equiv u_y \equiv v_x \equiv v_y \equiv 0$$

in D. Therefore, we follow from Theorem 1.10 that both u and v are constant in D. In particular, f is constant in D. This completes the proof of the problem. $\blacksquare$

Problem 3.6

Bak and Newman Chapter 3 Exercise 6.

Proof. Suppose that f is analytic in the region $D \subseteq \mathbb{C}$. By Definition 3.3, f is differentiable in D. By Problem 3.3, we establish that f^2 is also differentiable in D so that Definition 3.3 again implies that f^2 is analytic in D. Furthermore, we get from Problem 3.3 that

$$[f^2(z)]' = 2f(z)f'(z) \equiv 0$$

in D. By Problem 3.5, $[f(z)]^2$ is constant. Let $[f(z)]^2 = A$. Then we take the modulus to both sides to get

$$|f(z)|^2 = |A|$$

in D. In other words, $|f|$ is constant in D. By Proposition 3.7, f is constant in D, completing the proof of the problem. $\blacksquare$

Problem 3.7

Bak and Newman Chapter 3 Exercise 7.

Proof. Let $f : D \to \mathbb{C}$ be a nonconstant analytic function in the region D. Let $f = u + iv$. If f mapped D into a straight line $y = ax + b$ with $a, b \in \mathbb{R}$, then the point $f(x + iy)$ as a point in $\mathbb{C}$ (or in $\mathbb{R}^2$), we read

$$v(x, y) = au(x, y) + b$$

in D. Since f is analytic in D, it is differentiable there and Proposition 3.1 says that

$$u_x = v_y = au_y \quad \text{and} \quad u_y = -v_x = -au_x \qquad (3.4)$$

in D. These two equations (3.4) imply that

$$(1 + a^2)u_x = 0$$

in D. Since a is real, we have $u_x = 0$ and then the second equation (3.4) gives $u_y = 0$ in D. By Theorem 1.10, u is constant so that Proposition 3.6 implies that f is constant which contradicts to the hypothesis.

Next, assume that f mapped D into a circular arc of the circle $|w - w_0| = r$ centred at $w_0 \in \mathbb{C}$ and radius $r > 0$, i.e., $|f(z) - w_0| = r$. Consider the map $g(z) = f(z) - w_0$ which is also nonconstant analytic in D. Now

$$|g(z)| = |f(z) - w_0| = r$$

in D. Hence Proposition 3.7 ensures that g and then f are both constants, a contradiction again. This completes the proof of the problem. $\blacksquare$

Problem 3.8

Bak and Newman Chapter 3 Exercise 8.

Proof. Suppose that $f(x, y) = (x^2 - y^2) + iv$ is analytic at z. Then it must be differentiable in a neighborhood $D(z; r)$ of z and the Cauchy-Riemann equations give $u_x = v_y$ and $u_y = -v_x$ in $D(z; r)$. Since $u_x = 2x$ and $u_y = -2y$, we have $v_x = 2y$ and $v_y = 2x$ so that

$$v(x, y) = 2xy + g(y) \quad \text{and} \quad v(x, y) = 2xy + h(x) \tag{3.5}$$

in $D(z; r)$. Therefore, the equations (3.5) imply that $h(x) = g(y) = C$ for some constant C in $D(z; r)$. In conclusion, f must take the form

$$f(z) = f(x, y) = (x^2 - y^2) + i(2xy + C) = (x + iy)^2 + iC = z^2 + iC$$

for some constant C. This completes the proof of the problem.

Problem 3.9

Bak and Newman Chapter 3 Exercise 9.

Proof. Assume that f was an analytic function satisfying the requirement. By Proposition 3.1, it satisfies the Cauchy-Riemann equations, i.e., $v_y = u_x = 2x$ and $v_x = -u_y = -2y$. Therefore, we have

$$v(x, y) = 2xy + F(x) \quad \text{and} \quad v(x, y) = -2xy + G(y)$$

for some functions $F(x)$ and $G(y)$. Clearly, they imply that

$$F(x) - G(y) = -4xy. \tag{3.6}$$

Put $y = 0$ in the equation (3.6) to get

$$F(x) = G(0)$$

for all x in the domain of F. In other words, $F(x)$ is a constant function. Similarly, it can be shown that $G(y)$ is also a constant function. However, the equation (3.6) means that $-4xy$ is a constant *for all* x and y, a contradiction and then no such analytic function exits. This completes the proof of the problem.

Problem 3.10

Bak and Newman Chapter 3 Exercise 10.

Proof. Since f is entire, Proposition 3.1 implies that

$$u_x = v_y. \tag{3.7}$$

By the given form, u_x and v_y are functions of x and y respectively. These facts and the equation (3.7) force that $u_x = v_y = A$ for some real constant A. Then simple integration gives

$$u(x) = Ax + B \quad \text{and} \quad v(y) = Ay + C$$

for some $B, C \in \mathbb{R}$. Consequently, we have

$$f(z) = Ax + B + i(Ay + C) = A(x + iy) + B + iC = Az + B + iC.$$

This completes the proof of the problem.

> **Problem 3.11**
>
> *Bak and Newman Chapter 3 Exercise 11.*

Proof. By the definition, we have $e^z = e^x \cos y + i e^x \sin y$ so that $u = e^x \cos y$ and $v = e^x \sin y$.

(a) Direct computation gives $u_x = e^x \cos y$, $u_y = -e^x \sin y$, $v_x = e^x \sin y$ and $v_y = e^x \cos y$. It is clear that all u_x, u_y, v_x and v_y exist and continuous everywhere. Furthermore, since $u_x = v_y$ and $u_y = -v_x$ hold everywhere, Proposition 3.2 ensures that e^z is differentiable everywhere, i.e., e^z is entire.

(b) If $z_1 = a + ib$ and $z_2 = c + id$, then we see that

$$
\begin{aligned}
e^{z_1 + z_2} &= e^{(a+c) + i(b+d)} \\
&= e^{a+c}[\cos(b+d) + i\sin(b+d)] \\
&= e^{a+c}[\cos b \cos d - \sin b \sin d + i(\sin b \cos d + \cos b \sin d)] \\
&= e^{a+c}(\cos b + i\sin b) \times (\cos d + i\sin d) \\
&= e^{z_1} e^{z_2}.
\end{aligned}
$$

This completes the proof of the problem. ∎

> **Problem 3.12**
>
> *Bak and Newman Chapter 3 Exercise 12.*

Proof. Let $z = x + iy$. Then it follows from the definition of the modulus of z (see p. 5) that

$$
|e^z| = |e^{x+iy}| = |e^x| \cdot |e^{iy}| = e^x \cdot |\cos y + i\sin y| = e^x \cdot \sqrt{\cos^2 t + \sin^2 t} = e^x,
$$

completing the proof of the problem. ∎

> **Problem 3.13**
>
> *Bak and Newman Chapter 3 Exercise 13.*

Proof. When $z = x + iy$, we notice that

$$
e^z = e^x e^{iy}. \tag{3.8}
$$

Suppose that z is a ray lying in the right half-plane $\{z \in \mathbb{C} \mid \operatorname{Re}(z) = x > 0\}$, i.e., $x \to +\infty$ and $y \to L$ for some finite L or $L = \pm\infty$. Then it follows from the expression (3.8) that

$$
|e^z| = e^x \to \infty.
$$

Similarly, if z is a ray lying in the left half-plane $\{z \in \mathbb{C} \mid \operatorname{Re}(z) = x < 0\}$, i.e., $x \to -\infty$ and $y \to L$ for some finite L or $L = \pm\infty$. Then the expression (3.8) implies that

$$
|e^z| = e^x \to 0.
$$

Finally, if the ray lies on the imaginary axis, then $x = 0$ and $z = iy$ so that

$$
|e^z| = |e^{iy}| = 1.
$$

Since $y \to +\infty$ or $y \to -\infty$, e^z moves along the unit circle anticlockwise and clockwise infinitely many times respectively. We have completed the proof of the problem. ∎

Problem 3.14

Bak and Newman Chapter 3 Exercise 14.

Proof.

(a) Since $e^0 = 1$, $e^z = e^0$ implies that $z = 2k\pi i$ for all $k \in \mathbb{Z}$.

(b) Since $e^{\frac{\pi}{2}i} = i$, $e^z = e^{\frac{\pi}{2}i}$ implies that $z = \frac{\pi i}{2} + 2k\pi i = (\frac{\pi}{2} + 2k\pi)i$ for all $k \in \mathbb{Z}$.

(c) If we write $z = x + iy$, then $e^z = -3$ will give $e^x e^{iy} = 3e^{\pi i}$. Then its modulus implies that $e^x = 3$ or equivalently, $x = \ln 3$. In addition, $e^{iy} = e^{\pi i}$ leads to us that $y = \pi + 2k\pi$ for all $k \in \mathbb{Z}$. Hence the solutions of the equation have the form

$$z = x + iy = \ln 3 + (2k+1)\pi i,$$

where $k \in \mathbb{Z}$.

(d) Similar to part (c), if we write $z = x + iy$, then $e^z = 1 + i$ implies that

$$e^x e^{iy} = \sqrt{2} \cdot \left(\frac{1}{\sqrt{2}} + \frac{i}{\sqrt{2}} \right) = \sqrt{2} e^{\frac{\pi}{4}i}.$$

Thus $e^x = \sqrt{2}$ and then $x = \ln\sqrt{2} = \frac{1}{2}\ln 2$. Besides, it is easy to see that $y = \frac{\pi}{4} + 2k\pi$ for all $k \in \mathbb{Z}$. In conclusion, the solutions of the equation take the form

$$z = x + iy = \frac{1}{2}\ln 2 + \left(2k + \frac{1}{4} \right)\pi i,$$

where $k \in \mathbb{Z}$.

This completes the proof of the problem. $\blacksquare$

Problem 3.15

Bak and Newman Chapter 3 Exercise 15.

Proof.

(a) Recall the formulas of $\sin z$ and $\cos z$, we obtain immediately that

$$2\sin z \cos z = 2 \times \frac{1}{2i}(e^{iz} - e^{-iz}) \times \frac{1}{2}(e^{iz} + e^{-iz}) = \frac{1}{2i}(e^{2iz} - e^{-2iz}) = \sin 2z.$$

(b) Direct computation implies

$$\cos^2 z + \sin^2 z = \left[\frac{1}{2}(e^{iz} + e^{-iz}) \right]^2 + \left[\frac{1}{2i}(e^{iz} - e^{-iz}) \right]^2$$
$$= \frac{1}{4}(e^{2iz} + 2 + e^{-2iz}) - \frac{1}{4}(e^{2iz} - 2 + e^{-2iz})$$
$$= 1.$$

(c) Using the property $(e^z)' = e^z$ and Problem 3.3, we have

$$(\sin z)' = \frac{\mathrm{d}}{\mathrm{d}z}\left[\frac{1}{2i}(e^{iz} - e^{-iz}) \right] = \frac{1}{2i}(ie^{iz} + ie^{-z}) = \frac{1}{2}(e^{iz} + e^{-iz}) = \cos z.$$

We end the proof of the problem. $\blacksquare$

Proof. Recall the definitions of hyperbolic sine and cosine as follows:

$$\sinh x = \frac{e^x - e^{-x}}{2} \quad \text{and} \quad \cosh x = \frac{e^x + e^{-x}}{2}, \tag{3.9}$$

where $x \in \mathbb{R}$.

(a) By the definition of $\sin z$, we know that

$$\sin\left(\frac{\pi}{2} + iy\right) = \frac{1}{2i}[e^{i(\frac{\pi}{2}+iy)} - e^{-i(\frac{\pi}{2}+iy)}] = \frac{1}{2i}(e^{-y+i\frac{\pi}{2}} - e^{y-\frac{\pi}{2}i}) = \frac{1}{2}(e^y + e^{-y}) = \cosh y.$$

(b) Using the result of Problem 3.20 in advance, we have

$$
\begin{aligned}
|\sin z| &= \sqrt{(\sin x \cosh y)^2 + (\cos x \sinh y)^2} \\
&= \sqrt{\sin^2 x \cdot \left(\frac{e^{2y} + 2 + e^{-2y}}{4}\right) + \cos^2 x \cdot \left(\frac{e^{2y} - 2 + e^{-2y}}{4}\right)} \\
&= \sqrt{\frac{e^{2y} + 2 + e^{-2y} - 4\cos^2 x}{4}} \\
&= \sqrt{\left(\frac{e^y + e^{-y}}{2}\right)^2 - \cos^2 x} \\
&= \sqrt{\cosh^2 y - \cos^2 x}. \tag{3.10}
\end{aligned}
$$

On the left side or the right side of the square, we have $z = \pm(N + \frac{1}{2})\pi + iy$, where $-(N + \frac{1}{2})\pi \le y \le (N + \frac{1}{2})\pi$ so that $\cos \pm(N + \frac{1}{2})\pi = 0$ and then the expression (3.10) implies

$$|\sin z| = \cosh y.$$

By the A.M. $\ge$ G.M., we know that $\cosh y \ge 1$. Hence we have $|\sin z| \ge 1$ in this case.

Next, on the top side or the bottom side of the square, we see that $z = x \pm i(N + \frac{1}{2})\pi$, where $-(N + \frac{1}{2})\pi \le x \le (N + \frac{1}{2})\pi$. By the definition (3.9), $\cosh y$ is clearly an even increasing function of y, so we have $\cosh \pm(N + \frac{1}{2})\pi = \cosh(N + \frac{1}{2})\pi \ge \cosh \frac{\pi}{2} \ge 2$ and then

$$\cosh^2 y - \cos^2 x \ge 4 - 1 = 3.$$

Hence we deduce from the expression (3.10) that $|\sin z| \ge \sqrt{3} \ge 1$.

(c) If $y \to \pm\infty$, then it follows from the definition (3.9) that $\cosh y \to \infty$. Since $0 \le \cos^2 x \le 1$ for all $x \in \mathbb{R}$, we conclude from the expression (3.10) that $|\sin z| \to \infty$ as $\operatorname{Im} z = y \to \pm\infty$.

This completes the proof of the problem. ∎

Proof. Since $\cos z = \frac{1}{2}(e^{iz} + e^{-iz})$ and $\sin z = \frac{1}{2i}(e^{iz} - e^{-iz})$, we obtain from the Chain Rule that

$$\begin{aligned}
(\cos z)' &= \frac{1}{2}\left[\frac{d}{dz}(e^{iz}) + \frac{d}{dz}(e^{-iz})\right] \\
&= \frac{1}{2}\left[i\frac{d}{d(iz)}(e^{iz}) - i\frac{d}{d(-iz)}(e^{-iz})\right] \\
&= \frac{-1}{2i}(e^{iz} - e^{-iz}) \\
&= -\sin z,
\end{aligned}$$

completing the proof of the problem. ∎

Problem 3.18

Bak and Newman Chapter 3 Exercise 18.

Proof. By the definition of $\sin z$, we have

$$\frac{e^{iz} - e^{-iz}}{2i} = 2$$

$$e^{iz} - e^{-iz} = 4i. \tag{3.11}$$

Set $w = e^{iz}$. Then the equation (3.11) becomes $w^2 - 4iw - 1 = 0$ and then the quadratic formula [4, Eqn. (1), p. 4] gives

$$w = \frac{4i \pm \sqrt{(4i)^2 - 4 \times (-1)}}{2} = \frac{4i \pm 2\sqrt{3}i}{2} = (2 \pm \sqrt{3})i.$$

Thus $e^{iz} = (2 \pm \sqrt{3})i$ and then we get

$$\begin{aligned}
iz &= \ln(2 \pm \sqrt{3}) + \left(\frac{\pi}{2} + 2k\pi\right)i \\
z &= \left(\frac{\pi}{2} + 2k\pi\right) - i\ln(2 \pm \sqrt{3}),
\end{aligned}$$

where $k \in \mathbb{Z}$. This ends the proof of the problem. ∎

Problem 3.19

Bak and Newman Chapter 3 Exercise 19.[*]

Proof. Since $e^{e^z} = 1 = e^0$, we have

$$e^z = 2k\pi i, \tag{3.12}$$

where $k \in \mathbb{Z}$. If $k = 0$, then $e^z = 0$ which is impossible. Thus we suppose that $k \neq 0$.

If $k > 0$, then $2k\pi i = e^{\ln(2k\pi) + i\frac{\pi}{2}}$ and the equation (3.12) gives

$$e^z = e^{\ln(2k\pi) + i\frac{\pi}{2}}.$$

Consequently, we obtain

$$z = \ln(2k\pi) + i\frac{\pi}{2} + 2n\pi i = \ln(2k\pi) + i\left(2n\pi + \frac{\pi}{2}\right),$$

where $n \in \mathbb{Z}$.

Next, if $k < 0$, then we have $2k\pi i = (-2k\pi i)(-i) = e^{\ln(-2k\pi)+i\frac{3\pi}{2}}$. Similarly, the equation (3.12) becomes

$$e^z = e^{\ln(-2k\pi)+i\frac{3\pi}{2}}$$

which implies

$$z = \ln(-2k\pi) + i\frac{3\pi}{2} + 2n\pi i = \ln(-2k\pi) + i\left(2n\pi + \frac{3\pi}{2}\right),$$

where $n \in \mathbb{Z}$. Thus we complete the proof of the problem. $\blacksquare$

> **Problem 3.20**
>
> Bak and Newman Chapter 3 Exercise 20.

Proof. By the definition of $\sin z$, we have

$$\begin{aligned}
\sin(x + iy) &= \frac{e^{i(x+iy)} - e^{-i(x+iy)}}{2i} \\
&= \frac{e^{ix-y} - e^{-ix+y}}{2} \\
&= \frac{e^{-y}(\cos x + i\sin x) - e^{y}(\cos x - i\sin x)}{2i} \\
&= \frac{e^{y} + e^{-y}}{2i} \times (i\sin x) + \frac{e^{-y} - e^{y}}{2i}\cos x \\
&= \frac{e^{y} + e^{-y}}{2} \times (\sin x) + i \cdot \frac{e^{y} - e^{-y}}{2}\cos x \\
&= \sin x \cosh y + i\cos x \sinh y,
\end{aligned}$$

completing the proof of the problem. $\blacksquare$

> **Problem 3.21**
>
> Bak and Newman Chapter 3 Exercise 21.

Proof. Using similar attack as in the proof of Problem 2.14(a), the radius of convergence of the series of $f(z)$ is ∞. Furthermore, it is clear that the series converges absolutely, so Problem 2.17 says that

$$f(z)f(w) = \sum_{n=0}^{\infty}\frac{z^n}{n!}\sum_{n=0}^{\infty}\frac{w^n}{n!} = \sum_{n=0}^{\infty}\left[\sum_{k=0}^{n}\frac{z^k}{k!}\cdot\frac{w^{n-k}}{(n-k)!}\right] = \sum_{n=0}^{\infty}\frac{1}{n!}\left(\sum_{k=0}^{n}C_k^n z^k \cdot w^{n-k}\right). \tag{3.13}$$

By the binomial theorem, the bracket in the formula (3.13) is exactly $\frac{(z+w)^n}{n!}$, so the formula (3.13) becomes

$$f(z)f(w) = \sum_{n=0}^{\infty}\frac{(z + w)^n}{n!} = f(z + w). \tag{3.14}$$

Let $x \in \mathbb{R}$. By the power series representation of e^x [25, Definition 4.5.1, p. 90], it is easy to see that

$$f(x) = e^x. \tag{3.15}$$

Recall that the series converges absolutely so that rearrangement is permitted. Thus we have

$$f(iy) = 1 + iy - \frac{y^2}{2} - i\frac{y^3}{3!} + \cdots = \left(1 - \frac{y^2}{2} + \frac{y^4}{4} - \cdots\right) + i\left(y - \frac{y^3}{3!} + \cdots\right). \tag{3.16}$$

By the power series representation of $\sin y$ and $\cos y$ [25, p. 102]), the equation (3.16) can be written as

$$f(iy) = \cos y + i\sin y. \tag{3.17}$$

Hence we deduce immediately from the formulas (3.14), (3.15) and (3.17) that

$$f(z) = f(x + iy) = f(x)f(iy) = e^x(\cos y + i\sin y) = e^{x+iy} = e^z$$

as desired. This completes the proof of the problem. $\blacksquare$

Problem 3.22

Bak and Newman Chapter 3 Exercise 22.

Proof. By Problem 3.21, we have

$$e^{iz} = \sum_{n=0}^{\infty} \frac{(iz)^n}{n!} \tag{3.18}$$

and

$$e^{-iz} = \sum_{n=0}^{\infty} \frac{(-iz)^n}{n!}. \tag{3.19}$$

The difference of the two expressions (3.18) and (3.19) establishes

$$e^{iz} - e^{-iz} = 2i\left(z - \frac{z^3}{3!} + \frac{z^5}{5!} - \cdots\right) = 2ig(z).$$

Hence it follows from the definition of $\sin z$ that $g(z) = \sin z$, completing the proof of the problem. $\blacksquare$

Problem 3.23

Bak and Newman Chapter 3 Exercise 23.

Proof. Since $\sin z$ is entire, it follows from Theorem 2.9 and Problem 3.22 that

$$\cos z = (\sin z)' = g'(z) = 1 - \frac{z^2}{2!} + \frac{z^4}{4!} - \cdots.$$

We end the proof of the problem. $\blacksquare$

CHAPTER **4**

Line Integrals and Entire Functions

Problem 4.1

Bak and Newman Chapter 4 Exercise 1.

Proof. We say that $z \sim \omega$ if there exists an one-to-one C^1 mapping λ satisfying Definition 4.4. Now we are going to check the definition of an equivalence relation directly.

- **Reflexivity:** The identity $i : [a,b] \to [a,b]$ is clearly an one-to-one C^1 mapping, $i'(t) > 0$ and $z(i(t)) = z(t)$ for all $t \in [a,b]$. Thus we have $z \sim z$.

- **Symmetry:** Suppose that $z \sim \omega$. By the definition, there exists an one-to-one C^1 mapping $\lambda : [c,d] \to [a,b]$ such that $\lambda(c) = a$, $\lambda(d) = b$, $\lambda'(t) > 0$ and

$$\omega(t) = z(\lambda(t)) \tag{4.1}$$

for all $t \in [c,d]$.

We need a modified exercise from Rudin as follows:[a]

Suppose that $f'(x) > 0$ on $[a,b]$. Then f is strictly increasing in $[a,b]$ and if g denotes its inverse function, then g is differentiable on $[f(a), f(b)]$ and

$$g'(f(x)) = \frac{1}{f'(x)}$$

on $[a,b]$.

Since $\lambda'(t) > 0$ on $[c,d]$, Lemma 4.1 implies that its inverse function $\lambda^{-1} : [a,b] \to [c,d]$ is differentiable and

$$(\lambda^{-1})'(\lambda(t)) = \frac{1}{\lambda'(t)} > 0 \tag{4.2}$$

for all $t \in [c,d]$. Furthermore, the formula (4.2) ensures that λ^{-1} is also an one-to-one C^1 mapping on $[a,b]$.

---

[a] In fact, Lemma 4.1 is [22, Exercise 2, p. 114] with the open interval (a,b) replaced by closed interval $[a,b]$. A proof of this exercise can be found in [26, pp. 85, 86] and a proof of Lemma 4.1 can be modified from it once one-sided limits are considered.

To continue our proof, if $s = \lambda(t)$, then $t = \lambda^{-1}(s)$ and the expression (4.1) gives immediately that

$$z(s) = \omega(\lambda^{-1}(s))$$

which means that $\omega \sim z$.

- **Transitivity:** Suppose that $z \sim \omega$ and $\omega \sim \zeta$. Then there are one-to-one C^1 mappings $\lambda_1 : [c, d] \to [a, b]$ and $\lambda_2 : [e, f] \to [c, d]$ such that $\lambda_1(c) = a$, $\lambda_1(d) = b$, $\lambda_2(e) = c$, $\lambda_2(f) = d$, $\lambda_1'(t) > 0$ for all $t \in [c, d]$, $\lambda_2'(s) > 0$ for all $s \in [e, f]$ and

$$\omega(t) = z(\lambda_1(t)) \quad \text{and} \quad \zeta(s) = \omega(\lambda_2(s)). \tag{4.3}$$

Now the composition function $\lambda = \lambda_1 \circ \lambda_2 : [e, f] \to [a, b]$ satisfies $\lambda(e) = a$ and $\lambda(f) = b$. Furthermore, Problem 3.3 guarantees that

$$\lambda'(s) = (\lambda_1 \circ \lambda_2)'(s) = \lambda_1'(\lambda_2(s)) \cdot \lambda_2'(s) > 0$$

for all $s \in [e, f]$. Finally, the two equations (4.3) show that

$$z(\lambda(s)) = z(\lambda_1(\lambda_2(s))) = \omega(\lambda_2(s)) = \zeta(s)$$

for all $s \in [e, f]$. By Definition 4.4, $z \sim \zeta$.

Hence these prove that the concept of "smooth curves" is an equivalence relation and this completes the proof of the problem.

> **Problem 4.2**
>
> Bak and Newman Chapter 4 Exercise 2.

Proof. Note that $\dot{z}(t) = 2t + 2it$. By Definition 4.3, we see that

$$\begin{aligned}
\int_C f(z)\,\mathrm{d}z &= \int_0^1 f(t^2 + it^2)(2t + 2it)\,\mathrm{d}t \\
&= \int_0^1 (t^4 + it^4)(2t + 2it)\,\mathrm{d}t \\
&= 2(1 + i)^2 \int_0^1 t^5\,\mathrm{d}t \\
&= \frac{2i}{3},
\end{aligned}$$

completing the proof of the problem.

> **Problem 4.3**
>
> Bak and Newman Chapter 4 Exercise 3.

Proof. In this case, we have $\dot{z}(t) = \cos t - i \sin t$ so that Definition 4.3 asserts

$$\int_C f(z)\,\mathrm{d}z = \int_0^{2\pi} (\sin t - i \cos t)(\cos t - i \sin t)\,\mathrm{d}t = -i \int_0^{2\pi} \mathrm{d}t = -2\pi i. \tag{4.4}$$

Our result is different from that in Example 2 because our curve is in the opposite direction of that in Example 2.[b] We have ended the proof of the problem.

---

[b] We remark that the result (4.4) is an immediate consequence of Proposition 4.7.

> **Problem 4.4**
>
> *Bak and Newman Chapter 4 Exercise 4.*

Proof. We follow the given hint. Suppose that $f = u_1 + iv_1$, $g = u_2 + iv_2$ and $\alpha = A + iB$, where u_1, u_2, v_1, v_2 are real-valued functions and $A, B \in \mathbb{R}$. Suppose, further, that C is a smooth curve given by $z(t)$, where $a \leq t \leq b$. Then we follow from Definitions 4.1 and 4.3 that

$$
\begin{aligned}
\int_C [f(z) + g(z)] \, \mathrm{d}z &= \int_a^b [f(z(t)) + g(z(t))] \cdot \dot{z}(t) \, \mathrm{d}t \\
&= \int_a^b \left\{ [u_1(z(t)) + u_2(z(t))] \cdot \dot{z}(t) + i[v_1(z(t)) + v_2(z(t))] \cdot \dot{z}(t) \right\} \mathrm{d}t \\
&= \int_a^b [u_1(z(t)) + u_2(z(t))] \cdot \dot{z}(t) \, \mathrm{d}t + i \int_a^b [v_1(z(t)) + v_2(z(t))] \cdot \dot{z}(t) \, \mathrm{d}t \\
&= \left[\int_a^b u_1(z(t)) \cdot \dot{z}(t) \, \mathrm{d}t + i \int_a^b v_1(z(t)) \cdot \dot{z}(t) \, \mathrm{d}t \right] \\
&\quad + \left[\int_a^b u_2(z(t)) \cdot \dot{z}(t) \, \mathrm{d}t + i \int_a^b v_2(z(t)) \cdot \dot{z}(t) \, \mathrm{d}t \right] \\
&= \int_a^b [u_1(z(t)) + iv_1(z(t))] \cdot \dot{z}(t) \, \mathrm{d}t + \int_a^b [u_2(z(t)) + iv_2(z(t))] \cdot \dot{z}(t) \, \mathrm{d}t \\
&= \int_C f(z) \, \mathrm{d}z + \int_C g(z) \, \mathrm{d}z.
\end{aligned}
$$

Similarly, we have

$$
\begin{aligned}
\int_C \alpha f(z) \, \mathrm{d}z &= \int_a^b (A + iB)[u_1(z(t)) + iv_1(z(t))] \cdot \dot{z}(t) \, \mathrm{d}t \\
&= \int_a^b \left\{ [Au_1(z(t)) - Bv_1(z(t))] + i[Av_1(z(t)) + Bu_1(z(t))] \right\} \cdot \dot{z}(t) \, \mathrm{d}t \\
&= \int_a^b [Au_1(z(t)) - Bv_1(z(t))] \cdot \dot{z}(t) \, \mathrm{d}t + i \int_a^b [Av_1(z(t)) + Bu_1(z(t))] \cdot \dot{z}(t) \, \mathrm{d}t \\
&= (A + iB) \left[\int_a^b u_1(z(t)) \cdot \dot{z}(t) \, \mathrm{d}t + i \int_a^b v_1(z(t)) \cdot \dot{z}(t) \, \mathrm{d}t \right] \\
&= (A + iB) \int_a^b [u_1(z(t)) + iv_1(z(t))] \cdot \dot{z}(t) \, \mathrm{d}t \\
&= \alpha \int_C f(z) \, \mathrm{d}z.
\end{aligned}
$$

This completes the proof of the problem. $\blacksquare$

> **Problem 4.5**
>
> *Bak and Newman Chapter 4 Exercise 5.*

Proof. Fix $a \in \mathbb{C}$. Let $b \in \mathbb{C}$ with $a \neq b$ and C be a smooth curve connecting a and b. Furthermore, we let $z : [0, 1] \to \mathbb{C}$ be a parametrization such that $z(0) = a$ and $z(1) = b$. Since F is analytic on C, Proposition 4.12 implies that

$$
F(b) - F(a) = F(z(0)) - F(z(1)) = \int_C F'(z) \, \mathrm{d}z = 0.
$$

In other words, $F(b) = F(a)$. Since b is arbitrary, F is a constant, completing the proof of the problem.

> ### Problem 4.6
>
> Bak and Newman Chapter 4 Exercise 6.

Proof. It is clear that $z(t) = e^{it}$ is a parametrization of the smooth curve $|z| = 1$. By Definition 4.3, we know that

$$\int_{|z|=1} f(z)\,dz = \int_0^{2\pi} f(e^{it})ie^{it}\,dt.$$

By the hypothesis, $f(e^{it})ie^{it}$ is a continuous complex-valued function of t so that the proof of Lemma 4.9 gives

$$\int_{|z|=1} f(z)\,dz = Re^{i\theta}$$

for some $R \geq 0$ and $\theta \in [0, 2\pi]$.[c] Since R and $f(z)$ are real for all $|z| = 1$, we have

$$\left| \int_{|z|=1} f(z)\,dz \right| = R$$

$$= e^{-i\theta} \int_{|z|=1} f(z)\,dz$$

$$= \int_0^{2\pi} f(e^{it})ie^{i(t-\theta)}\,dt$$

$$= \operatorname{Re} \left\{ \int_0^{2\pi} f(e^{it})[-\sin(t-\theta) + i\cos(t-\theta)]\,dt \right\}$$

$$= -\int_0^{2\pi} f(e^{it})\sin(t-\theta)\,dt. \tag{4.5}$$

Since $-f(e^{it})\sin(t-\theta) \leq |f(e^{it})\sin(t-\theta)|$ for all $t \in [0, 2\pi]$, the equation (4.5) reduces to

$$\left| \int_{|z|=1} f(z)\,dz \right| \leq \int_0^{2\pi} |f(e^{it})\sin(t-\theta)|\,dt \leq \int_0^{2\pi} |\sin(t-\theta)|\,dt = \int_0^{2\pi} |\sin t|\,dt. \tag{4.6}$$

Direct computation shows

$$\int_0^{2\pi} |\sin t|\,dt = \int_0^{\pi} \sin t\,dt - \int_\pi^{2\pi} \sin t\,dt = 4.$$

We conclude from the inequality (4.6) that

$$\left| \int_{|z|=1} f(z)\,dz \right| \leq 4,$$

ending the proof of the problem.

> ### Problem 4.7
>
> Bak and Newman Chapter 4 Exercise 7.

---

[c] Since the integral is a complex number, we can express it in the polar form.

Proof. Suppose that $f(z) = \alpha + \beta z$, $a \le b$ and $c \le d$. Let $z_1 = a + ic$, $z_2 = b + ic$, $z_3 = b + id$ and $z_4 = a + id$. Now the line segment Γ_1 from z_1 to z_2 can be parameterized by

$$\gamma_1(t) = z_1 + t(z_2 - z_1)$$

for $t \in [0, 1]$. Then Definition 4.3 shows that

$$\int_{\Gamma_1} \mathrm{d}z = \int_0^1 1 \cdot (z_2 - z_1)\,\mathrm{d}t = z_2 - z_1 \tag{4.7}$$

and

$$\int_{\Gamma_1} z\,\mathrm{d}z = \int_0^1 [z_1 + t(z_2 - z_1)](z_2 - z_1)\,\mathrm{d}t = z_1(z_2 - z_1) + \frac{(z_2 - z_1)^2}{2} = \frac{z_2^2 - z_1^2}{2}. \tag{4.8}$$

Similarly, if Γ_2, Γ_3 and Γ_4 are the line segments from z_2 to z_3, z_3 to z_4 and z_4 to z_1 respectively, then we have

$$\int_{\Gamma_2} \mathrm{d}z = z_3 - z_2, \quad \int_{\Gamma_3} \mathrm{d}z = z_4 - z_3 \quad \text{and} \quad \int_{\Gamma_4} \mathrm{d}z = z_1 - z_4 \tag{4.9}$$

as well as

$$\int_{\Gamma_2} z\,\mathrm{d}z = \frac{z_3^2 - z_2^2}{2}, \quad \int_{\Gamma_3} z\,\mathrm{d}z = \frac{z_4^2 - z_3^2}{2} \quad \text{and} \quad \int_{\Gamma_4} z\,\mathrm{d}z = \frac{z_1^2 - z_4^2}{2}. \tag{4.10}$$

Since

$$\int_\Gamma f(z)\,\mathrm{d}z = \sum_{k=1}^4 \int_{\Gamma_k} f(z)\,\mathrm{d}z,$$

the integrals (4.7), (4.8), (4.9), (4.10) and Proposition 4.8 together imply that

$$\int_\Gamma f(z)\,\mathrm{d}z = \sum_{k=1}^4 \left[\alpha \int_{\Gamma_k} \mathrm{d}z + \beta \int_{\Gamma_k} z\,\mathrm{d}z \right] = \alpha \sum_{k=1}^4 \int_{\Gamma_k} \mathrm{d}z + \beta \sum_{k=1}^4 \int_{\Gamma_k} z\,\mathrm{d}z = 0$$

which completes the proof of the problem. ∎

Problem 4.8

Bak and Newman Chapter 4 Exercise 8.

Proof.

(a) Clearly, $\dot{z}(\theta) = Rie^{i\theta} \ne 0$ for all $t \in [0, 2\pi]$, so C is a smooth curve by Definition 4.2. Since $k \ne -1$, it must be true that

$$\left(\frac{z^{k+1}}{k+1} \right)' = z^k.$$

Since $\frac{z^{k+1}}{k+1}$ is analytic on the smooth curve C, Proposition 4.12 implies that

$$\int_C z^k\,\mathrm{d}z = \frac{(Re^{2\pi i})^{k+1}}{k+1} - \frac{(Re^0)^{k+1}}{k+1} = 0.$$

(b) By Definition 4.2, we get

$$\int_C f(z)\,\mathrm{d}z = \int_0^{2\pi} (Re^{i\theta})^k \cdot Rie^{i\theta}\,\mathrm{d}\theta = iR^{k+1} \int_0^{2\pi} e^{i(k+1)\theta}\,\mathrm{d}\theta = \frac{R^{k+1}}{k+1} e^{i(k+1)\theta} \Big|_0^{2\pi} = 0.$$

This completes the analysis of the problem. ∎

Problem 4.9

Bak and Newman Chapter 4 Exercise 9.

Proof.

(a) Let $f(z) = z - i$ and $F(z) = \frac{z^2}{2} - iz$. Since $\dot{z}(t) = 1 + 2it \neq 0$, C is a smooth curve. In addition, F is analytic on C and $F'(z) = f(z)$, so we apply Proposition 4.12 to conclude that

$$\int_C (z - i)\, dz = F(z(1)) - F(z(-1)) = \left[\frac{(1+i)^2}{2} - i(1+i)\right] - \left[\frac{(-1+i)^2}{2} - i(-1+i)\right] = 0.$$

(b) If Γ is the line segment from $-1 + i$ to $1 + i$ which is parameterized by

$$\gamma(t) = (-1 + i) + t[1 + i - (-1 + i)] = (2t - 1) + i$$

for $t \in [0, 1]$, then we have

$$\int_\Gamma (z - i)\, dz = \int_0^1 (2t - 1) \cdot 2\, dt = 0. \tag{4.11}$$

Since the sum $C + \Gamma$ is a smooth closed curve and $f(z) = z - i$ is entire, Theorem 4.16 (The Closed Curve Theorem) implies that

$$\int_C (z - i)\, dz + \int_\Gamma (z - i)\, dz = \int_{C+\Gamma} (z - i)\, dz = 0. \tag{4.12}$$

Combining the two results (4.11) and (4.12), we conclude that

$$\int_C (z - i)\, dz = 0.$$

We have completed the proof of the problem.

Problem 4.10

*Bak and Newman Chapter 4 Exercise 10.**

Proof.

(a) Recall that $(e^z)' = e^z$. If $z(t) = it$ for $t \in [0, 1]$ which is the parametrization of a smooth curve connecting 0 and i, then it follows from Proposition 4.12 that

$$\int_0^i e^z\, dz = e^{z(1)} - e^{z(0)} = e^i - 1.$$

(b) Recall that $\left(\frac{1}{2} \sin 2z\right)' = \cos 2z$. If $z(t) = \frac{\pi}{2} + it$ for $t \in [0, 1]$ which is the parametrization of a smooth curve connecting $\frac{\pi}{2}$ and $\frac{\pi}{2} + i$, then it follows from Proposition 4.12 that

$$\int_{\frac{\pi}{2}}^{\frac{\pi}{2}+i} \cos 2z\, dz = \frac{\sin 2z(1)}{2} - \frac{\sin 2z(0)}{2} = \frac{\sin(\pi + 2i)}{2} - \frac{\sin \pi}{2} = \frac{\sin(\pi + 2i)}{2}.$$

By Problem 3.20, we see that $\sin(\pi + 2i) = -i \sinh 2$, so we conclude that

$$\int_{\frac{\pi}{2}}^{\frac{\pi}{2}+i} \cos 2z\, dz = \frac{-i \sinh 2}{2}.$$

This completes the proof of the problem.

Problem 4.11

Bak and Newman Chapter 4 Exercise 11.

Proof. Let $a, b \in D$. Define the curve C connecting a and b by $z(t) = (1-t)a + tb$ for $t \in [0, 1]$. Since D is convex, we see that $C \subseteq D$. Thus Proposition 4.12 is applicable to f' along the curve C to get

$$f(b) - f(a) = \int_C f'(z)\,\mathrm{d}z. \tag{4.13}$$

Since $|f'(z)| \le 1$ on D and C is a line segment of length $b-a$, Theorem 4.10 (The *M-L* Formula) reduces the expression (4.13) to

$$|f(b) - f(a)| = \left| \int_C f'(z)\,\mathrm{d}z \right| \le 1 \cdot |b-a| = |b-a|$$

which is our desired result. This ends the proof of the problem. ∎

Problem 4.12

Bak and Newman Chapter 4 Exercise 12.

Proof. If $a = b$, then we have $\mathrm{e}^a - \mathrm{e}^b = 0 = a - b$. Therefore, we suppose that $a \ne b$ in the following discussion. Let $f(z) = \mathrm{e}^z$ and $D = \{z \in \mathbb{C} \,|\, \mathrm{Re}\, z < 0\}$. Then D is a convex region containing a and b and f is analytic in D. Furthermore, if $z = x + iy \in D$, then $x < 0$ so that

$$|f'(z)| = |\mathrm{e}^z| = |\mathrm{e}^x \cdot \mathrm{e}^{iy}| = \mathrm{e}^x \le 1. \tag{4.14}$$

Hence we may apply Problem 4.11 to conclude that

$$|\mathrm{e}^a - \mathrm{e}^b| \le |a - b|. \tag{4.15}$$

We note that the inequality in Problem 4.11 can be strict provided that we have $|f'(z)| < 1$ in D. Since $x < 0$, we know that the inequality (4.14) is in fact strict and this implies that the inequality (4.15) there is strict. We complete the proof of the problem. ∎

CHAPTER 5

Properties of Entire Functions

Problem 5.1

Bak and Newman Chapter 5 Exercise 1.

Proof. It is clear that f is entire. Since $f'(z) = 2z$, $f''(z) = 2$ and $f^{(k)}(z) = 0$ for all $k \geq 3$, it follows from Corollary 5.7 that

$$f(z) = 4 + 4(z - 2) + 2(z - 2)^2.$$

This completes the proof of the problem. ∎

Problem 5.2

Bak and Newman Chapter 5 Exercise 2.

Proof. It is obvious that f is entire. Since $f^{(k)}(z) = \mathrm{e}^z$ for all $k \geq 1$, it follows from Corollary 5.7 that

$$f(z) = \mathrm{e}^a + \mathrm{e}^a(z - a) + \frac{\mathrm{e}^a}{2!}(z - a)^2 + \cdots = \mathrm{e}^a \sum_{k=0}^{\infty} \frac{(z - a)^k}{k!},$$

completing the proof of the problem. ∎

Problem 5.3

Bak and Newman Chapter 5 Exercise 3.

Proof.

(a) Using Problem 3.3 repeatedly, we have

$$f^{(k)}(z) = (-1)^{k+1} f^{(k)}(-z)$$

for all $k \in \mathbb{N}$. In other words, the derivative of an odd (resp. even) function is an even (resp.) function. Furthermore, $f(z) = -f(-z)$ implies that $f(0) = 0$. These two facts combine to give

$$f^{(2n)}(0) = 0$$

55

for all $n = 0, 1, 2, \ldots$. Hence we obtain from Corollary 5.7 that

$$f(z) = f'(0)z + \frac{f^{(3)}(0)}{3!}z^3 + \cdots = \sum_{k=1}^{\infty} \frac{f^{(2k-1)}(0)}{(2k-1)!} z^{2k-1}.$$

(b) By the analysis in part (a), we have $f^{(2k-1)}(0) = 0$ for all $k \in \mathbb{N}$, so we follow again from Corollary 5.7 that

$$f(z) = f(0) + \frac{f''(0)}{2!}z^2 + \cdots = \sum_{k=1}^{\infty} \frac{f^{(2k)}(0)}{(2k)!} z^{2k}.$$

We have ended the proof of the problem. $\blacksquare$

Problem 5.4

Bak and Newman Chapter 5 Exercise 4.

Proof. Suppose that

$$f(z) = \sum_{k=0}^{\infty} C_k z^k.$$

On the one hand, using [4, Eqn. (1), p. 63], we have

$$C_k = \frac{1}{2\pi i} \int_C \frac{f(\omega)}{\omega^{k+1}} \, d\omega, \tag{5.1}$$

where $k = 0, 1, 2, \ldots$. On the other hand, Corollary 2.11 implies that

$$C_k = \frac{f^{(k)}(0)}{k!} \tag{5.2}$$

for all $k = 0, 1, 2, \ldots$. Hence our desired formula follows immediately by comparing the different expressions (5.1) and (5.2). We complete the proof of the problem. $\blacksquare$

Problem 5.5

Bak and Newman Chapter 5 Exercise 5.

Proof. Let $g(z) = f(z + a)$. Then g is entire because f is entire. It is easy to check that $g^{(k)}(z) = f^{(k)}(z+a)$ so that $g^{(k)}(0) = f^{(k)}(a)$ for all $k = 0, 1, 2, \ldots$. Now it follows from Problem 5.4 that

$$f^{(k)}(a) = g^{(k)}(0) = \frac{k!}{2\pi i} \int_C \frac{g(\zeta)}{\zeta^{k+1}} \, d\zeta = \frac{k!}{2\pi i} \int_C \frac{f(\zeta + a)}{\zeta^{k+1}} \, d\zeta, \tag{5.3}$$

where C is any circle surrounding the origin. If $\zeta(t) = Re^{it}$ for some $R > 0$ and $0 \le t \le 2\pi$, then Definition 4.3 shows that the integral in the expression (5.3) becomes

$$\int_C \frac{f(\zeta + a)}{\zeta^{k+1}} \, d\zeta = \int_0^{2\pi} \frac{f(Re^{it} + a)}{(Re^{it})^{k+1}} \cdot Rie^{it} \, dt. \tag{5.4}$$

Let $\omega = \zeta + a$. Then the locus of ω is the circle centred at a with radius R. Besides, we have $Re^{it} = \zeta(t) = \omega(t) - a$ and $\dot{\omega}(t) = Rie^{it}$. Therefore, we observe from the integral (5.4) that

$$
\begin{aligned}
\int_C \frac{f(\zeta + a)}{\zeta^{k+1}}\, d\zeta &= \int_0^{2\pi} \frac{f(Re^{it} + a)}{(Re^{it})^{k+1}} \cdot Rie^{it}\, dt \\
&= \int_0^{2\pi} \frac{f(\omega(t))}{[\omega(t) - a]^{k+1}} \dot{\omega}(t)\, dt \\
&= \int_{C'} \frac{f(\omega)}{(\omega - a)^{k+1}}\, d\omega,
\end{aligned}
\tag{5.5}
$$

where C' is the circle surrounding the point a. Combining the two formulas (5.3) and (5.5), we conclude that

$$
f^{(k)}(a) = \frac{k!}{2\pi i} \int_{C'} \frac{f(\omega)}{(\omega - a)^{k+1}}\, d\omega.
$$

This completes the proof of the problem.

Problem 5.6

Bak and Newman Chapter 5 Exercise 6.

Proof.

(a) By the expression (5.1), we have

$$
C_k = \frac{f^{(k)}(0)}{k!} = \frac{1}{2\pi i} \int_C \frac{f(\omega)}{\omega^{k+1}}\, d\omega,
$$

where C is the circle centred at the origin with radius R. Since f is entire, $\frac{f(\omega)}{\omega^{k+1}}$ is continuous on the smooth curve C. Now the length of C is $2\pi R$ and

$$
\left| \frac{f(\omega)}{\omega^{k+1}} \right| = \frac{|f(\omega)|}{R^{k+1}} \le \frac{M}{R^{k+1}}
$$

on C, so Theorem 4.10 (The M-L Formula) implies that

$$
|C_k| = \frac{1}{2\pi} \left| \int_C \frac{f(\omega)}{\omega^{k+1}}\, d\omega \right| \le \frac{1}{2\pi} \cdot \frac{M}{R^{k+1}} \cdot (2\pi) = \frac{M}{R^{k+1}},
$$

as desired.

(b) Apply part (a) with $R = M = 1$ to get $|C_k| \le 1$ for all $k = 0, 1, 2, \ldots$.

We have completed the proof of the problem.

Problem 5.7

Bak and Newman Chapter 5 Exercise 7.

Proof. Let $R > 0$. On the circle $|z| = R$, we have $|f(z)| \le A + BR^k$. By Problem 5.6(a), we know that

$$
|C_j| \le \frac{A + BR^k}{R^j},
\tag{5.6}
$$

where $j = 0, 1, 2, \ldots$. If $j > k$, then $j - k > 0$ and the inequality (5.6) implies that

$$\lim_{R \to +\infty} |C_j| \le \lim_{R \to +\infty} \left(\frac{A}{R^j} + \frac{B}{R^{j-k}} \right) = 0.$$

Hence $C_j = 0$ for all $j = k + 1, k + 2, \ldots$ and Theorem 5.11 (The Extended Liouville Theorem) follows, completing the proof of the problem. ▨

Problem 5.8

Bak and Newman Chapter 5 Exercise 8.

Proof. On the circle $|z| = R > 0$, the hypothesis implies that

$$|f(z)| \le A + BR^{\frac{3}{2}}. \tag{5.7}$$

Thus we deduce from the bound (5.7), Problem 5.6(a) and Theorem 5.5 (The Taylor Expansion of an Entire Function) that

$$|f^{(k)}(0)| = |k!C_k| \le \frac{k!(A + BR^{\frac{3}{2}})}{R^k}. \tag{5.8}$$

Notice that the inequality (5.8) holds for every $R > 0$, so if $k \ge 2$, then we conclude that

$$\lim_{R \to +\infty} |f^{(k)}(0)| \le \lim_{R \to +\infty} \frac{k!(A + BR^{\frac{3}{2}})}{R^k} = 0.$$

Hence $f^{(k)}(0) = 0$ for all $k = 2, 3, \ldots$ and then f is a linear polynomial. We have completed the proof of the problem. ▨

Problem 5.9

Bak and Newman Chapter 5 Exercise 9.

Proof. Since f is entire, f' is everywhere differentiable by Corollary 5.6. By Definition 3.3, f' is also entire. By Theorem 5.11 (The Extended Liouville Theorem), $f'(z)$ is a polynomial of degree at most 1 and then we apply Theorem 5.5 (The Taylor Expansion of an Entire Function) to f' to get

$$f^{(k)}(0) = 0 \tag{5.9}$$

for $k = 3, 4, \ldots$. Furthermore, we have $f'(0) = 0$ so that Theorem 5.5 (The Taylor Expansion of an Entire Function) and the values (5.9) ensure that

$$f(z) = f(0) + f'(0)z + \frac{f''(0)}{2!} z^2 = a + bz^2,$$

where $a - f(0)$ and $b = \frac{f''(0)}{2}$.

Next, we recall from Problem 5.4 that

$$f''(0) = \frac{1}{2\pi i} \int_C \frac{f'(\omega)}{\omega^2} \, d\omega, \tag{5.10}$$

where C is a circle surrounding the origin. In particular, we take C to be the unit circle $|\omega| = 1$ so that $\frac{f'(\omega)}{\omega^2} \ll 1$ on C. Therefore, we apply Theorem 4.10 (The M-L Formula) to the integral (5.10) to yield

$$|f''(0)| \le \frac{1}{2\pi} \cdot 1 \cdot 2\pi = 1$$

and hence

$$|b| = \frac{|f''(0)|}{2} \le \frac{1}{2}.$$

This completes the proof of the problem. $\blacksquare$

Problem 5.10

Bak and Newman Chapter 5 Exercise 10.

Proof. Assume that f was a nonconstant entire function satisfying the two equations. By repeated applications of the two equations, we can show that f satisfies the following equations

$$f(z) = f(z+n) \quad \text{and} \quad f(z) = f(z+in) \tag{5.11}$$

for all $n \in \mathbb{Z}$.

Consider the unit square $D = \{a + ib \,|\, a, b \in [0,1]\}$ which is compact. Since f is entire, it modulus is continuous on $\mathbb{C}$ and thus $|f(D)|$ is a compact subset of $\mathbb{R}$. By the Heine-Borel Theorem [27, p. 30], $|f(z)|$ is bounded on D.

Let $z = x + iy \in \mathbb{C}$. Recall that x can be written as the sum of $[x]$ and $\{x\}$, where $[x]$ and $\{x\}$ are the greatest integer function and the fractional part of the real number x respectively. We note that $[x] \in \mathbb{Z}$ and $0 \le \{x\} < 1$. Similarly, we have $y = [y] + \{y\}$. Therefore, we obtain

$$z = x + iy = ([x] + i[y]) + \omega,$$

where $\omega = \{x\} + i\{y\}$. It is easily seen that $\omega \in D$. By the equations (5.11), we see that

$$f(z) = f((\omega + i[y]) + [x]) = f(\omega + i[y]) = f(\omega).$$

In other words, f is bounded on $\mathbb{C}$ and Theorem 5.10 (Liouville's Theorem) implies that f is constant, a contradiction. Hence no such function exists and we have completed the proof of the problem. $\blacksquare$

Problem 5.11

Bak and Newman Chapter 5 Exercise 11.

Proof. Let $P(z)$ be a real polynomial of odd degree. By Problem 1.5, we know that the complex zeros of a real polynomial appear in conjugate pairs. Consequently, the number of complex zeros of $P(z)$ must be even. Since $\deg P(z)$ is odd, Theorem 5.12 (The Fundamental Theorem of Algebra) asserts that the number of zeros (counting multiplicity) of $P(z)$ must be odd. Hence $P(z)$ has *at least* one real zero. This ends the analysis of the problem. $\blacksquare$

Problem 5.12

Bak and Newman Chapter 5 Exercise 12.

Proof. Let $P(z)$ be a real polynomial of degree n. By Theorem 5.12 (The Fundamental Theorem of Algebra), we can express $P(z)$ as

$$P(z) = a_n(z - z_1)(z - z_2) \cdots (z - z_n)$$

for some $a_n \in \mathbb{C}$ and $z_1, z_2, \ldots, z_n$ are the zeros of $P(z)$. By Problem 1.5, z_i is a zero of $P(z)$ if and only if $\overline{z_i}$ is a zero of $P(z)$. There are two cases:

- **Case (i):** z_i **is real.** Then $\overline{z_i}$ is also real and so $P(z)$ has $z - z_i$ as its factor.

- **Case (ii):** z_i **is non-real.** Then we have

$$(z - z_i)(z - \overline{z_i}) = z^2 - 2\operatorname{Re}(z_i)z + |z_i|^2$$

  which is a real quadratic polynomial.

Hence $P(z)$ is a product of real linear and quadratic polynomials, completing the proof of the problem. ∎

Problem 5.13

Bak and Newman Chapter 5 Exercise 13.

Proof. We prove the result in several steps.

- **Step 1:** $v(x,y) \equiv 0$ **if and only if** $y = 0$. If we suppose that $P(x+iy) = u(x,y)+iv(x,y)$, where u and v are real-valued functions, then the hypothesis says that $v(x,y) = 0$ if and only if $y = 0$.

- **Step 2: Either** $v(x,y) > 0$ **or** $v(x,y) < 0$ **in the upper half-plane.** Since P is a polynomial, it is differentiable everywhere and thus v is continuous on $\mathbb{C}$. Fix a point x first, if $v(x,y_1) > 0$ and $v(x,y_2) < 0$ with $0 < y_1 < y_2$, then the Intermediate Value Theorem [27, p. 101] asserts that there exists a $y_0 \in (y_1, y_2)$ such that

$$v(x, y_0) = 0$$

which contradicts **Step 1**. This means that either $v(x,y) > 0$ or $v(x,y) < 0$ for all $y > 0$. Next, we suppose that x_1 is a number such that $v(x_1,y) > 0$ for all $y > 0$. Fix a $y > 0$. If $v(x_2,y) < 0$, where $x_2 \neq x_1$, then we apply the Intermediate Value Theorem to conclude that

$$v(x_0, y) = 0$$

for some $x_0 \in (x_1, x_2)$ or $x_0 \in (x_2, x_1)$ which is a contradiction again. Thus we have the desired result.

- **Step 3: Either** $v_y(x,0) \geq 0$ **for all** $x \in \mathbb{R}$ **or** $v_y(x,0) \leq 0$ **for all** $x \in \mathbb{R}$. Assume that $v_y(x,0)$ changed sign on $\mathbb{R}$, i.e., there exist $x_1 < x_2$ such that $v_y(x_1,0) > 0$ and $v_y(x_2,0) < 0$. Define $G_1, G_2 : \mathbb{R} \to \mathbb{R}$ by

$$G_1(y) = v(x_1, y) \quad \text{and} \quad G_2(y) = v(x_2, y).$$

The assumption $v_y(x_1,0) > 0$ implies that G_1 is strictly increasing in $(-\delta_1, \delta_1)$ for some $\delta_1 > 0$. Similarly, G_2 is strictly decreasing in $(-\delta_2, \delta_2)$ for some $\delta_2 > 0$. Therefore, by using **Step 1**, one can find a $\epsilon > 0$ such that

$$v(x_1, \epsilon) > v(x_1, 0) = 0 \quad \text{and} \quad v(x_2, \epsilon) < v(x_2, 0) = 0. \tag{5.12}$$

Now if we define $F(x) = v(x, \epsilon)$ which is a continuous function with respect to x, then we follow from the Intermediate Value Theorem and the inequalities (5.12) that

$$F(x_3) = v(x_3, \epsilon) = 0$$

for some $x_3 \in (x_1, x_2)$, a contradiction to **Step 1**.

- **Step 4: $u(x, 0)$ is monotonic on $\mathbb{R}$.** Since $P(z)$ is entire, Proposition 2.3 or 3.1 implies that

$$u_x(x, 0) = v_y(x, 0)$$

so that either $u_x(x, 0) \geq 0$ or $u_x(x, 0) \leq 0$ for all $x \in \mathbb{R}$ by **Step 3**. Hence $u(x, 0)$ is monotonic on $\mathbb{R}$.

- **Step 5: $P(z) = \alpha$ has only one solution for all $\alpha \in \mathbb{R}$.** Let $\deg P(z) = n$. By Theorem 5.12 (The Fundamental Theorem of Algebra), the polynomial $P(z)$ has exactly n roots $z_1, z_2, \ldots, z_n \in \mathbb{C}$, counted with multiplicity. Let $z_k = x_k + iy_k$, where $k = 1, 2, \ldots, n$. Then we get

$$P(z) = A(z - z_1)(z - z_2) \cdots (z - z_n) \tag{5.13}$$

for some $A \in \mathbb{C}$.

Recall that $P = u + iv$, so $P(z_k) = 0$ implies that $u(x_k, y_k) = v(x_k, y_k) = 0$. By **Step 1**, we must have $y_k = 0$ for all $k = 1, 2, \ldots, n$. Hence we obtain from the expression (5.13) that

$$P(z) = A(z - x_1)(z - x_2) \cdots (z - x_n). \tag{5.14}$$

Without loss of generality, we may assume that $x_1 \leq x_2 \leq \cdots \leq x_n$. Now our hypothesis forces that A must be real. Since $P(x, 0) = u(x, 0)$, we get from the expression (5.14) that

$$u(x, 0) = A(x - x_1)(x - x_2) \cdots (x - x_n).$$

Assume that $P(x, 0)$ had *at least* two distinct zeros, say $x_1 < x_2$. Since $P(x, 0)$ is a continuous function of x, $P(x, 0) = u(x, 0)$ *cannot* be monotonic, but this contradicts **Step 4**. Therefore, we conclude that $x_1 = x_2 = \cdots = x_n$ and the representation (5.14) reduces to

$$P(z) = A(z - x_1)^n. \tag{5.15}$$

Next, if $n > 1$, then $z_1 = x_1 + e^{\frac{\pi i}{n}}$ is a non-real complex number and

$$P(z_1) = A(z_1 - x_1)^n = Ae^{\pi i} = -A. \tag{5.16}$$

Recall that A is real, so the result (5.16) contradicts to our hypothesis that P is real if and only if z is real. Consequently, we have $n = 1$ and so the expression (5.15) gives

$$P(z) = Az - Az_1,$$

as desired.

We complete the analysis of the problem. $\blacksquare$

Problem 5.14

Bak and Newman Chapter 5 Exercise 14.

Proof. If $P(\alpha) = P'(\alpha) = \cdots = P^{(k-1)}(\alpha) = 0$ and $P^{(k)}(\alpha) \neq 0$, then Corollary 5.7 implies that

$$P(z) = \frac{P^{(k)}(\alpha)}{k!}(z - \alpha)^k + \frac{P^{(k+1)}(\alpha)}{(k + 1)!}(z - \alpha)^{k+1} + \cdots = (z - \alpha)^k Q(z),$$

where

$$Q(z) = \frac{P^{(k)}(\alpha)}{k!} + \frac{P^{(k+1)}(\alpha)}{(k + 1)!}(z - \alpha) + \cdots$$

and $Q(\alpha) = \frac{P^{(k)}(\alpha)}{k!} \neq 0$.

Conversely, suppose that $P(z) = (z - \alpha)^k Q_0(z)$, where $Q_0(\alpha) \neq 0$. Then we have

$$P'(z) = (z - \alpha)^{k-1} Q_1(z),$$

where $Q_1(z) = (z - \alpha)Q_0'(z) + kQ_0(z)$. Therefore, it is easily seen that

$$P^{(n)}(z) = (z - \alpha)^{k-n} Q_n(z), \tag{5.17}$$

where $Q_n(z) = (z-\alpha)Q_{n-1}'(z) + kQ_{n-1}(z)$ which satisfies $Q_n(\alpha) = kQ_{n-1}(\alpha)$ for $n = 1, 2, \ldots, k$. Thus the expression (5.17) implies immediately that

$$P^{(n)}(\alpha) = \begin{cases} 0, & \text{if } 1 \leq n \leq k - 1; \\ k!Q_0(\alpha), & \text{if } k = n. \end{cases}$$

Hence we have completed the analysis of the problem. $\blacksquare$

> ### Problem 5.15
>
> *Bak and Newman Chapter 5 Exercise 15.*

Proof. Fix a point $z \in \mathbb{C}$. Consider the line segment connecting 0 and z which is parameterized by $\{tz \mid t \in [0, 1]\}$. Suppose that $K_z = \{t \in [0, 1] \mid |f(tz)| \leq 1\}$ which is a bounded set. Define

$$d = \begin{cases} \sup K_z, & \text{if } K_z \neq \varnothing; \\ 0, & \text{otherwise.} \end{cases} \tag{5.18}$$

Obviously, we have $0 \leq d \leq 1$. Let $z_d = dz$. We claim that

$$|f(z_d)| \leq \max\{1, |f(0)|\}. \tag{5.19}$$

On the one hand, if $d = 0$, then $z_d = 0$ so that

$$|f(z_d)| = |f(0)|. \tag{5.20}$$

On the other hand, suppose that $d > 0$ and $\{t_n\}$ is a sequence of K_z with limit point t_0. Since f is continuous and $|f(t_n z)| \leq 1$ for all $n \in \mathbb{N}$, we have

$$|f(t_0 z)| = \lim_{n \to \infty} |f(t_n z)| \leq 1.$$

In other words, $t_0 \in K_z$ and thus K_z is closed in $[0, 1]$. By the Heine-Borel Theorem, K_z is compact. Furthermore, we recall from [27, Problem 4.27, pp. 44, 45] that $d \in K_z$ which implies

$$|f(z_d)| = |f(dz)| \leq 1. \tag{5.21}$$

Hence our claim (5.19) follows immediately from the results (5.20) and (5.21).

By the definition (5.18), we have $|f(tz)| > 1$ for all $t \in [d, 1]$ and the hypothesis implies that $|f'(tz)| \le 1$ for all $t \in [d, 1]$, i.e., $|f'(\omega)| \le 1$ along the line segment connecting z_d and z. Combining this fact and Proposition 4.12, we have[a]

$$|z| \ge |z - z_d| \ge \int_{z_d}^{z} |f'(\omega)| \, \mathrm{d}\omega \ge \left| \int_{z_d}^{z} f'(\omega) \, \mathrm{d}\omega \right| = |f(z) - f(z_d)| \ge |f(z)| - |f(z_d)|$$

and so

$$|f(z)| \le |f(z_d)| + |z| \le \max\{1, |f(0)|\} + |z|,$$

where $z \in \mathbb{C}$. Consequently, it follows from Theorem 5.11 (The Extended Liouville Theorem) that f is a linear polynomial. This completes the proof of the problem. ◼

Problem 5.16

*Bak and Newman Chapter 5 Exercise 16.**

Proof. Let $P(z) = a_n z^n + a_{n-1} z^{n-1} + \cdots + a_n$ be a polynomial having zeros $z_1, z_2, \ldots, z_n$, where $a_n \ne 0$. By [4, Eqn. (4), p. 67], the centroid of zeros of $P(z)$ is

$$\frac{z_1 + z_2 + \cdots + z_n}{n} = -\frac{a_{n-1}}{n a_n}. \tag{5.22}$$

Now we let $\omega_1, \omega_2, \ldots, \omega_{n-1}$ be the zeros of $P'(z) = n a_n z^{n-1} + (n-1) a_{n-1} z^{n-2} + \cdots + a_2 z + a_1$. In other words, we have

$$n a_n z^{n-1} + (n-1) a_{n-1} z^{n-2} + \cdots + a_2 z + a_1 = n a_n (z - \omega_1)(z - \omega_2) \cdots (z - \omega_{n-1}). \tag{5.23}$$

By comparing coefficients of z^{n-2} on both sides of the expression (5.23), we see immediately that

$$(n-1) a_{n-1} = -n a_n (\omega_1 + \omega_2 + \cdots + \omega_{n-1})$$
$$\frac{\omega_1 + \omega_2 + \cdots + \omega_{n-1}}{n-1} = -\frac{a_{n-1}}{n a_n}. \tag{5.24}$$

Hence the result follows if we combine the two expressions (5.22) and (5.24). This completes the proof of the problem. ◼

Problem 5.17

*Bak and Newman Chapter 5 Exercise 17.**

Proof. Let S be a convex set. The statement for $n = 1$ and $n = 2$ are trivial. Assume that the statement is true for $n = k \in \mathbb{N}$, i.e., if $z_1, z_2, \ldots, z_k \in S$, then

$$a_1 z_1 + a_2 z_2 + \cdots + a_k z_k \in S, \tag{5.25}$$

where $a_j \ge 0$ for all $1 \le j \le k$ and $\sum_{j=1}^{k} a_j = 1$. Now let $z_1, z_2, \ldots, z_k, z_{k+1} \in S$, $a_j \ge 0$ for all $1 \le j \le k+1$ and $\sum_{j=1}^{k+1} a_j = 1$.

---

[a] In fact, we also apply Definition 4.3 and Lemma 4.9 to show the third inequality holds.

Define $A = \sum_{j=1}^{k} a_i$. Then the definition gives $1 - A = a_{k+1}$ which implies that

$$(a_1 z_1 + a_2 z_2 + \cdots + a_k z_k) + a_{k+1} z_{k+1} = A\left(\frac{a_1}{A} z_1 + \frac{a_2}{A} z_2 + \cdots + \frac{a_k}{A} z_k\right) + (1 - A)z_{k+1}. \quad (5.26)$$

Since $\sum_{j=1}^{k} \frac{a_j}{A} = 1$, the induction assumption (5.25) ensures that the bracket on the right-hand side of the expression (5.26) is an element of S. Therefore, the convexity of S further implies that the right-hand side of the expression (5.26) belongs to S. Hence our required result follows from induction and it ends the proof of the problem.

Problem 5.18

*Bak and Newman Chapter 5 Exercise 18.**

Proof.

(a) Let $z_{1,k}, z_{2,k}, \ldots, z_{k,k}$ be the zeros of $P_k(z)$, where $k \in \mathbb{N}$. Applying [4, Eqn. (4), p. 67], we have

$$\frac{z_{1,k} + z_{2,k} + \cdots + z_{k,k}}{k} = \frac{1}{k} \times \left[-\frac{\frac{1}{(k-1)!}}{\frac{1}{k!}}\right] = -1.$$

(b) Without loss of generality, we may assume that $z_{k,k}$ is a zero of $P_k(z)$ with maximal possible modulus. We note that

$$P'_{k+1}(z) = 1 + z + \cdots + \frac{z^k}{k!} = P_k(z)$$

for every $k = 0, 1, 2, \ldots$. By the hypothesis, we know that

$$|z_{j,k+1}| \le |z_{k+1,k+1}|$$

for all $j = 1, 2 \ldots, k$. In other words, we get

$$z_{j,k+1} \in \overline{D(0; |z_{k+1,k+1}|)}$$

for all $j = 1, 2 \ldots, k + 1$. Since $\overline{D(0; |z_{k+1,k+1}|)}$ is convex, it contains the convex hull of $z_{1,k+1}, z_{2,k+1}, \ldots, z_{k+1,k+1}$ by Definition 5.13. Finally, Theorem 5.14 (The Gauss-Lucas Theorem) implies that all the zeros of $P'_{k+1}(z) = P_k(z)$ lie within the convex hull of the zeros of $P_{k+1}(z)$. Particularly, we obtain

$$|z_{k,k}| \le |z_{k+1,k+1}|$$

and hence $\{|z_{k,k}|\}$ is an increasing sequence.

We end the proof of the problem.

Problem 5.19

*Bak and Newman Chapter 5 Exercise 19.**

Proof. Let $Q(z) = 1 + z + z^2 + \cdots + z^n$. Then it is easy to check that $Q'(z) = P(z)$. Note that

$$Q(z) = \frac{z^{n+1} - 1}{z - 1},$$

so the zeros of $Q(z)$ lie on the unit circle.

Let $S = \{z_1, z_2, \ldots, z_{n+1}\}$ be set of the zeros of $Q(z)$ and Conv (S) the convex hull of S. Since $S \subseteq \overline{D(0;1)}$ and $\overline{D(0;1)}$ is a convex set, Definition 5.13 ensures that Conv $(S) \subseteq \overline{D(0;1)}$. Hence we deduce from Theorem 5.14 (The Gauss-Lucas Theorem) that the zeros of $P(z)$ lie within Conv $(S) \subseteq \overline{D(0;1)}$. This completes the proof of the problem. $\blacksquare$

Remark 5.1

In fact, the convex hull of a finite set $S = \{z_1, z_2, \ldots, z_{n+1}\}$ is given by

$$\text{Conv}\,(S) = \Big\{ a_1 z_1 + a_2 z_2 + \cdots + a_{n+1} z_{n+1} \Big| a_j \ge 0 \text{ for } 1 \le j \le n+1 \text{ and } \sum_{j=1}^{n+1} a_j \Big\}.$$

See, for example, [26, Eqn. (10.33), p. 268].

Problem 5.20

*Bak and Newman Chapter 5 Exercise 20.**

Proof. Using the method of Chapter 1, we see that the solutions of the polynomial equation $z^2 = i$ are

$$z = \pm \frac{1}{\sqrt{2}} (1 + i).$$

Therefore, we have $\sqrt{i} = \frac{1}{\sqrt{2}}(1 + i)$.

To find the estimates of $\sqrt{i}$ by Newton's method, we first let

$$f(z) = z^2 - i \quad \text{and} \quad g(z) = z - \frac{f(z)}{f'(z)} = z - \frac{z^2 - i}{2z}.$$

Take $z_0 = 1$. Then we have

$$z_1 = g(z_0) = 1 - \frac{1 - i}{2} = \frac{1 + i}{2},$$

$$z_2 = g(z_1) = \frac{1 + i}{2} - \frac{-\frac{i}{2}}{1 + i} = \frac{1 + i}{2} + \frac{1 + i}{4} = \frac{3}{4}(1 + i),$$

$$z_3 = g(z_2) = \frac{17}{24}(1 + i),$$

$$z_4 = g(z_3) = \frac{865}{1224}(1 + i).$$

This completes the proof of the problem. $\blacksquare$

CHAPTER **6**

Properties of Analytic Functions

Problem 6.1

Bak and Newman Chapter 6 Exercise 1.

Proof. It is easily seen that $f(z) = \frac{1}{z}$ is analytic everywhere except at $z = 0$. Therefore, if $z \in D(1 + i; \sqrt{2})$, then we obtain

$$
\begin{aligned}
\frac{1}{z} &= \frac{1}{(1 + i) + [z - (1 - i)]} \\
&= \frac{1}{1 + i} \cdot \frac{1}{1 + \frac{z - (1+i)}{1+i}} \\
&= \frac{1}{1 + i} \cdot \left[1 - \frac{z - (1 + i)}{1 + i} + \frac{[z - (1 + i)]^2}{(1 + i)^2} - \cdots \right] \\
&= \sum_{k=0}^{\infty} \frac{(-1)^k (z - 1 - i)^k}{(1 + i)^{k+1}}.
\end{aligned}
$$

This completes the proof of the problem. ∎

Problem 6.2

*Bak and Newman Chapter 6 Exercise 2.**

Proof. By partial fractions, we have

$$
f(z) = \frac{1}{1 - z - 2z^2} = \frac{1}{(1 + z)(1 - 2z)} = \frac{2}{3} \cdot \frac{1}{1 + z} + \frac{1}{3} \cdot \frac{1}{1 - 2z}. \tag{6.1}
$$

We know that

$$
\frac{1}{1 + z} = \sum_{n=0}^{\infty} (-z)^n \quad \text{and} \quad \frac{1}{1 - 2z} = \sum_{n=0}^{\infty} (2z)^n \tag{6.2}
$$

for $|z| < 1$ and for $|z| < \frac{1}{2}$ respectively. Hence, for $|z| < \frac{1}{2}$, by putting the identities (6.2) into the expression (6.1), we see that

$$
f(z) = \sum_{n=0}^{\infty} \left[\frac{2(-1)^n}{3} + \frac{2^n}{3} \right] z^n,
$$

ending the proof of the problem. ∎

Problem 6.3

Bak and Newman Chapter 6 Exercise 3.

Proof. We note that the series $\sum_{n=0}^{\infty} z^n$ converges for $|z| < 1$, so Theorem 2.9 implies that

$$\left(\sum_{n=0}^{\infty} z^n\right)' = \sum_{n=0}^{\infty} n z^{n-1} \tag{6.3}$$

throughout $|z| < 1$. Therefore, we observe from the formula (6.3) that

$$\sum_{n=0}^{\infty} n z^n = z \sum_{n=0}^{\infty} n z^{n-1} = z \left(\sum_{n=0}^{\infty} z^n\right)' = z \cdot \left(\frac{1}{1-z}\right)' = \frac{z}{(1-z)^2} \tag{6.4}$$

throughout $|z| < 1$.

Similarly, further application of Theorem 2.9 gives

$$\left(\sum_{n=0}^{\infty} z^n\right)'' = \sum_{n=0}^{\infty} n(n-1) z^{n-2}. \tag{6.5}$$

throughout $|z| < 1$. Hence we deduce from the formulas (6.4) and (6.5) that

$$\sum_{n=0}^{\infty} n^2 z^n = \sum_{n=0}^{\infty} n(n-1) z^n + \sum_{n=0}^{\infty} n z^n$$

$$= z^2 \sum_{n=0}^{\infty} n(n-1) z^{n-2} + \frac{z}{(1-z)^2}$$

$$= z^2 \left(\sum_{n=0}^{\infty} z^n\right)'' + \frac{z}{(1-z)^2}$$

$$= z^2 \left(\frac{1}{1-z}\right)'' + \frac{z}{(1-z)^2}$$

$$= \frac{2z^2}{(1-z)^3} + \frac{z}{(1-z)^2}$$

$$= \frac{z(1+z)}{(1-z)^3}.$$

We have completed the proof of the problem.

Problem 6.4

Bak and Newman Chapter 6 Exercise 4.

Proof. Assume that $f(\frac{1}{n}) = \frac{1}{n+1}$ for all $n \in \mathbb{N}$. Consider $g(z) = f(z) - \frac{z}{1+z}$. Then we have

$$g\left(\frac{1}{n}\right) = f\left(\frac{1}{n}\right) - \frac{\frac{1}{n}}{\frac{1}{n}+1} = \frac{1}{n+1} - \frac{1}{n+1} = 0$$

for all $n \in \mathbb{N}$. Since $\{\frac{1}{n}\}$ is a sequence of distinct points and $\frac{1}{n} \to 0 \in D(0;1)$, Theorem 6.9 (The Uniqueness Theorem) implies that

$$f(z) = \frac{z}{1+z} \tag{6.6}$$

in $D(0;1)$. Since f is analytic in $\overline{D(0;1)}$, it must be continuous there. This means that

$$\lim_{n\to\infty} f(z_n) = f(-1) \tag{6.7}$$

holds for every sequence $\{z_n\} \subseteq \overline{D(0;1)}$ such that $z_n \to -1$ as $n \to \infty$ and $z_n \neq -1$ for all $n \in \mathbb{N}$. However, the representation (6.6) guarantees that the limit (6.7) is impossible. Hence no such analytic function exists and we complete the proof of the problem. ▪

Problem 6.5

Bak and Newman Chapter 6 Exercise 5.

Proof. We fix $z_2 \in \mathbb{R}$. Let $f(z) = \sin(z + z_2)$ and $g(z) = \sin z \cos z_2 + \cos z \sin z_2$. Then f and g are entire with respect to z. Since

$$f\left(\frac{1}{n}\right) = g\left(\frac{1}{n}\right)$$

for all $n \in \mathbb{N}$, it follows from Corollary 6.10 that $f \equiv g$ in $\mathbb{C}$ (with respect to z).

Next, we fix $z_1 \in \mathbb{C}$ and consider $F(\omega) = \sin(z_1 + \omega)$ and $G(\omega) = \sin z_1 \cos \omega + \cos z_1 \sin \omega$. Obviously, they are entire with respect to ω. The previous paragraph ensures that

$$F\left(\frac{1}{n}\right) = \sin\left(z_1 + \frac{1}{n}\right) = f(z_1) = g(z_1) = \sin z_1 \cos\left(\frac{1}{n}\right) + \cos z_1 \sin\left(\frac{1}{n}\right) = G\left(\frac{1}{n}\right)$$

for all $n \in \mathbb{N}$. Thus Corollary 6.10 shows that $F \equiv G$ in $\mathbb{C}$ (with respect to ω) and hence the formula

$$\sin(z_1 + z_2) = \sin z_1 \cos z_2 + \cos z_1 \sin z_2$$

holds for all $z_1, z_2 \in \mathbb{C}$. We complete the proof of the problem. ▪

Problem 6.6

Bak and Newman Chapter 6 Exercise 6.

Proof. Let $D = \mathbb{C} \setminus \{(k + \frac{1}{2})\pi \mid k \in \mathbb{Z}\}$. Since π is irrational, we have $\frac{1}{n} \in D$ for all $n \in \mathbb{N}$. Since $f(z)$ agrees with the analytic function $\tan z$ on the set $\{\frac{1}{n}\}$ and $\frac{1}{n} \to 0$ in D as $n \to \infty$, Corollary 6.10 implies that

$$f(z) = \tan z$$

in D. Thus $f(z) = i$ if and only if $\tan z = i$. Using the identities for $\sin z$ and $\cos z$ on [4, p. 41], we have

$$\frac{e^{iz} - e^{-iz}}{e^{iz} + e^{-iz}} = -1$$
$$e^{iz} - e^{-iz} = -e^{iz} - e^{-iz}$$
$$e^{iz} = 0$$

which is impossible.

Furthermore, if f is entire, then f is continuous in $\mathbb{C}$. Thus using similar argument as in the proof of Problem 6.4, it is impossible that f is not entire. This completes the proof of the problem. ▪

Proof. We notice that $N \in \mathbb{N}$. Since f is entire and $|f(z)| \geq |z|^N$, $f(z) \to \infty$ as $z \to \infty$. By Theorem 6.11, f is a polynomial. Let $f(z) = a_n z^n + a_{n-1} z^{n-1} + \cdots + a_1 z + a_0$, where a_n is a nonzero constant. Then the triangle inequality implies that

$$|f(z)| \leq |a_n| \cdot |z|^n + |a_{n-1}| \cdot |z|^{n-1} + \cdots + |a_1| \cdot |z| + |a_0|$$
$$\leq |z|^n \cdot \left(|a_n| + \frac{|a_{n-1}|}{|z|} + \cdots + \frac{|a_0|}{|z|^n} \right)$$
$$\leq 2|a_n| \cdot |z|^n \tag{6.8}$$

for sufficiently large z. Assume that $n < N$. Now the hypothesis and the inequality (6.8) give

$$|z|^N \leq 2|a_n| \cdot |z|^n$$

for sufficiently large enough z, but this means that $a_n = \infty$, a contradiction. Hence $\deg f(z) \geq N$ and we complete the proof of the problem.   ▨

Proof. Suppose that f is constant. Since $|f(z)| \leq 2$ for all $|z| = 1$ with $\operatorname{Im} z \geq 0$, it is actually true for all $|z| \leq 1$. In particular, we have $|f(0)| \leq 2 < \sqrt{6}$.

Without loss of generality, we may assume that f is nonconstant. Define $F(z) = f(z)f(-z)$. By the definition on [4, p. 86], F is analytic in $D(0;1)$ and continuous on the compact set $\overline{D(0;1)}$ so that $\max |F(z)|$ exits in $\overline{D(0;1)}$ by the Extreme Value Theorem. Now Theorem 6.13 (The Maximum Modulus Theorem) implies that this maximum must lie on $|z| = 1$. Hence our hypotheses assert that

$$|F(z)| \leq \max_{|z|=1} F(z) = \max_{|z|=1} (|f(z)| \cdot |f(-z)|) \leq 2 \times 3 = 6$$

for all $|z| \leq 1$. Particularly, if $z = 0$, we have

$$|f(0)|^2 = |F(0)| \leq 6$$

which implies $|f(0)| \leq \sqrt{6}$, completing the proof of the problem.   ▨

Proof. Let $z = x + iy$ and $K \subseteq \mathbb{C}$ be compact. Since e^z is continuous on K, the Extreme Value Theorem ensures that the function $|e^z|$ attains its maximum and minimum in K. Let K° be the interior of K and $z_0 = x_0 + iy_0 \in K^\circ$. Assume that z_0 was an extreme of the $|e^z|$. We know that K° is open, so there exists a $\delta > 0$ such that $D(z_0; \delta) \subseteq K^\circ$. It is obvious that $z_{\pm} = z_0 \pm \frac{\delta}{2} \in D(z_0; \delta)$ so that

$$|e^{z_+}| = e^{x_0 + \frac{\delta}{2}} > e^{x_0} = |e^{z_0}| \quad \text{and} \quad |e^{z_-}| = e^{x_0 - \frac{\delta}{2}} < e^{x_0} = |e^{z_0}|$$

which imply contradictions. Hence the modulus of e^z can only have an extreme on ∂K which completes the proof of the problem. ■

Problem 6.10

Bak and Newman Chapter 6 Exercise 10.

Proof. Let $f(z) = z^2 - z = z(z - 1)$ which is clearly nonconstant and analytic in $\overline{D(0;1)}$. On the one hand, since $f(0) = 0$, we have

$$\min_{|z|\leq 1} |f(z)| = 0.$$

On the other hand, the Extreme Value Theorem and Theorem 6.13 (The Maximum Modulus Theorem) ensure that $\max\limits_{|z|\leq 1} |f(z)|$ exists on $|z| = 1$. To find this maximum, if $|z| = 1$, then $z = re^{i\theta}$ so that

$$|f(z)| = |z - 1| = \sqrt{(\cos\theta - 1)^2 + \sin^2\theta} = \sqrt{2 - 2\cos\theta}.$$

Therefore, f takes its maximum when $\theta = \pi$ or equivalently at $z = -1$ and furthermore, we have

$$\max_{|z|\leq 1} |f(z)| = 2.$$

This completes the proof of the problem. ■

Problem 6.11

*Bak and Newman Chapter 6 Exercise 11.**

Proof. Let $C = \partial D(\alpha;r)$ for some $\alpha \in \mathbb{C}$ and $r > 0$. Since f is analytic in the compact set $\overline{D(\alpha;r)}$, it is analytic in an open set D containing $\overline{D(\alpha;r)}$ by Definition 3.3. Consequently, there exists a $R > r$ such that

$$\overline{D(\alpha;r)} \subseteq D(\alpha;R) \subseteq D. \tag{6.9}$$

Otherwise, for each $n \in \mathbb{N}$, one can find a $\omega_n \in D(\alpha;r+\frac{1}{n})$ but $\omega_n \notin D$. Since $\{\omega_n\} \subseteq D(\alpha;r+1)$ and $|\omega_n| \leq |\omega_n - \alpha| + |\alpha|$, $\{|\omega_n|\}$ is a bounded sequence. Thus the Bolzano-Weierstrass Theorem [27, Problem 5.25] ensures that there is a convergent subsequence $\{\omega_{n_k}\}$. Let $\omega_{n_k} \to \omega_0$ as $k \to \infty$. Next, we note that $\{\omega_{n_k}\} \subseteq \mathbb{C} \setminus D$. Since D is open in $\mathbb{C}$, $\mathbb{C} \setminus D$ is closed in $\mathbb{C}$ so that

$$\omega_0 \in \mathbb{C} \setminus D. \tag{6.10}$$

However, since $|\omega_{n_k} - \alpha| < r + \frac{1}{n_k}$ for all $k \in \mathbb{N}$, we have

$$|\omega_0 - \alpha| \leq r,$$

i.e., $\omega_0 \in \overline{D(\alpha;r)}$ which is contrary to the set relation (6.10). Hence the set relation (6.9) holds for some $R > r$.

Let $z_0 \in D(\alpha;r)$ and $n \in \mathbb{N}$. Using Theorem 6.4 (The Cauchy Integral Formula) to the analytic function f^n to get

$$f^n(z_0) = \frac{1}{2\pi i} \int_C \frac{f^n(z)}{z - z_0} \, dz. \tag{6.11}$$

Now we want to apply Theorem 4.10 (The M-L Formula) to the expression (6.11), so we check the hypotheses. Since $z_0 \notin C$, the function $\frac{f^n(z)}{z - z_0}$ is continuous on C. Let $\rho_C(z_0) = \inf\limits_{z \in C} |z - z_0|$. Recall that $\rho_C(z_0) = 0$ if and only if $z_0 \in C$. Therefore, $\rho_C(z_0) > 0$. Since $\rho_C(z_0) \le |z - z_0|$ for all $z \in C$, it must be true that

$$\left| \frac{f^n(z)}{z - z_0} \right| \le \frac{M^n}{\rho_C(z_0)}. \tag{6.12}$$

Consequently, it follows from Theorem 4.10 (The M-L Formula) and the estimate (6.12) that

$$|f^n(z_0)| = \left| \frac{1}{2\pi i} \int_C \frac{f^n(z)}{z - z_0} \, \mathrm{d}z \right| \le \frac{1}{2\pi} \cdot \frac{M^n}{\rho_C(z_0)} \cdot (2\pi r) = \frac{M^n r}{\rho_C(z_0)}$$

which gives

$$|f(z_0)| \le \left(\frac{r}{\rho_C(z_0)} \right)^{\frac{1}{n}} \cdot M. \tag{6.13}$$

Take $n \to \infty$, the inequality (6.13) implies that $|f(z_0)| \le M$ for all $z_0 \in D(\alpha; r)$. This proves Theorem 6.13 (The Maximum-Modulus Theorem) and we complete the proof of the problem. ▧

> **Remark 6.1**
>
> For Problem 6.11, please also see the book of Pólya and Szegő [21, p. 29].

> **Problem 6.12**
>
> *Bak and Newman Chapter 6 Exercise 12.*

Proof. If both f and g are constant, then there is nothing to prove. Thus, without loss of generality, we may assume that *at least* one of f and g is nonconstant. Let z_0 be a maximum of $|f(z)| + |g(z)|$ in D. Furthermore, we let $f(z_0) = R_1 \mathrm{e}^{-i\alpha}$ and $g(z_0) = R_2 \mathrm{e}^{-i\beta}$ for some $\alpha, \beta \in [0, 2\pi]$. We consider the function $h : \mathbb{C} \to \mathbb{C}$ defined by

$$h(z) = f(z)\mathrm{e}^{i\alpha} + g(z)\mathrm{e}^{i\beta}.$$

Since f and g are analytic in D and at least one of them is nonconstant, h is nonconstant and analytic in D. Therefore, the Extreme Value Theorem and Theorem 6.13 (The Maximum Modulus Theorem) show that the modulus $|h(z)|$ attains its maximum on ∂D.

By the definition, we have

$$|h(z)| \le |\mathrm{e}^{i\alpha}| \cdot |f(z)| + |\mathrm{e}^{i\beta}| \cdot |g(z)| = |f(z)| + |g(z)| \tag{6.14}$$

for all $z \in D$. Besides, we also have

$$|h(z_0)| = |f(z_0)\mathrm{e}^{i\alpha} + g(z_0)\mathrm{e}^{i\beta}| = R_1 + R_2 = |f(z_0)| + |g(z_0)|. \tag{6.15}$$

Now these two facts (6.14) and (6.15) combine to say that the modulus $|h(z)|$ attains its maximum, which is exactly $|f(z_0)| + |g(z_0)|$, at z_0. Hence we conclude from the previous paragraph that $|f(z)| + |g(z)|$ takes its maximum on ∂D. This completes the proof of the problem. ▧

> **Problem 6.13**
>
> *Bak and Newman Chapter 6 Exercise 13.*

Proof. Suppose that $P(z)$ is a nonconstant polynomial and $R > 0$. Assume that $P(z) \neq 0$ in the region $D(0; R)$. Now Theorem 6.14 (The Minimum Modulus Theorem) says that the minimum modulus of $P(z)$ in $D(0; R)$ exists on $|z| = R$, i.e., there exists a z_0 such that $|z_0| = R$ and

$$|P(z_0)| < |P(z)|$$

holds for all $|z| < R$. Particularly, we have

$$|P(z_0)| < |P(0)| \tag{6.16}$$

However, we recall from Problem 1.26 that $|P(z)| \geq M|z|^n$ for large $|z|$, where M is a positive constant and $n = \deg P$. So if R is large enough, then we have

$$|P(z_0)| > 2|P(0)|$$

which definitely contradicts the inequality (6.16). Hence $P(\omega) = 0$ for some $\omega \in \mathbb{C}$ which proves the Fundamental Theorem of Algebra. This ends the analysis of the problem. $\blacksquare$

Problem 6.14

Bak and Newman Chapter 6 Exercise 14.

Proof. If $P(z)$ is constant, then $P(z) = a_0$ with $|a_0| \leq 1$ and thus there is nothing to prove. We suppose that $P(z)$ is nonconstant. By Problem 5.6(b), we have

$$|a_k| \leq 1 \tag{6.17}$$

for all $k = 0, 1, 2, \ldots, n$. Take $R > 1$ and consider

$$f(z) = \frac{P(z)}{z^n} = \frac{a_0}{z^n} + \frac{a_1}{z^{n-1}} + \cdots + a_n \tag{6.18}$$

defined in the compact annulus $D = \{z \in \mathbb{C} \,|\, 1 \leq |z| \leq R\}$. Since f is C-analytic in the region $\{z \in \mathbb{C} \,|\, 1 < |z| < R\}$, the Extreme Value Theorem and Theorem 6.13 (The Maximum Modulus Theorem) guarantee that the maximum modulus of f in D occurs on ∂D. Thus if $|z| = 1$, since $|P(z)| \leq 1$ for all $|z| \leq 1$, then $|f(z)| \leq 1$. Next, if $|z| = R$, then we obtain from the inequality (6.17) and the expression (6.18) that

$$|f(z)| \leq \frac{|a_0|}{|z|^n} + \frac{|a_1|}{|z|^{n-1}} + \cdots + |a_n| \leq 1 + \frac{1}{R} + \cdots + \frac{1}{R^n} = \left(1 - \frac{1}{R^{n+1}}\right) \cdot \frac{R}{R-1} \to 1$$

as $R \to \infty$. In other words, this means that $|f(z)| \leq 1$ for all $|z| \geq 1$ and hence

$$|P(z)| \leq |z|^n$$

for all $|z| \geq 1$. This completes the proof of the problem. $\blacksquare$

Problem 6.15

*Bak and Newman Chapter 6 Exercise 15.**

Proof. We notice that any line passing through $z = 2$ can be parameterized as

$$z(t) = 2 + te^{i\theta}, \tag{6.19}$$

where $t \in \mathbb{R}$ and $\theta \in [0, 2\pi]$ is the angle between the line and the positive real axis. Therefore, we observe that

$$\begin{aligned}
f(z(t)) &= (1 + te^{i\theta}) \cdot (-2 + te^{i\theta})^2 \\
&= 4 - 3t^2 e^{2i\theta} + t^3 e^{3i\theta} \\
&= (4 - 3t^2 \cos 2\theta + t^3 \cos 3\theta) + i(-3t^2 \sin 2\theta + t^3 \sin 3\theta).
\end{aligned}$$

Direct computation gives

$$\begin{aligned}
g(t) &= |f(z(t))|^2 \\
&= (4 - 3t^2 \cos 2\theta + t^3 \cos 3\theta)^2 + (-3t^2 \sin 2\theta + t^3 \sin 3\theta)^2 \\
&= t^6 + 9t^4 + 16 - 6t^5 \cos \theta - 24t^2 \cos 2\theta + 8t^3 \cos 3\theta
\end{aligned}$$

so that

$$\begin{aligned}
g'(t) &= 6t^5 + 36t^3 - 30t^4 \cos \theta - 48t \cos 2\theta + 24t^2 \cos 3\theta \\
&= 6t(t^4 - 5t^3 \cos \theta + 6t^2 + 4t \cos 3\theta - 8 \cos 2\theta)
\end{aligned} \tag{6.20}$$

and

$$g''(t) = 30t^4 + 108t^2 - 120t^3 \cos \theta - 48 \cos 2\theta + 48t \cos 3\theta.$$

Since the problem requires that $|f(z)|$ has a relative extremum at $z = 2$, it follows from Fermat's Theorem [27, Theorem 8.6, p. 129] that $g'(0) = 0$ which is obvious by the expression (6.20). It is evident that

$$g''(0) = -48 \cos 2\theta,$$

so we have $g''(0) > 0$ if and only if $\frac{\pi}{4} < \theta < \frac{3\pi}{4}$ and $\frac{5\pi}{4} < \theta < \frac{7\pi}{4}$. Similarly, $g''(0) < 0$ if and only if $0 \le \theta < \frac{\pi}{4}$, $\frac{3\pi}{4} < \theta < \frac{5\pi}{4}$ and $\frac{7\pi}{4} < \theta \le 2\pi$. In other words, it follows from the Second Derivative Test [2, §4.17, pp. 188, 189] that $|f(z)|$ has a local minimum at $z = 2$ on any line (6.19) with $\theta \in (\frac{\pi}{4}, \frac{3\pi}{4}) \cup (\frac{5\pi}{4}, \frac{7\pi}{4})$ and $|f(z)|$ has a relative maximum at $z = 2$ through any line (6.19) with $\theta \in [0, \frac{\pi}{4}) \cup (\frac{3\pi}{4}, \frac{5\pi}{4}) \cup (\frac{7\pi}{4}, 2\pi]$.

Now it remains to test the behaviour of $|f|$ along the line (6.19) for

$$\theta = \frac{\pi}{4}, \frac{3\pi}{4}, \frac{5\pi}{4} \quad \text{and} \quad \frac{7\pi}{4}.$$

In fact, it follows easily from the First Derivative Test [2, Theorem 4.8, p. 188] that $|f|$ attains its local maximum at $z = 2$ when $\theta = \frac{\pi}{4}, \frac{7\pi}{4}$ and $|f|$ has a local minimum at $z = 2$ when $\theta = \frac{3\pi}{4}, \frac{5\pi}{4}$. This completes the proof of the problem. $\blacksquare$

Problem 6.16

*Bak and Newman Chapter 6 Exercise 16.**

Proof. It is clear that f is analytic in $\mathbb{C} \setminus \{0\}$. Note that

$$f'(z) = \frac{z^2 - 1}{z^2} = \frac{(z - 1)(z + 1)}{z^2}$$

so that $f'(z) = 0$ if and only if $z = \pm 1$. Since $f(1) \neq 0$, Theorem 6.17 ensues that $z = 1$ is a saddle point of f. Since $f(-1) = 0$, the modulus $|f|$ definitely has the absolute minimum at $z = -1$ and thus $z = -1$ is not a saddle point of f by Definition 6.16. In conclusion, $z = 1$ is the *only* saddle point of f.

Suppose that

$$z(t) = 1 + te^{i\theta} \tag{6.21}$$

where $t \in \mathbb{R}$ and $\theta \in [0, 2\pi]$ is the angle between the line and the positive real axis. Therefore, we observe that

$$
\begin{aligned}
f(z(t)) &= \frac{(2 + te^{i\theta})^2}{1 + te^{i\theta}} \\
&= \frac{(4 + 4te^{i\theta} + t^2 e^{2i\theta})(1 + te^{-i\theta})}{1 + 2t\cos\theta + t^2} \\
&= \frac{4(1 + t^2) + t(4 + t^2)e^{i\theta} + 4te^{-i\theta} + t^2 e^{2i\theta}}{1 + 2t\cos\theta + t^2} \\
&= \frac{4(1 + t^2) + t(8 + t^2)\cos\theta + t^2\cos 2\theta}{1 + 2t\cos\theta + t^2} + i\frac{t^3\sin\theta + t^2\sin 2\theta}{1 + 2t\cos\theta + t^2}
\end{aligned}
$$

so that

$$
\begin{aligned}
g(t) &= |f(z(t))|^2 \\
&= \left[\frac{4(1 + t^2) + t(8 + t^2)\cos\theta + t^2\cos 2\theta}{1 + 2t\cos\theta + t^2}\right]^2 + \left(\frac{t^3\sin\theta + t^2\sin 2\theta}{1 + 2t\cos\theta + t^2}\right)^2 \\
&= \frac{h_1^2(t) + h_2^2(t)}{h_3^2(t)},
\end{aligned} \tag{6.22}
$$

where

$$
\begin{aligned}
h_1(t) &= 4(1 + t^2) + t(8 + t^2)\cos\theta + t^2\cos 2\theta, \\
h_2(t) &= t^3\sin\theta + t^2\sin 2\theta, \\
h_3(t) &= 1 + 2t\cos\theta + t^2.
\end{aligned}
$$

Direct computation shows

$$
\begin{aligned}
h_1(0) &= 4, & h_1'(0) &= 8\cos\theta, & h_1''(0) &= 8 + 2\cos 2\theta, \\
h_2(0) &= 0, & h_2'(0) &= 0, & h_2''(0) &= 2\sin 2\theta, \\
h_3(0) &= 1, & h_3'(0) &= 2\cos\theta, & h_3''(0) &= 2.
\end{aligned} \tag{6.23}
$$

Now it yields from the expression (6.22) that

$$g' = 2 \cdot \frac{h_3(h_1 h_1' + h_2 h_2') - (h_1^2 + h_2^2)h_3'}{h_3^3}$$

and then the values (6.23) imply that

$$g'(0) = 0$$

for every $\theta \in [0, 2\pi]$. Next, further differentiation gives

$$
\begin{aligned}
g'' &= 2 \cdot \frac{h_3^3[h_3(h_1 h_1' + h_2 h_2') - (h_1^2 + h_2^2)h_3']' - [h_3(h_1 h_1' + h_2 h_2') - (h_1^2 + h_2^2)h_3'] \cdot (3h_3^2 h_3')}{h_3^6} \\
&= 2 \cdot \frac{h_3[h_3(h_1 h_1' + h_2 h_2') - (h_1^2 + h_2^2)h_3']' - [h_3(h_1 h_1' + h_2 h_2') - (h_1^2 + h_2^2)h_3'] \cdot (3h_3')}{h_3^4}
\end{aligned}
$$

and then

$$g''(0) = 2 \cdot \frac{[h_3(h_1 h_1' + h_2 h_2')]' - [(h_1^2 + h_2^2)h_3']'}{h_3^3}\bigg|_{t=0}. \tag{6.24}$$

On the one hand, $[h_3(h_1 h_1' + h_2 h_2')]' = h_3'(h_1 h_1' + h_2 h_2') + h_3[h_1 h_1'' + (h_1')^2 + h_2 h_2'' + (h_2')^2]$, so the values (6.23) give

$$\frac{\mathrm{d}}{\mathrm{d}t}[h_3(h_1 h_1' + h_2 h_2')]\bigg|_{t=0} = 2\cos\theta(32\cos\theta) + [4(8 + 2\cos 2\theta) + 64\cos^2\theta]$$

$$= 128\cos^2\theta + 8(4 + \cos 2\theta). \tag{6.25}$$

On the other hand, since $[(h_1^2 + h_2^2)h_3']' = (2h_1 h_1' + 2h_2 h_2')h_3' + (h_1^2 + h_2^2)h_3''$, we get from the values (6.23) that

$$\frac{\mathrm{d}}{\mathrm{d}t}[(h_1^2 + h_2^2)h_3']\bigg|_{t=0} = 32 + 128\cos^2\theta. \tag{6.26}$$

Putting the results (6.25) and (6.26) into the expression (6.24), we observe that

$$g''(0) = 2[128\cos^2\theta + 8(4 + \cos 2\theta) - 32 - 128\cos^2\theta] = 16\cos 2\theta$$

Hence the argument of the proof of Problem 6.15 can be applied here to yield that $|f|$ has a local minimum at $z = 1$ on any line (6.21) with $\theta \in [0, \frac{\pi}{4}) \cup (\frac{3\pi}{4}, \frac{5\pi}{4}) \cup (\frac{7\pi}{4}, 2\pi]$. Similarly, $|f|$ has a local maximum at $z = 1$ through any line (6.21) with $\theta \in (\frac{\pi}{4}, \frac{3\pi}{4}) \cup (\frac{5\pi}{4}, \frac{7\pi}{4})$. Now the remaining unclear points are

$$\theta = \frac{\pi}{4}, \frac{3\pi}{4}, \frac{5\pi}{4} \quad \text{and} \quad \frac{7\pi}{4}. \tag{6.27}$$

However, we can check from the graphs of $|f|$ that the point $z = 1$ is in fact an inflection point[a] along the lines corresponding to these values (6.27).

Figure 6.1 to Figure 6.3 below are some typical examples for the above discussion. In fact, the orange line, the light red line and the blue line correspond to the graphs of the function $\sqrt{g(t)} = |f(z(t))|$, $g'(t)$ and $g''(t)$ respectively. They show us that $|f|$ has a local maximum, a local minimum and an inflection point at $z = 1$ on the line $1 + t\exp(\frac{\pi}{2})$, $1 + t\exp(\frac{33\pi}{18})$ and $1 + t\exp(\frac{3\pi}{4})$ respectively.

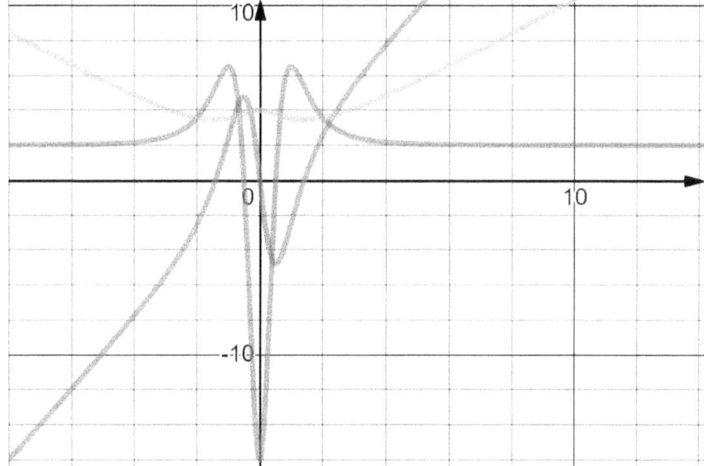

Figure 6.1: The function $|f(z(t))|$ has a local maximum at $z = 1$ on the line $1 + t\exp(\frac{\pi}{2})$.

---

[a]Points where the concavity changes direction, see [2, p. 191].

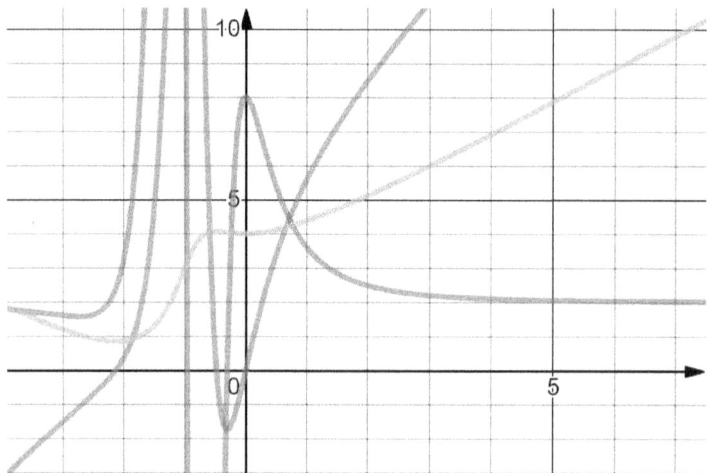

Figure 6.2: The function $|f(z(t))|$ has a local minimum at $z = 1$ on the line $1 + t\exp(\frac{33\pi}{18})$.

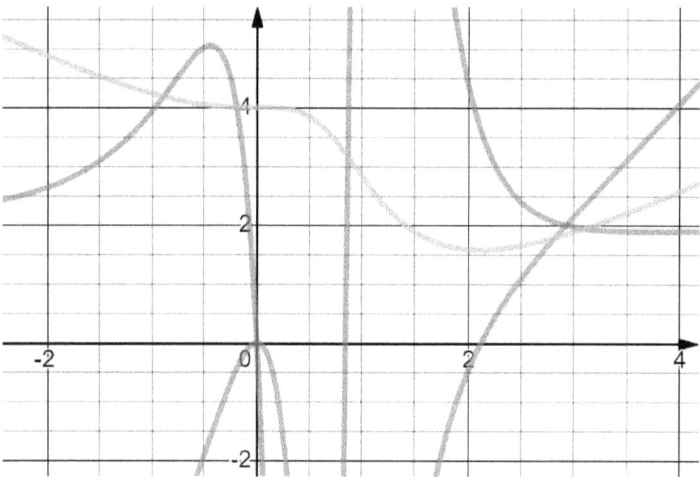

Figure 6.3: The function $|f(z(t))|$ has an inflection point at $z = 1$ on the line $1 + t\exp(\frac{3\pi}{4})$.

This completes the proof of the problem.

Problem 6.17

Bak and Newman Chapter 6 Exercise 17.

Proof.

(a) By direction computation, we have

$$f'(z) = \frac{z^3 + z - 2}{z^3} = \frac{(z-1)(z^2 + z + 2)}{z^3}.$$

So $f'(z) = 0$ if and only if $z = 1, \frac{-1 \pm i\sqrt{7}}{2}$. By Theorem 6.17, the saddle points of $f(z)$ are exactly

$$z_1 = \frac{-1 + i\sqrt{7}}{2} \quad \text{and} \quad z_2 = \frac{-1 - i\sqrt{7}}{2}.$$

(b) We divide the proof into several steps:

- **Step 1: Finding possible relative extrema of** $|f(z)|$**.** Note that $|z_1| = |z_2| = \sqrt{2}$. Let $z = \sqrt{2}e^{i\theta}$, where $\theta \in [0, 2\pi]$. Then we have

$$|f(z)| = |f(\sqrt{2}e^{i\theta})|$$

$$= \frac{\sqrt{3 - 2\sqrt{2}\cos\theta} \cdot \sqrt{5 - 4\cos 2\theta}}{2} \tag{6.28}$$

$$= \frac{\sqrt{15 - 12\cos 2\theta - 10\sqrt{2}\cos\theta + 8\sqrt{2}\cos\theta\cos 2\theta}}{2}. \tag{6.29}$$

Suppose that $g(\theta)$ is the function inside the square root of the expression (6.29). It is clear from the expression (6.29) that it suffices to find the extrema of $g(\theta)$. Simple differentiation shows that

$$g'(\theta) = 24\sin 2\theta + 10\sqrt{2}\sin\theta - 8\sqrt{2}\sin\theta\cos 2\theta - 16\sqrt{2}\cos\theta\sin 2\theta$$

$$= 24\sin 2\theta + 10\sqrt{2}\sin\theta - 8\sqrt{2}\sin\theta(1 - 2\sin^2\theta) - 32\sqrt{2}(1 - \sin^2\theta)\sin\theta$$

$$= 48\sqrt{2}\sin^3\theta - 30\sqrt{2}\sin\theta + 24\sin 2\theta.$$

Therefore, $g'(\theta) = 0$ if and only if

$$48\sqrt{2}\sin^3\theta - 30\sqrt{2}\sin\theta + 24\sin 2\theta = 0$$

$$\sin\theta(8\sqrt{2}\sin^2\theta - 5\sqrt{2} + 8\cos\theta) = 0$$

$$\sin\theta(-8\sqrt{2}\cos^2\theta + 8\cos\theta + 3\sqrt{2}) = 0$$

if and only if

$$\sin\theta = 0 \quad \text{or} \quad \cos\theta = \frac{3\sqrt{2}}{4}, -\frac{\sqrt{2}}{4}.$$

Note that $\cos\theta = \frac{3\sqrt{2}}{4}$ is impossible because of the definition. Consequently, Fermat's Theorem ensues that $g(\theta)$ takes its possible relative maxima or minima when $\sin\theta = 0$ or $\cos\theta = -\frac{\sqrt{2}}{4}$.

- **Step 2: Determination of the relative minima of** $|f(z)|$**.** We know that

$$g''(\theta) = 144\sqrt{2}\sin^2\theta\cos\theta - 30\sqrt{2}\cos\theta + 48\cos 2\theta.$$

If $\sin\theta = 0$, then $\cos\theta = \pm 1$ and $\cos 2\theta = 1$ so that

$$g''(\theta) = 48 \mp 30\sqrt{2} > 0.$$

By the Second Derivative Test, $g(\theta)$ takes its relative minima when $\sin\theta = 0$ and we follow from the expression (6.28) that

$$|f(\sqrt{2})| = \frac{\sqrt{3 - 2\sqrt{2}}}{2} \quad \text{or} \quad |f(-\sqrt{2})| = \frac{\sqrt{3 + 2\sqrt{2}}}{2}$$

are relative minima of $|f(z)|$.

- **Step 3: Determination of the relative maxima of $|f(z)|$.** If $\cos\theta = -\frac{\sqrt{2}}{4}$, then $\sin\theta = \pm\frac{\sqrt{14}}{4}$ and $\cos 2\theta = -\frac{3}{4}$. In this case, we have

$$g''(\theta) = -75 < 0,$$

so the Second Derivative Test implies that $g(\theta)$ and thus $|f(z)|$ takes its relative maxima when $\cos\theta = -\frac{\sqrt{2}}{4}$.

Next, the pair of values $\cos\theta = -\frac{\sqrt{2}}{4}$ and $\sin\theta = \frac{\sqrt{14}}{4}$ corresponds to

$$z = \sqrt{2}e^{i\theta} = -\frac{1}{2} + i\frac{\sqrt{28}}{4} = \frac{-1 + i\sqrt{7}}{2} = z_1.$$

Similarly, the pair of values $\cos\theta = -\frac{\sqrt{2}}{4}$ and $\sin\theta = -\frac{\sqrt{14}}{4}$ corresponds to

$$z = \sqrt{2}e^{i\theta} = -\frac{1}{2} - i\frac{\sqrt{28}}{4} = \frac{-1 - i\sqrt{7}}{2} = z_2.$$

Hence we conclude immediately that

$$|f(z_i)| = \max |f(z)|$$

on $|z| = \sqrt{2}$, where $i = 1, 2$.

(c) Rewrite f as

$$f(z) = \frac{(z-1)^2(z+1)}{z^2} = \left(1 - \frac{\bar{z}}{|z|^2}\right)^2(1+z) = \left(z - \frac{\bar{z}}{|z|^2}\right)\left(1 - \frac{\bar{z}}{|z|^2}\right) = z - 1 + \frac{1 - \bar{z}}{|z|^2}. \quad (6.30)$$

We apply the idea of the proof of Problem 6.16.

- **Case (i): Lines through z_1.** Consider

$$z_1(t) = z_1 + te^{i\theta},$$

where $t \in \mathbb{R}$ and $\theta \in [0, 2\pi]$ is the angle between the line and the positive real axis. Since $z_1(t) = -\frac{1}{2} + t\cos\theta + i(\frac{\sqrt{7}}{2} + t\sin\theta)$, we have

$$|z_1(t)|^2 = 2 + (\sqrt{7}\sin\theta - \cos\theta)t + t^2.$$

Since $\overline{z_1} = -\frac{1}{2} + t\cos\theta - i(\frac{\sqrt{7}}{2} + t\sin\theta)$, the expression (6.30) becomes

$$f(z_1(t)) = -\frac{3}{2} + t\cos\theta + i\left(\frac{\sqrt{7}}{2} + t\sin\theta\right) + \frac{\frac{3}{2} - t\cos\theta + i(\frac{\sqrt{7}}{2} + t\sin\theta)}{2 + (\sqrt{7}\sin\theta - \cos\theta)t + t^2}$$

$$= -\frac{3}{2} + t\cos\theta + \frac{\frac{3}{2} - t\cos\theta}{2 + (\sqrt{7}\sin\theta - \cos\theta)t + t^2}$$

$$i\left[\frac{\sqrt{7}}{2} + t\sin\theta + \frac{\frac{\sqrt{7}}{2} + t\sin\theta}{2 + (\sqrt{7}\sin\theta - \cos\theta)t + t^2}\right]$$

$$= \left(\frac{3}{2} - t\cos\theta\right) \cdot \frac{-1 - (\sqrt{7}\sin\theta - \cos\theta)t - t^2}{2 + (\sqrt{7}\sin\theta - \cos\theta)t + t^2}$$

$$+ i\left(\frac{\sqrt{7}}{2} + t\sin\theta\right) \cdot \frac{3 + (\sqrt{7}\sin\theta - \cos\theta)t + t^2}{2 + (\sqrt{7}\sin\theta - \cos\theta)t + t^2}$$

and then

$$g(t) = |f(z_1(t))|^2 = \frac{h_1^2(t) + h_2^2(t)}{h_3^2(t)},$$

where

$$h_1(t) = \left(\frac{3}{2} - t\cos\theta\right)[1 + (\sqrt{7}\sin\theta - \cos\theta)t + t^2],$$

$$h_2(t) = \left(\frac{\sqrt{7}}{2} + t\sin\theta\right)[3 + (\sqrt{7}\sin\theta - \cos\theta)t + t^2],$$

$$h_3(t) = 2 + (\sqrt{7}\sin\theta - \cos\theta)t + t^2.$$

Direct computation shows that

$$h_1(0) = \frac{3}{2}, \quad h_1'(0) = \frac{1}{2}(3\sqrt{7}\sin\theta - 5\cos\theta),$$

$$h_2(0) = \frac{3\sqrt{7}}{2}, \quad h_2'(0) = \frac{1}{2}(13\sin\theta - \sqrt{7}\cos\theta),$$

$$h_3(0) = 2, \quad h_3'(0) = \sqrt{7}\sin\theta - \cos\theta.$$

Therefore, we obtain

$$g'(0) = 2 \cdot \left.\frac{h_3(h_1 h_1' + h_2 h_2') - (h_1^2 + h_2^2)h_3'}{h_3^3}\right|_{t=0}$$

$$= \frac{2[\frac{3}{4}(3\sqrt{7}\sin\theta - 5\cos\theta) + \frac{3\sqrt{7}}{4}(13\sin\theta - \sqrt{7}\cos\theta)] - (\frac{9}{4} + \frac{63}{4})(\sqrt{7}\sin\theta - \cos\theta)}{4}$$

$$= \frac{6(4\sqrt{7}\sin\theta - 3\cos\theta) - 18(\sqrt{7}\sin\theta - \cos\theta)}{4}$$

$$= \frac{3\sqrt{7}}{2}\sin\theta$$

so that $g'(0) = 0$ if and only if $\theta = 0, \pi$. By the First Derivative Test, it is easily seen that $|f|$ has local minima at $z = z_1$ through the lines $z_1 \pm t$.

- **Case (ii): Lines through z_2.** Next, we consider

$$z_2(t) = z_2 + te^{i\theta}.$$

Since $z_2(t) = -\frac{1}{2} + t\cos\theta + i(-\frac{\sqrt{7}}{2} + t\sin\theta)$, we have

$$|z_2(t)|^2 = 2 - (\sqrt{7}\sin\theta + \cos\theta)t + t^2.$$

Since $\overline{z_2} = -\frac{1}{2} + t\cos\theta - i(-\frac{\sqrt{7}}{2} + t\sin\theta)$, the expression (6.30) gives

$$f(z_2(t)) = -\frac{3}{2} + t\cos\theta + i\left(-\frac{\sqrt{7}}{2} + t\sin\theta\right) + \frac{\frac{3}{2} - t\cos\theta + i(\frac{-\sqrt{7}}{2} + t\sin\theta)}{2 - (\sqrt{7}\sin\theta + \cos\theta)t + t^2}$$

$$= -\frac{3}{2} + t\cos\theta + \frac{\frac{3}{2} - t\cos\theta}{2 - (\sqrt{7}\sin\theta + \cos\theta)t + t^2}$$

$$i\left[-\frac{\sqrt{7}}{2} + t\sin\theta + \frac{\frac{-\sqrt{7}}{2} + t\sin\theta}{2 - (\sqrt{7}\sin\theta + \cos\theta)t + t^2}\right]$$

$$= \left(\frac{3}{2} - t\cos\theta\right) \cdot \frac{-1 + (\sqrt{7}\sin\theta + \cos\theta)t - t^2}{2 - (\sqrt{7}\sin\theta + \cos\theta)t + t^2}$$

$$+ i\left(-\frac{\sqrt{7}}{2} + t\sin\theta\right) \cdot \frac{3 - (\sqrt{7}\sin\theta + \cos\theta)t + t^2}{2 - (\sqrt{7}\sin\theta + \cos\theta)t + t^2}.$$

Thus we get

$$G(t) = |f(z_2(t))|^2 = \frac{H_1^2(t) + H_2^2(t)}{H_3^2(t)},$$

where

$$H_1(t) = \left(\frac{3}{2} - t\cos\theta\right)\left[-1 + (\sqrt{7}\sin\theta + \cos\theta)t - t^2\right],$$

$$H_2(t) = \left(-\frac{\sqrt{7}}{2} + t\sin\theta\right)\left[3 - (\sqrt{7}\sin\theta + \cos\theta)t + t^2\right],$$

$$H_3(t) = 2 - (\sqrt{7}\sin\theta + \cos\theta)t + t^2.$$

Simple differentiation yields that

$$H_1(0) = -\frac{3}{2}, \quad H_1'(0) = \frac{1}{2}(3\sqrt{7}\sin\theta + 5\cos\theta),$$

$$H_2(0) = -\frac{3\sqrt{7}}{2}, \quad H_2'(0) = \frac{1}{2}(13\sin\theta + \sqrt{7}\cos\theta),$$

$$H_3(0) = 2, \quad H_3'(0) = -(\sqrt{7}\sin\theta + \cos\theta).$$

Clearly, these values imply that

$$\begin{aligned}
G'(0) &= 2 \cdot \left.\frac{H_3(H_1 H_1' + H_2 H_2') - (H_1^2 + H_2^2)H_3'}{H_3^3}\right|_{t=0} \\
&= \frac{2\left[-\frac{3}{4}(3\sqrt{7}\sin\theta + 5\cos\theta) - \frac{3\sqrt{7}}{4}(13\sin\theta + \sqrt{7}\cos\theta)\right] + 18(\sqrt{7}\sin\theta + \cos\theta)}{4} \\
&= \frac{2(-12\sqrt{7}\sin\theta - 9\cos\theta) + 18(\sqrt{7}\sin\theta + \cos\theta)}{4} \\
&= -\frac{3\sqrt{7}}{2}\sin\theta.
\end{aligned}$$

Hence $G'(0) = 0$ if and only if $\theta = 0, \pi$. Similar to **Case (i)**, $|f|$ has local maxima at $z = z_2$ through the lines $z_2 \pm t$.

We complete the analysis of the problem.

CHAPTER **7**

Further Properties of Analytic Functions

Bak and Newman Chapter 7 Exercise 1.

Proof. Let K be the compact domain of f. Consider $e^{f(z)}$ and $e^{-if(z)}$. Since f nonconstant analytic on K, both $e^{f(z)}$ and $e^{-if(z)}$ are nonconstant analytic on K. Furthermore, we know that $e^{f(z)} \neq 0$ and $e^{-if(z)} \neq 0$ on K. Thus the Extreme Value Theorem, Theorems 6.13 (The Maximum Modulus Theorem) and 6.14 (The Minimum Modulus Theorem) together assert that $\max |e^{f(z)}|, \max |e^{if(z)}|, \min |e^{f(z)}|$ and $\min |e^{-if(z)}|$ assume on the boundary ∂K. Recall that e^x is an increasing function on $\mathbb{R}$, so the facts

$$|e^{f(z)}| = e^{\operatorname{Re} f(z)} \quad \text{and} \quad |e^{-if(z)}| = e^{\operatorname{Im} f(z)}$$

imply that maxima/minima of $e^{\operatorname{Re} f(z)}$ correspond to maxima/minima of $\operatorname{Re} f(z)$ and similarly, maxima/minima of $e^{\operatorname{Im} f(z)}$ correspond to maxima/minima of $\operatorname{Im} f(z)$. Hence $\operatorname{Re} f$ and $\operatorname{Im} f$ assume their maxima and minima on ∂K, completing the proof of the problem. ∎

Bak and Newman Chapter 7 Exercise 2.

Proof. Let f be a nonconstant analytic function on the region D. By Theorem 7.1 (The Open Mapping Theorem), $f(D)$ is open in $\mathbb{C}$. Since f is continuous and D is connected, $f(D)$ is also connected. By Definition 1.6, $f(D)$ is also a region. This completes the proof of the problem. ∎

Bak and Newman Chapter 7 Exercise 3.

Proof.

(a) Assume that z was an interior point of S. Then there exists a $\delta > 0$ such that $D(z; \delta) \subseteq S$. By Theorem 7.1 (The Open Mapping Theorem), $f(D(z; \delta))$ is an open set in $\mathbb{C}$ containing $f(z)$, i.e., $f(z)$ is an interior point of $f(S) = T$ which contradicts the hypothesis.

83

(b) The graph of the set S is shown in Figure 7.1 below. By the definition of $f(z) = z^2$,

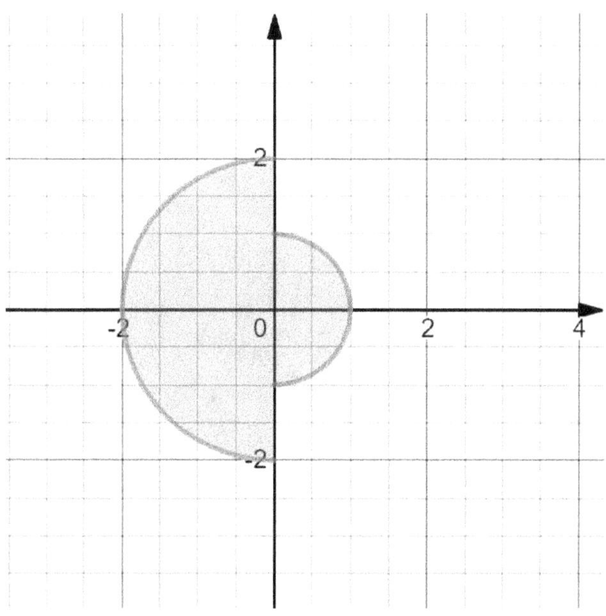

Figure 7.1: The graph of the set S.

we see that $f(S) \subseteq \overline{D(0;4)}$. We claim that $f(S) = \overline{D(0;4)}$. Let $u + iv \in \overline{D(0;4)}$ and $f(x + iy) = u + iv$. Then we have $u = x^2 - y^2$ and $v = 2xy$. Solving these equations give

$$4x^4 - 4ux^2 - v^2 = 0$$

and it implies that

$$x^2 = \frac{u \pm \sqrt{u^2 + v^2}}{2}.$$

If we take

$$x = -\sqrt{\frac{u + \sqrt{u^2 + v^2}}{2}} < 0,$$

then we have $z = x + iy \in S_1 \subset S$ and we have the claim. Next, the points $z_1 = 1$ and $z_2 = \frac{1}{2} + i\frac{\sqrt{3}}{2}$ are clearly boundary points on S. Since $f(z_1) = 1$ and $f(z_2) = -\frac{1}{2} + i\frac{\sqrt{3}}{2}$, they are clearly interior points of $\overline{D(0;4)}$.

We complete the proof of the problem.

Problem 7.4

Bak and Newman Chapter 7 Exercise 4.

Proof. Suppose that f is nonconstant and $D = D(0;1)$. We want to verify that

$$f(D) = D. \tag{7.1}$$

If $\zeta \in D \setminus f(D)$, then the map $g : \overline{D} \to \overline{D}$ defined by

$$g(z) = \frac{1}{f(z) - \zeta}$$

is nonconstant C-analytic in D. Now the Extreme Value Theorem and Theorem 6.13 (The Maximum Modulus Theorem) assert that $|g(z)| \le \max\limits_{w \in C(0;1)} |g(w)|$ and then

$$\left| \frac{1}{f(z) - \zeta} \right| \le \max_{w \in C(0;1)} \left| \frac{1}{f(w) - \zeta} \right| \le \max_{w \in C(0;1)} \left| \frac{1}{|f(w)| - |\zeta|} \right| = \frac{1}{1 - |\zeta|}$$

for all $z \in D$. Thus we have $0 < 1 - |\zeta| \le |f(z) - \zeta|$ which means that the distance between the point ζ and the set $f(D)^{\mathrm{a}}$ is at least $1 - |\zeta|$. Therefore, we have

$$D\left(\zeta, \frac{1 - |\zeta|}{2}\right) \subseteq D \setminus f(D)$$

and so the set $D \setminus f(D)$ is open in $\mathbb{C}$. Next, we follow from Theorem 7.1 (The Open Mapping Theorem) that $f(D)$ is open.

Recall the set relation

$$D = (D \setminus f(D)) \cup f(D). \tag{7.2}$$

Since $D \setminus f(D)$ and $f(D)$ are disjoint open subsets of D, the relation (7.2) implies that D is disconnected by the definition, but this is definitely a contradiction. In other words, we have either $f(D) = D$ or $f(D) = \varnothing$. However, the latter is impossible by the hypothesis and then we have the result (7.1), completing the analysis of the problem. ▪

Problem 7.5

Bak and Newman Chapter 7 Exercise 5.

Proof. Let $D = D(0;1)$. If $f(z) = 1$, then we are done, so we suppose that f is nonconstant. We claim that the number of zeros of f inside $\overline{D}$ is finite. To this end, let $\{\alpha_n\}$ be the sequence of zeros (counted with their multiplicities) of f inside $\overline{D}$. Since $|f(z)| = 1$ on $|z| = 1$, we actually have $\{\alpha_n\} \subseteq D$. As a consequence of the Bolzano-Weierstrass Theorem, $\{\alpha_n\}$ contains a convergent subsequence $\{\alpha_{n_k}\}$. Let $\alpha_{n_k} \to \alpha_0$. Since f is continuous in $\overline{D}$, we have

$$0 = \lim_{k \to \infty} f(\alpha_{n_k}) = f(\alpha_0)$$

which implies that $\alpha_0 \notin C(0;1)$, i.e., $\alpha_0 \in D$. Since D is a region, Theorem 6.9 (The Uniqueness Theorem) says that $f \equiv 0$ in D but it contradicts the continuity of f in $\overline{D}$. This proves our claim.

Next, we define the function

$$g(z) = \frac{f(z)}{B_{\alpha_1}(z) B_{\alpha_2}(z) \cdots B_{\alpha_n}(z)}, \tag{7.3}$$

where $B_\alpha(z)$ is the Möbius transformation considered on [4, p. 95]. There are two cases.

- **Case (i): g is constant.** Then it means that

$$f(z) = c B_{\alpha_1}(z) B_{\alpha_2}(z) \cdots B_{\alpha_n}(z) \tag{7.4}$$

for some $c \in \mathbb{C}$. If $\alpha_j \ne 0$, then f is undefined at $z = \frac{1}{\overline{\alpha_j}}$ which contradicts the hypothesis that f is entire. Consequently, we have $\alpha_j = 0$ for all $j = 1, 2, \ldots, n$ so that the expression (7.4) reduces to

$$f(z) = c z^n$$

as required.

---

[a] That is $\rho_{f(D)}(\zeta) = \inf\limits_{z \in D} |f(z) - \zeta|$.

- **Case (ii): g is nonconstant.** We claim that g is analytic in D. In fact, it is clear for $z \in D \setminus \{\alpha_1, \alpha_2, \ldots, \alpha_n\}$. In the case $z = \alpha_j$, we need the following lemma:[b]

> **Lemma 7.1**
>
> If f is analytic in a region G and $f \not\equiv 0$ on G, then for each $\alpha \in G$ with $f(\alpha) = 0$ there exists an integer $n \geq 1$ and an analytic function $g : G \to \mathbb{C}$ such that
> $$f(z) = (z - \alpha)^n g(z) \quad \text{and} \quad g(\alpha) \neq 0.$$

Apply Lemma 7.1 to the function $f(z)$, the expression (7.3) can be rewritten as

$$g(z) = \frac{(z - \alpha_1)(z - \alpha_2) \cdots (z - \alpha_n) h(z)}{\displaystyle\prod_{j=1}^{n} \frac{z - \alpha_j}{1 - \overline{\alpha_j} z}} = h(z) \prod_{j=1}^{n} (1 - \overline{\alpha_j} z), \qquad (7.5)$$

where h is analytic in D and $h(\alpha_j) \neq 0$ for all $j = 1, 2, \ldots, n$. Thus the representation (7.5) implies that $g(z)$ is analytic at $z = \alpha_j$ for all $j = 1, 2, \ldots, n$.

Since g is nonconstant analytic in D, we apply the Extreme Value Theorem and Theorem 6.13 (The Maximum Modulus Theorem) to the expression (7.3) to get

$$|g(z)| \leq \max_{|\zeta|=1} |g(\zeta)| = \max_{|\zeta|=1} \frac{|f(\zeta)|}{\left| \displaystyle\prod_{j=1}^{n} \frac{\zeta - \alpha_j}{1 - \overline{\alpha_j} \zeta} \right|} = \max_{|\zeta|=1} |f(\zeta)| = 1 \qquad (7.6)$$

for all $z \in \overline{D}$. Recall that $|\alpha_j| < 1$ so that $|\overline{\alpha_j}|^{-1} > 1$. By the representation (7.5) and the property of h, we see that $g(z) \neq 0$ for all $z \in D$. Now it follows from Theorem 6.14 (The Minimum Modulus Theorem) that

$$1 = \min_{|\zeta|=1} |f(\zeta)| = \min_{|\zeta|=1} |g(\zeta)| \leq |g(z)| \qquad (7.7)$$

for all $z \in \overline{D}$. Combining the inequalities (7.6) and (7.7), we have $|g(z)| = 1$ for all $z \in \overline{D}$, i.e., $g(z) = e^{i\theta}$ for some θ so that f has the form (7.4) and then our result follows in this case.

We have completed the proof of the problem.

> **Problem 7.6**
>
> *Bak and Newman Chapter 7 Exercise 6.*[*]

Proof. Suppose that $D = D(0; 1)$ and $f(z)$ has poles at $\{\alpha_1, \alpha_2, \ldots, \alpha_n\} \subseteq D$. Let $f(z) = \frac{P(z)}{Q(z)}$, where $P(z)$ and $Q(z)$ are polynomials. Then we have

$$f(z) - \frac{P(z)}{\displaystyle\prod_{j=1}^{n}(z - \alpha_j)}.$$

---

[b]Read [7, Corollary 3.9, p. 79] for a proof of it. The case for polynomials has been given on [4, p. 67].

Define

$$g(z) = f(z)B_{\alpha_j}(z) = \frac{P(z)}{\prod\limits_{j=1}^{n}(z - \alpha_j)} \cdot \prod_{j=1}^{n} \frac{z - \alpha_j}{1 - \overline{\alpha_j}z} = \frac{P(z)}{\prod\limits_{j=1}^{n}(1 - \overline{\alpha_j}z)} \tag{7.8}$$

which is rational in $\overline{D}$. Since $|\alpha_j| < 1$, we have $|\overline{\alpha_j}|^{-1} > 1$ and then the expression (7.8) ensures that $g(z)$ has no poles in D. Furthermore, the representation (7.8) also implies that, on $|z| = 1$,

$$|g(z)| = |f(z)| \cdot |B_{\alpha_j}(z)| = |f(z)|.$$

This completes the proof of the problem. ∎

Problem 7.7

Bak and Newman Chapter 7 Exercise 7.[*]

Proof.

(a) Let $D = D(0; 1)$ and $|a| < 1$. Recall that $B_a : \overline{D} \to \overline{D}$ given by

$$B_a(\omega) = \frac{\omega - a}{1 - \overline{a} \cdot \omega} \tag{7.9}$$

is analytic and bounded in $\overline{D}$ by 1. Put $a = \frac{\alpha}{R}$ and $\omega = \frac{z}{R}$ into the representation (7.9) to get

$$B_{\frac{\alpha}{R}}\left(\frac{z}{R}\right) = \frac{\frac{z}{R} - \frac{\alpha}{R}}{1 - \frac{\overline{\alpha}}{R} \cdot \frac{z}{R}} = \frac{R(z - \alpha)}{R^2 - \overline{\alpha} \cdot z}$$

which is analytic in $\overline{D(0; R)}$. Note that $|\alpha| < R$. Following the idea on [4, p. 95], on $|z| = R$, we have

$$\left|B_{\frac{\alpha}{R}}\left(\frac{z}{R}\right)\right|^2 = \frac{R(z - \alpha)}{R^2 - \overline{\alpha} \cdot z} \cdot \frac{R(\overline{z} - \overline{\alpha})}{R^2 - \alpha \cdot \overline{z}}$$

$$= \frac{R^2(|z|^2 - \alpha\overline{z} - \overline{\alpha}z + |\alpha|^2)}{R^4 - R^2\alpha\overline{z} - R^2\overline{\alpha}z + |\alpha|^2 \cdot |z|^2}$$

$$= \frac{R^2 - \alpha\overline{z} - \overline{\alpha}z + |\alpha|^2}{R^2 - \alpha\overline{z} - \overline{\alpha}z + |\alpha|^2}$$

$$= 1.$$

In other words, the function $B_{\frac{\alpha}{R}}\left(\frac{z}{R}\right)$ maps the circle $|z| = R$ into $C(0; 1)$.

(b) Let $f(z) = (z - \alpha_1)(z - \alpha_2)\cdots(z - \alpha_n)$ and $g(z) = \dfrac{1}{R^n}\prod\limits_{k=1}^{n}(R^2 - \overline{\alpha_k}z)$. It is clear that both f and g are analytic throughout $\overline{D(0; R)}$ and part (a) ensures that they have the same magnitude on $|z| = R$, i.e.,

$$|f(z)| = |g(z)| \tag{7.10}$$

on $|z| = R$. Since we have

$$g(0) = \frac{1}{R^n} \cdot R^{2n} = R^n \neq 0$$

Theorem 6.13 (The Maximum Module Theorem) and Theorem 6.14 (The Minimum Module Theorem) imply that there exist points $|z_1| = |z_2| = R$ such that

$$|g(z_1)| > R^n \quad \text{and} \quad |g(z_2)| < R^n. \tag{7.11}$$

Consequently, we obtain from the results (7.10) and (7.11) that

$$\sqrt[n]{|z_1 - \alpha_1| \cdot |z_1 - \alpha_2| \cdots |z_1 - \alpha_n|} = \sqrt[n]{|f(z_1)|} > R$$

and

$$\sqrt[n]{|z_2 - \alpha_1| \cdot |z_2 - \alpha_2| \cdots |z_2 - \alpha_n|} = \sqrt[n]{|f(z_2)|} < R.$$

We end the proof of the problem.

> ## Problem 7.8
>
> *Bak and Newman Chapter 7 Exercise 8.*

Proof. If f is constant, then there is nothing to prove. Thus we may assume that f is nonconstant. Since f is analytic in the annulus $1 \leq |z| \leq 2$, if we define

$$g(z) = \frac{f(z)}{z^2},$$

then g is also nonconstant analytic in the annulus. By Theorem 6.13 (The Maximum Modulus Theorem), $\max |g(\omega)|$ occurs on $|\omega| = 1$ or $|\omega| = 2$ so that we obtain

$$\frac{|f(z)|}{|z|^2} = |g(z)| \leq \max |g(\omega)| = \max \frac{|f(\omega)|}{|\omega|^2} \leq 1$$

throughout the annulus. Hence we have $|f(z)| \leq |z|^2$ throughout the annulus which ends the proof of the problem.

> ## Problem 7.9
>
> *Bak and Newman Chapter 7 Exercise 9.*

Proof. Define $F(z) = \frac{1}{10} f(2z)$. Then F is analytic in $D(0; 1)$. Furthermore, we have $|F(z)| \leq 1$ in $D(0; 1)$ and $F(\frac{1}{2}) = 0$. It is easily seen that our F is exactly the function f considered in [4, Example 1, p. 95]. Therefore, we have

$$|F(z)| \leq \left| \frac{z - \frac{1}{2}}{1 - \frac{1}{2}z} \right|$$

or equivalently,

$$|f(2z)| \leq 10 \left| \frac{z - \frac{1}{2}}{1 - \frac{1}{2}z} \right|. \tag{7.12}$$

Put $z = \frac{1}{4}$ into the inequality (7.12), we get

$$\left| f\left(\frac{1}{2}\right) \right| \leq \frac{20}{7},$$

completing the proof of the problem.

> ## Problem 7.10
>
> *Bak and Newman Chapter 7 Exercise 10.*

Proof. Let $D = D(0; 1)$. We define $g : D \to \mathbb{C}$ by

$$g = B_{f(\alpha)} \circ f$$

where B_α is the Möbius transformation

$$B_\alpha(z) = \frac{z - \alpha}{1 - \overline{\alpha}z}.$$

Since f and $B_{f(\alpha)}$ are analytic in D, g is also analytic in D. Furthermore, since f and $B_{f(\alpha)}$ are bounded by 1 in D, g is also bounded by 1 in D. By direct computation, we know that

$$B_\alpha'(\alpha) = \frac{1}{1 - |\alpha|^2}. \tag{7.13}$$

Using Problem 3.3 and the fact (7.13), we obtain

$$g'(\alpha) = B_{f(\alpha)}'(f(\alpha)) \times f'(\alpha) = \frac{f'(\alpha)}{1 - |f(\alpha)|^2} \tag{7.14}$$

Since $f(\alpha) \neq 0$, we have $1 - |f(\alpha)|^2 < 1$. Furthermore, it is impossible that $|f(\alpha)| = 1$ by Theorem 6.13 (The Maximum Modulus Theorem). Hence the expression (7.14) implies that

$$|f'(\alpha)| < |g'(\alpha)|.$$

This completes the proof of the problem. $\blacksquare$

Problem 7.11

Bak and Newman Chapter 7 Exercise 11.

Proof. By Problem 7.10, we may assume that $f(\alpha) = 0$ so that

$$f'(\alpha) = \lim_{z \to \alpha} \frac{f(z) - f(\alpha)}{z - \alpha} = \lim_{z \to \alpha} \frac{f(z)}{z - \alpha}. \tag{7.15}$$

Again the assumption and [4, Example 1, p. 95] give

$$|f(z)| \leq |B_\alpha(z)| \tag{7.16}$$

throughout $D(0; 1)$. Therefore, the limit (7.15) and the estimate (7.16) combine to imply

$$f'(\alpha) \ll \lim_{z \to \alpha} \frac{B_\alpha(z)}{z - \alpha} = \lim_{z \to \alpha} \frac{B_\alpha(z) - B_\alpha(\alpha)}{z - \alpha} = B_\alpha'(\alpha).$$

Since f is arbitrary, we conclude that

$$\max_f |f'(\alpha)| = B_\alpha'(\alpha).$$

This ends the proof of the problem. $\blacksquare$

Problem 7.12

Bak and Newman Chapter 7 Exercise 12.

Proof. We imitate the proof of Proposition 7.3. In fact, we introduce the auxiliary function

$$g(z) = (z^2 + R^2)^2 f(z).$$

For any z with $|z| = R$ and $\operatorname{Im} z \geq 0$, we have

$$|(z - iR)^2 f(z)| \leq \left(\frac{|z - iR|}{|\operatorname{Re} z|}\right)^2 = \sec^2 \theta$$

for some $\theta \in [0, \frac{\pi}{4}]$ so that

$$|(z - iR)^2 f(z)| \leq 2. \tag{7.17}$$

Similarly, if $|z| = R$ and $\operatorname{Im} z \leq 0$, then we obtain

$$|(z + iR)^2 f(z)| \leq 2. \tag{7.18}$$

Combining the estimates (7.17) and (7.18), for all z with $|z| = R$, we get

$$|g(z)| = |(z + iR)^2| \cdot |(z - iR)^2| \cdot |f(z)| \leq 2 \times (|z| + |R|)^2 = 8R^2.$$

Since g is nonconstant analytic in $\overline{D(0; R)}$, Theorem 6.13 (The Maximum Modulus Theorem) implies that

$$|g(z)| \leq 8R^2$$

in $D(0; R)$ and so

$$|f(z)| \leq \frac{8R^2}{|z^2 + R^2|^2} \tag{7.19}$$

for all $z \in D(0; R)$. If we take $R \to \infty$, then we deduce from the inequality (7.19) that $f(z) = 0$ for all $z \in \mathbb{C}$, completing the proof of the problem. ∎

> **Problem 7.13**
>
> *Bak and Newman Chapter 7 Exercise 13.*

Proof.

(a) Let Γ be the boundary of a rectangle R in $\mathbb{C}$ and $z \in \mathbb{C}$. We want to apply Theorem 7.4 (Morera's Theorem), so we check its hypotheses into four steps.

- **Step 1: The continuity of an auxiliary function.** Define $\varphi : \mathbb{C} \times [0, 1] \to \mathbb{C}$ by

$$\varphi(z, t) = \begin{cases} \dfrac{\sin zt}{t}, & \text{if } t \neq 0; \\[2ex] z, & \text{if } t = 0. \end{cases} \tag{7.20}$$

We claim that φ is continuous on $\mathbb{C} \times [0, 1]$. To this end, we need the following lemma:

> **Lemma 7.2**
>
> Suppose that ζ is complex. Then we have
>
> $$\lim_{\zeta \to 0} \frac{\sin \zeta}{\zeta} = 1.$$

Proof of Lemma 7.2. Notice that

$$\lim_{\zeta \to 0} \frac{\sin \zeta}{\zeta} = \lim_{\zeta \to 0} \frac{\sin \zeta - \sin 0}{\zeta - 0} = \cos 0 = 1$$

as required. ∎

It is clear from the definition (7.20) that $\varphi(z, t)$ is continuous for every $z \in \mathbb{C}$ and $t \in (0, 1]$, so it suffices to check its continuity at $t = 0$. In fact, for every fixed $z \in \mathbb{C}$, we get from Lemma 7.2 that

$$\lim_{\substack{t \to 0 \\ t \neq 0}} \varphi(z, t) = z \lim_{\substack{zt \to 0 \\ t \neq 0}} \frac{\sin zt}{zt} = z = \varphi(z, 0).$$

This shows that φ is continuous on $\mathbb{C} \times [0, 1]$ as desired.

- **Step 2: f is continuous on $\mathbb{C}$.** With the aid of the auxiliary function (7.20), we can write

$$f(z) = \int_0^1 \varphi(z, t) \, \mathrm{d}t$$

so that

$$|f(z) - f(\omega)| \leq \int_0^1 |\varphi(z, t) - \varphi(\omega, t)| \, \mathrm{d}t. \tag{7.21}$$

Given $\epsilon > 0$. By **Step 1**, there exists a $\delta > 0$ such that $|(z, t) - (\omega, s)| < \delta$ implies

$$|\varphi(z, t) - \varphi(\omega, s)| < \epsilon.$$

Particularly, if we take $t = s$, then $|z - \omega| < \delta$ implies

$$|\varphi(z, t) - \varphi(\omega, t)| < \epsilon$$

and then the inequality (7.21) can be reduced to

$$|f(z) - f(\omega)| < \epsilon.$$

In other words, f is continuous on $\mathbb{C}$.

- **Step 3: Absolute Convergence of the Double Integral.** To evaluate the integral

$$\int_\Gamma f(z) \, \mathrm{d}z = \int_\Gamma \int_0^1 \frac{\sin zt}{t} \, \mathrm{d}t \, \mathrm{d}z, \tag{7.22}$$

we need to interchange the order of integration of the double integral (7.22). As suggested by the example given in [4, pp. 98, 99], we have to establish the absolute convergence of the double integral (7.22). To this end, we fix $z \in \mathbb{C}$. Now Lemma 7.2 ensures that there exists a $M > 0$ such that

$$\left| \frac{\sin \zeta}{\zeta} \right| \leq M \tag{7.23}$$

for all $0 < |\zeta| \leq |z|$. If $\zeta = tz$, where $t \in (0, 1]$, then the inequality (7.23) implies that

$$\left| \frac{\sin zt}{t} \right| \leq M|z|$$

which gives

$$\int_0^1 \left| \frac{\sin zt}{t} \right| \, \mathrm{d}t \leq M|z|. \tag{7.24}$$

Since Γ is a rectangle, it must be of finite length. Since the function $F(z) = |z|$ is continuous on the compact set Γ, F is bounded there. Hence we follow from the inequality (7.24) and Theorem 4.10 (The M-L Formula) that

$$\int_\Gamma \int_0^1 \left| \frac{\sin zt}{t} \right| \mathrm{d}t\, \mathrm{d}z \le M \int_\Gamma |z|\, \mathrm{d}z < \infty.$$

- **Step 4: The Application of Theorem 7.4 (Morera's Theorem).** For each fixed $t \in [0,1]$, since $\sin zt$ is entire, we follow from Theorem 4.14 (The Rectangle Theorem) that

$$\int_\Gamma \sin zt\, \mathrm{d}z = 0.$$

Thus we establish from this and **Step 3** that

$$\int_\Gamma f(z)\, \mathrm{d}z = \int_\Gamma \int_0^1 \frac{\sin zt}{t}\, \mathrm{d}t\, \mathrm{d}z = \int_0^1 \int_\Gamma \frac{\sin zt}{t}\, \mathrm{d}z\, \mathrm{d}t = \int_0^1 \frac{1}{t} \left(\int_\Gamma \sin zt\, \mathrm{d}z \right) \mathrm{d}t = 0.$$

Finally, we conclude that f is entire.

(b) Fix $z \in \mathbb{C}$ and let $R = |z|$. By Problem 3.22, we have

$$\frac{\sin zt}{t} = \sum_{n=0}^{\infty} \frac{(-1)^n z^{2n+1}}{(2n+1)!} t^{2n}, \tag{7.25}$$

where $t \in (0,1]$. It is clear that $tz \in \overline{D(0;R)}$, so we see that

$$\left| \frac{(-1)^n z^{2n+1}}{(2n+1)!} t^{2n} \right| \le \frac{R^{2n+1}}{(2n+1)!}.$$

Since

$$\limsup_{n\to\infty} \frac{R^{2n+3}}{(2n+3)!} \times \frac{(2n+1)!}{R^{2n+1}} = \limsup_{n\to\infty} \frac{R^2}{(2n+3)(2n+2)} = 0,$$

the Ratio Test [27, Theorem 6.8, p. 77] implies that the series

$$\sum_{n=0}^{\infty} \frac{R^{2n+1}}{(2n+1)!}$$

converges. Thus the Weierstrass M-test [28, Theorem 10.5, p. 3] asserts that the series (7.25) converges uniformly for all $t \in (0,1]$. Consequently, this fact allows us to interchange the integration and summation in the following calculation

$$\begin{aligned}
f(z) &= \int_0^1 \sum_{n=0}^{\infty} \frac{(-1)^n z^{2n+1}}{(2n+1)!} t^{2n}\, \mathrm{d}t \\
&= \sum_{n=0}^{\infty} \frac{(-1)^n z^{2n+1}}{(2n+1)!} \int_0^1 t^{2n}\, \mathrm{d}t \\
&= \sum_{n=0}^{\infty} \frac{(-1)^n z^{2n+1}}{(2n+1)(2n+1)!}. \tag{7.26}
\end{aligned}$$

Since z is arbitrary, the power series representation (7.25) of f holds throughout its domain of convergence. To determine the radius of convergence, since

$$\lim_{n\to\infty} \frac{(2n+1)(2n+1)!}{(2n+3)(2n+3)!} = \lim_{n\to\infty} \frac{2n+1}{(2n+3)^2(2n+2)} = 0,$$

we apply Problem 2.13 and Theorem 2.8 to conclude that the power series (7.26) converges for all $z \in \mathbb{C}$. In other words, f is entire.

We have completed the proof of the problem.

> **Remark 7.1**
>
> (a) In **Step 3** above, the reason why absolute convergence of the double integral implies the interchange of order of integration depends on the application of an advanced tool called **Fubini Theorem**, see [23, Theorem 8.8, pp. 164, 165].
>
> (b) We also remark that Problem 7.13(a) is actually a special case of the complex version of the so-called **Leibniz's Rule**: Suppose that $\varphi(z, t)$ is defined on $\Omega \times [0, 1]$, where Ω is open in $\mathbb{C}$. If $\varphi(z, t)$ is analytic in z for each t and φ is continuous on $\Omega \times [0, 1]$, then the function $f : \Omega \to \mathbb{C}$ given by
>
> $$f(z) = \int_0^1 \varphi(z, t) \, dt$$
>
> is analytic. This statement can be found in Conway [7, Exercise 2, pp. 73, 74] or Stein and Shakarchi [24, Theorem 5.4, p. 56]. The tools Stein and Shakarchi applied there are Riemann sums and Theorem 7.6. The corresponding real version can be found in [22, Theorem 9.42, pp. 236, 237].

> **Problem 7.14**
>
> *Bak and Newman Chapter 7 Exercise 14.*

Proof.

(a) Clearly, we have

$$f(z) = \int_0^1 \frac{\sin zt}{t} \, dt = \int_0^1 \int_0^z \cos \zeta t \, d\zeta \, dt = \int_0^z \left(\int_0^1 \cos \zeta t \, dt \right) d\zeta.$$

We let

$$g(\zeta) = \int_0^1 \cos \zeta t \, dt = \frac{\sin \zeta}{\zeta}$$

which is entire. Now by the proof of Theorem 4.15 (The Integral Theorem), we conclude immediately that

$$f'(z) = \frac{d}{dz} \int_0^z g(\zeta) \, d\zeta = g(z) = \int_0^1 \cos zt \, dt.$$

(b) Applying Theorem 2.9 to the power series (7.26) and then using Problem 3.22, we conclude that

$$f'(z) = \sum_{n=0}^{\infty} \frac{(-1)^n z^{2n}}{(2n+1)!} = \frac{\sin z}{z} = \int_0^1 \cos zt \, dt.$$

We end the proof of the problem. $\blacksquare$

> **Problem 7.15**
>
> *Bak and Newman Chapter 7 Exercise 15.*

Proof. It is reasonable to assume that $z_0 \neq z_1$. For every real x, we have

$$g(x) = z_0 + x e^{i\theta}$$

$$= z_0 + \frac{x|z_1 - z_0| \cdot e^{i\operatorname{Arg}(z_1 - z_0)}}{|z_1 - z_0|}$$

$$= z_0 + \frac{x(z_1 - z_0)}{|z_1 - z_0|}$$

$$= \left(1 - \frac{x}{|z_1 - z_0|}\right) z_0 + \frac{x}{|z_1 - z_0|} z_1. \tag{7.27}$$

Therefore, the formula (7.27) implies exactly that $g(\mathbb{R}) = L$. This completes the proof of the problem. ∎

Problem 7.16

Bak and Newman Chapter 7 Exercise 16.

Proof. Let $\Omega^+ = \{z \in \mathbb{C} \,|\, \operatorname{Im} z > 0\}$ and $\Omega^- = \{z \in \mathbb{C} \,|\, \operatorname{Im} z < 0\}$. Since f is analytic in $\Omega^+ \cup \mathbb{R}$ and f is real for real z, Theorem 7.8 (The Schwarz Reflection Principle)[c] implies that f can be extended to an entire function F such that $F(z) = f(z)$ on Ω^+. Since f is bounded in $\Omega^+ \cup \mathbb{R}$, F is bounded in $\mathbb{C}$. By Theorem 5.10 (Liouville's Theorem), F and then f is constant which completes the proof of the problem. ∎

Problem 7.17

Bak and Newman Chapter 7 Exercise 17.

Proof. Since f is entire, it deduces from Theorem 5.5 (The Taylor Expansion of an Entire Function) that

$$f(z) = \sum_{k=0}^{\infty} \frac{f^{(k)}(0)}{k!} z^k \tag{7.28}$$

for all $z \in \mathbb{C}$. Since $\mathbb{C}$ is certainly a region symmetric with respect to $\mathbb{R}$ and f is real on $\mathbb{R}$, Corollary 7.9 shows that

$$f(z) = \overline{f(\overline{z})} \tag{7.29}$$

for all $z \in \mathbb{C}$. Combining the two expressions (7.28) and (7.29), we obtain

$$\sum_{k=0}^{\infty} \frac{f^{(k)}(0)}{k!} z^k = \overline{\sum_{k=0}^{\infty} \frac{f^{(k)}(0)}{k!} \overline{z}^k} = \sum_{k=0}^{\infty} \frac{\overline{f^{(k)}(0)}}{k!} z^k$$

for all $z \in \mathbb{C}$. By Theorem 6.9 (The Uniqueness Theorem), we have

$$\frac{f^{(k)}(0)}{k!} = \frac{\overline{f^{(k)}(0)}}{k!} \tag{7.30}$$

for all $k = 0, 1, 2, \ldots$.

Next, we consider

$$g(z) = if(iz) = \sum_{k=0}^{\infty} i^{k+1} \cdot \frac{f^{(k)}(0)}{k!} z^k.$$

---

[c]Here we use the stronger form of the Schwarz Reflection Principle given in [7, §1, pp. 210 - 213] or [24, p. 60] because the form of the principle in our book is *only* applicable to "$\mathbb{C}$-analytic function f in a region D" whose definition assumes that D is a bounded region.

Since f is imaginary on the imaginary axis, g is real on the real axis. Thus the preceding paragraph shows that

$$i^{k+1}\frac{f^{(k)}(0)}{k!} = \overline{\frac{i^{k+1}f^{(k)}(0)}{k!}} = \overline{i^{k+1}}\cdot\overline{\frac{f^{(k)}(0)}{k!}}. \tag{7.31}$$

If k is even, then $k = 2m$ so that $\overline{i^{k+1}} = \overline{(-1)^m i} = -(-1)^m i = -i^{2m+1} = -i^{k+1}$ and the expression (7.31) reduces to

$$\frac{f^{(k)}(0)}{k!} = -\overline{\frac{f^{(k)}(0)}{k!}}. \tag{7.32}$$

Combining the results (7.30) and (7.32), we conclude that

$$\frac{f^{(k)}(0)}{k!} = 0$$

for even k. Hence the representation (7.28) can be written as

$$f(z) = \sum_{m=0}^{\infty} \frac{f^{(2m+1)}(0)}{(2m+1)!} z^{2m+1}$$

which is indeed an odd function. This completes the proof of the problem. $\blacksquare$

> **Problem 7.18**
>
> *Bak and Newman Chapter 7 Exercise 18.**

Proof. Let $z = x + iy$, $\omega = \gamma(z) = u + iv$ and $\omega^* = \gamma(\bar z)$. Since the line $u = v$ can be parameterized by $\gamma(t) = t + it$, where $t \in \mathbb{R}$, we have $z + iz = u + iv$ so that

$$(x + iy) + i(x + iy) = u + iv$$
$$(x - y) + i(x + y) = u + iv.$$

Thus we have $x = \frac{u+v}{2}$ and $y = \frac{v-u}{2}$ so that

$$\omega^* = \gamma(\bar z) = \gamma\left(\overline{\frac{u+v}{2} + i\frac{v-u}{2}}\right) = \frac{u+v}{2} - i\frac{v-u}{2} + i\left(\frac{u+v}{2} - i\frac{v-u}{2}\right) = v + iu$$

as required, completing the proof of the problem. $\blacksquare$

> **Problem 7.19**
>
> *Bak and Newman Chapter 7 Exercise 19.*

Proof. Let $D = \{z \in \mathbb{C} \,|\, |z| < 1 \text{ and } \operatorname{Im} z > 0\}$ and $D^* = \{z \in \mathbb{C} \,|\, |z| > 1 \text{ and } \operatorname{Im} z > 0\}$. By Problem 1.17, we know that the mapping

$$\omega = i\frac{z-1}{z+1} \tag{7.33}$$

maps the upper unit circle $C^+ = \{z \in \mathbb{C} \,|\, |z| = 1 \text{ and } \operatorname{Im} z > 0\}$ into the negative real axis $\mathbb{R}^-$. In fact, it is easy to show that this mapping is bijective. Therefore, its inverse

$$z = \frac{i+\omega}{i-\omega} \tag{7.34}$$

maps $\mathbb{R}^-$ one-one and onto C^+.

We claim that the mapping (7.34) maps the region $H = \{\omega \in \mathbb{C} \,|\, \mathrm{Re}\,\omega < 0 \text{ and } \mathrm{Im}\,\omega < 0\}$ onto D.[d] Let $\omega = a + ib$, where $a < 0$ and $b < 0$. Then direct computation gives

$$z = \frac{a + (b+1)i}{-a - (b-1)i} = \frac{1 - a^2 - b^2}{a^2 + (b-1)^2} - i\frac{2a}{a^2 + (b-1)^2}$$

and so

$$|z|^2 = \frac{(1 - a^2 - b^2)^2 + 4a^2}{[a^2 + (b-1)^2]^2} = \frac{1 + 2a^2 + a^4 - 2b^2 + 2a^2 b^2 + b^4}{[a^2 + (b-1)^2]^2}. \tag{7.35}$$

Further simplification of the expression (7.35) implies that $|z| < 1$ if and only if

$$b[a^2 + (b-1)^2] < 0$$

or equivalently, $b < 0$. This means that the mapping (7.34) maps H into D. Next, given $z = x + iy \in D$, where $y > 0$ and $x^2 + y^2 < 1$. Now the expression (7.33) implies that

$$\omega = i\frac{x + iy - 1}{x + iy + 1} = i\frac{x - 1 + iy}{x + 1 + iy} = -\frac{2y}{(x+1)^2 + y^2} - i\frac{1 - x^2 - y^2}{(x+1)^2 + y^2}. \tag{7.36}$$

Since $y > 0$ and $x^2 + y^2 < 1$, the expression (7.36) ensures that $\mathrm{Re}\,\omega < 0$ and $\mathrm{Im}\,\omega < 0$. Consequently, the mapping is onto and our claim follows.

Define $F : H \to D$ by

$$F(\omega) = f\!\left(\frac{i + \omega}{i - \omega}\right)$$

which is obviously analytic in H and is real for $\omega < 0$. Let $H^* = \{\omega \in \mathbb{C} \,|\, \mathrm{Re}\,\omega < 0 \text{ and } \mathrm{Im}\,\omega > 0\}$, the left upper half-plane. By Theorem 7.8 (The Schwarz Reflection Principle), F has an analytic extension $G : H \cup \mathbb{R}^- \cup H^* \to \mathbb{C}$ which is defined by

$$G(\omega) = \begin{cases} F(\omega), & \text{if } \omega \in H \cup \mathbb{R}^-; \\[2mm] \overline{F(\overline{\omega})}, & \text{if } \omega \in H^*. \end{cases} \tag{7.37}$$

By the expression (7.34), we know that[e]

$$\frac{1}{\overline{z}} = \frac{i + \overline{\omega}}{i - \overline{\omega}}.$$

Now the analysis of the above claim also shows that if $\omega \in H^*$, then $\mathrm{Im}\,z > 0$ and $|z| > 1$. Hence the analytic function (7.37) can be expressed as

$$g(z) = G\!\left(i\frac{z - 1}{z + 1}\right) = \begin{cases} f(z), & \text{if } z \in D \cup C^+; \\[2mm] \overline{f\!\left(\frac{1}{\overline{z}}\right)}, & \text{if } z \in D^*. \end{cases}$$

This concludes the proof of the problem.   ▪

Problem 7.20

Bak and Newman Chapter 7 Exercise 20.

---

[d]Actually, it is easily seen that H is the left lower half-plane. For example, if $\omega = -1 - i$, then $z = -\frac{1}{5} + \frac{2}{5}i$ which belongs to D.

[e]By the example on [4, pp. 102, 103], $\frac{1}{\overline{z}}$ is the reflection of the point z across the curve C^+.

Proof. Assume that $f : D(0;1) \to \mathbb{C}$ was a nonconstant analytic function in $D(0;1)$ and $\overline{f(z)}$ is real for all $z \in C(0;1)$. By Definition 3.3, f is continuous on the closed unit disc $\overline{D(0;1)}$. Similar to the proof of Problem 7.19, we can show that the mapping $T : \mathbb{C} \to \mathbb{C}$ defined by

$$T(z) = \frac{z - i}{z + i}$$

maps the sets $U = \{z \in \mathbb{C} \mid \text{Im } z > 0\}$ and $\mathbb{R}$ onto $D(0;1)$ and $C(0;1)$ respectively. Therefore, the composition $F = f \circ T : U \to \mathbb{C}$ is also nonconstant and continuous. Furthermore, F is analytic in U and $F(z)$ takes real values for real z. Thus Theorem 7.8 (The Schwarz Reflection Principle)[f] implies that F can be extended to a nonconstant entire function G defined by

$$G(z) = \begin{cases} F(z), & \text{if } z \in U \cup \mathbb{R}; \\[2mm] \overline{F(\overline{z})}, & \text{if } z \in U^*, \end{cases} \tag{7.38}$$

where $U^* = \{z \in \mathbb{C} \mid \overline{z} \in U\}$. Since $\overline{D(0;1)}$ is compact, we know that there exists a positive constant M such that $|F(z)| = |f(T(z))| \leq M$ for all $z \in U \cup \mathbb{R}$. Therefore, this fact and the expression (7.38) imply that

$$|G(z)| = |f(T(z))| \leq M$$

for all $z \in \mathbb{C}$. By Theorem 5.10 (Liouville's Theorem), G is constant and hence f is also constant, a contradiction. We complete the proof of the problem. $\blacksquare$

> ### Problem 7.21
>
> *Bak and Newman Chapter 7 Exercise 21.*

Proof. Assume that $f(x) = |x|$ for all $x \in \mathbb{R}$. Now, since f is analytic in the bounded region $\Omega^+ = \{z \in \mathbb{C} \mid |z| < 1 \text{ and Im } z > 0\}$ and f is continuous on the upper semi-disc, we observe from Theorem 7.8 (The Schwarz Reflection Principle) that f can be extended to an analytic function g in $D(0,1)$.

Since g is analytic in $D(0;1)$, we see that

$$\lim_{\substack{h \to 0 \\ h > 0}} \frac{g(h) - g(0)}{h} = \lim_{\substack{h \to 0 \\ h > 0}} \frac{f(h) - f(0)}{h} = \lim_{\substack{h \to 0 \\ h > 0}} \frac{h - 0}{h} = 1$$

but

$$\lim_{\substack{h \to 0 \\ h < 0}} \frac{g(h) - g(0)}{h} = \lim_{\substack{h \to 0 \\ h < 0}} \frac{f(h) - f(0)}{h} = \lim_{\substack{h \to 0 \\ h < 0}} \frac{-h - 0}{h} = -1.$$

Hence these mean that $g'(0)$ does not exist and this contradiction shows that $f(x) = |x|$ for all $x \in \mathbb{R}$ is impossible, completing the analysis of the problem. $\blacksquare$

> ### Problem 7.22
>
> *Bak and Newman Chapter 7 Exercise 22.*

---

[f]Of course, we also use the stronger version of the principle here.

Proof. Suppose that the two original horizontal lines are $\operatorname{Im} z = a$ and $\operatorname{Im} z = b$ and they are mapped by f to the horizontal lines $\operatorname{Im} z = c$ and $\operatorname{Im} z = d$ respectively, where $a, b, c, d \in \mathbb{R}$. In other words, we have

$$f(x + ia) = u + ic \quad \text{and} \quad f(x + ib) = v + id, \tag{7.39}$$

where $x, u, v \in \mathbb{R}$.

Of course, the two functions

$$g_1(z) = f(z + ia) - ic \quad \text{and} \quad g_2(z) = f(z + ib) - id$$

are entire and the expressions (7.39) give

$$g_1(x) = f(x + ia) - ic = u \quad \text{and} \quad g_2(x) = f(x + ib) - id = v.$$

In other words, g_1 and g_2 are real on the real axis. Now Corollary 7.9 leads to us that

$$f(z + ia) - ic = g_1(z) = \overline{g_1(\overline{z})} = \overline{f(\overline{z} + ia)} + ic \tag{7.40}$$

and

$$f(z + ib) - id = g_2(z) = \overline{g_2(\overline{z})} = \overline{f(\overline{z} + ib)} + id. \tag{7.41}$$

Let $z = x + iy$. If we apply the formula (7.40) to $f(z)$, then we have

$$f(z) = f(x + iy) = f(x + i(y - a) + ia) = \overline{f(x - i(y - a) + ia)} + 2ic \tag{7.42}$$

Next, we apply the formula (7.41) to the expression (7.42) and get

$$\begin{aligned}
f(z) &= \overline{f(x - i(y - 2a + b) + ib)} + 2ic \\
&= f(x + i(y - 2a + b) + ib) + 2ic - 2id \\
&= f(x + iy + 2i(b - a)) + 2i(c - d) \\
&= f(z + 2i(b - a)) + 2i(c - d). \tag{7.43}
\end{aligned}$$

By taking derivative with respect to z to both sides of the expression, we conclude that

$$f'(z + 2i(b - a)) = f'(z)$$

for all $z \in \mathbb{C}$, i.e., f' is periodic which completes the proof of the problem. $\blacksquare$

Problem 7.23

Bak and Newman Chapter 7 Exercise 23.[*]

Proof. Suppose that P and Q are two parallelograms and f maps P onto Q. Let L_1, L_2, L_3 and L_4 be the sides of P, where L_1 is parallel to L_2 and L_3 is parallel to L_4. Similarly, we let ℓ_1, ℓ_2, ℓ_3 and ℓ_4 be the sides of Q, where ℓ_1 is parallel to ℓ_2 and ℓ_3 is parallel to ℓ_4.

Suppose that $f(L_1) = \ell_1$. Generally speaking, if the lines L and ℓ make the angles θ and ϕ with the positive real axis respectively, then $L' = Le^{-i\theta}$ and $\ell' = \ell e^{-i\phi}$ are two horizontal lines. Next, it is easy to see that the function

$$F(z) = e^{-i\phi} f(ze^{-i\theta}) \tag{7.44}$$

is entire and satisfies

$$F(L') = \ell'.$$

In other words, without loss of generality, we may assume that

$$L_1 : \operatorname{Im} z = a \quad \text{and} \quad \ell_1 : \operatorname{Im} \zeta = c.$$

Since L_1 and ℓ_1 are parallel to L_2 and ℓ_2 respectively, we also have

$$L_2 : \operatorname{Im} z = b \quad \text{and} \quad \ell_2 : \operatorname{Im} \zeta = d.$$

In other words, F maps $L_1 : \operatorname{Im} z = a$ and $L_2 : \operatorname{Im} z = b$ onto $\ell_1 : \operatorname{Im} \zeta = c$ and $\ell_2 : \operatorname{Im} \zeta = d$ respectively. Thus it follows from Problem 7.22 that F' is periodic with period $\omega_1 = 2i(b - a)$. By the definition (7.44) (with $\theta = \phi = 0$), f' is periodic with period $2i(b - a)$.

Since P and Q are parallelograms, L_1 is *not* parallel to L_3 and ℓ_1 is *not* parallel to ℓ_3. Let $\alpha(\neq 0, \pi)$ and $\beta(\neq 0, \pi)$ be the angles which transform L_3 and ℓ_3 onto horizontal lines L_3' and ℓ_3' respectively. In other words, we may assume that

$$L_3' = L_3 \mathrm{e}^{-i\alpha} : \operatorname{Im} z = A \quad \text{and} \quad L_4 = L_4 \mathrm{e}^{-i\alpha} : \operatorname{Im} z = B,$$

where $A, B \in \mathbb{R}$. Similarly, we have

$$\ell_3' = \ell_3 \mathrm{e}^{-\beta} : \operatorname{Im} \zeta = C \quad \text{and} \quad \ell_4' = \ell_4 \mathrm{e}^{-\beta} : \operatorname{Im} \zeta = D,$$

where $C, D \in \mathbb{R}$. Now the function $G(z) = \mathrm{e}^{-i\beta} f(z\mathrm{e}^{-\alpha})$ is entire such that $G(L_3') = \ell_3'$ and $G(L_4') = \ell_4'$. By Problem 7.22 again, G' is period with period $\omega_2 = 2i(B - A)$ and so f' is periodic with period $2i(B - A)\mathrm{e}^{-\alpha}$. If $2i(b - a) = 2i(B - A)\mathrm{e}^{-\alpha}$, then we have $\alpha = 0$ or π which is impossible. Hence the two periods are linearly independent.

Since f' has two linearly independent periods ω_1 and ω_2, every $z \in \mathbb{C}$ can be expressed as $z = x\omega_1 + y\omega_2$ for some $x, y \in \mathbb{R}$. This implies that

$$f'(z) = f'(x\omega_1 + y\omega_2) = f'(\{x\}\omega_1 + \{y\}\omega_2 + [x]\omega_1 + [y]\omega_2) = f'(\{x\}\omega_1 + \{y\}\omega_2), \qquad (7.45)$$

where $0 \leq \{x\}, \{y\} < 1$. Since every $\{x\}\omega_1 + \{y\}\omega_2$ is a point of the (compact) parallelogram with vertices $0, \omega_1, \omega_2, \omega_1 + \omega_2$, f' is bounded on this parallelogram. Combining this fact with the representation (7.45), we conclude that f' is a bounded entire function. Hence we apply Theorem 5.10 (Liouville's Theorem) to f' to conclude that f' is constant. Finally, Problem 5.15 guarantees that f is a linear polynomial. This completes the proof of the problem. ∎

CHAPTER 8

Simply Connected Domains

Problem 8.1

Bak and Newman Chapter 8 Exercise 1.

Proof. Let S be a star-like region and $\alpha \in S$. Pick $z \in \mathbb{C} \setminus S$ and consider $\gamma(t) = tz + (1-t)\alpha$, where $t \geq 1$. Clearly, γ is a continuous curve, $\gamma(1) = z$. Since $z - \alpha \neq 0$, we must have $\gamma(t) = \alpha + t(z - \alpha)$ which gives

$$\lim_{t \to \infty} \gamma(t) = \infty.$$

Thus z is connected to ∞. We claim that $\gamma([1, \infty]) \subseteq \mathbb{C} \setminus S$. Otherwise, there exists a $t_1 > 1$ such that $z_1 = \gamma(t_1) \in S$. Since S is a star-like region and γ is a line segment, it must be true that

$$\gamma([1, t_1]) \subseteq S.$$

However, this implies that $z \in S$, a contradiction. Hence the point z_1 does not exist and our claim follows. By Definition 8.1, S is simply connected and it completes the proof of the problem. $\blacksquare$

Remark 8.1

(a) One may use the concept of **homotopy of paths** from topology to define simply connectivity and prove Problem 8.1, see [24, p. 96; Exercise 21, p. 107].

(b) Furthermore, we note that Burckel [6, p. 344] provides 15 equivalent definitions of a simply connected region. In fact, some are defined with respect to its analytic senses and the others are defined in the topological aspects of the region.

Problem 8.2

*Bak and Newman Chapter 8 Exercise 2.**

Proof. Let S be a nonempty convex region. Pick any point $\alpha \in S$. Since S is convex, the line segment connecting α and $z \in S$ is contained in S for every $z \in S$. By the definition, S is star-like. Since S is assumed to be a region, Problem 8.1 ensures immediately that S is simply connected, completing the proof of the problem. $\blacksquare$

> **Problem 8.3**
>
> *Bak and Newman Chapter 8 Exercise 3.*

Proof. Since S is open in $\mathbb{C}$, for each $z \in C$, there exists a $\delta_z > 0$ such that $D(z; 2\delta_z) \subseteq S$. Obviously, we have

$$C \subseteq \bigcup_{z \in C} D(z; \delta_z).$$

Since C is compact, there are finitely many points $z_1, z_2, \ldots, z_N$ in C such that

$$C \subseteq \bigcup_{k=1}^{N} D(z_k; \delta_{z_k}). \tag{8.1}$$

Let $\delta = \min(\delta_{z_1}, \delta_{z_2}, \ldots, \delta_{z_N}) > 0$. We claim that the annulus

$$B = \{\omega \in \mathbb{C} \mid r - \delta \le |\omega - \alpha| \le r + \delta\}$$

is contained in S. To see this, we let ω' be the projection of the point $\omega \in B$ on C. Now the set relation (8.1) ensures that $\omega' \in D(z_k; \delta_{z_k})$ for some $1 \le k \le N$, see Figure 8.1 below.

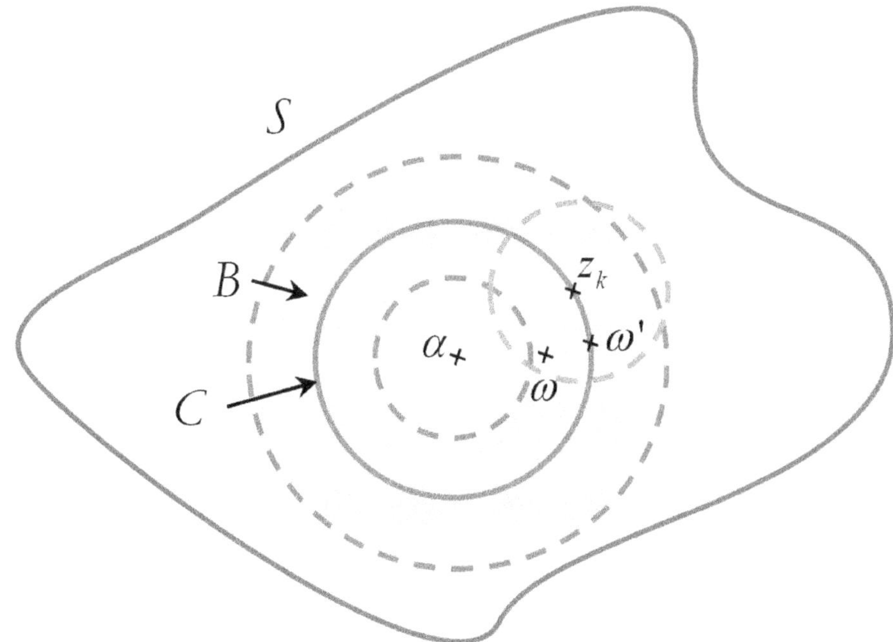

Figure 8.1: The geometry of the circle C and the annulus B.

Then the triangle inequality implies that

$$|\omega - z_k| \le |\omega - \omega'| + |\omega' - z_k| < \delta + \delta_{z_k} < 2\delta_{z_k}.$$

In other words, we have established that $\omega \in D(z; 2\delta_{z_k}) \subseteq S$ and thus $B \subseteq S$. We have completed the proof of the problem. ∎

> **Problem 8.4**
>
> *Bak and Newman Chapter 8 Exercise 4.*

Proof. Rewrite $\widetilde{S} = T \cup I$, where

$$T = \left\{ x + i \sin \frac{1}{x} \,\middle|\, 0 < x \le 1 \right\} \quad \text{and} \quad I = \{ iy \mid -1 \le y < \infty \}.$$

We note that the set T is part of the **Topologist's sine curve** studied in Problem 1.23. Since $\widetilde{S}$ contains all its limit points[a], it is closed in $\mathbb{C}$ and therefore, S is open in $\mathbb{C}$.

Given $x \in (0, 1]$ and $\epsilon > 0$. Let $n \in \mathbb{N}$ be large enough such that $\frac{1}{n\pi} < x$. Since

$$\lim_{t \to \frac{1}{n\pi}} \sin \frac{1}{t} = \sin n\pi = 0,$$

the point $x + i \sin \frac{1}{x} \in T \subseteq \widetilde{S}$ is connected to the point $\frac{1}{n\pi} \in T \subseteq \widetilde{S}$ by the continuous curve $\gamma_1 : [\frac{1}{n\pi}, x] \to T$ given by

$$\gamma_1(t) = \left(x + \frac{1}{n\pi} - t \right) + i \sin \frac{1}{x + \frac{1}{n\pi} - t}.$$

Recall the definition from [22, Exercise 20, p. 20] that

$$d(\gamma_1(t), \widetilde{S}) = \inf_{z \in \widetilde{S}} d(\gamma_1(t), z).$$

If we choose $n > \frac{1}{\pi \epsilon}$, then since $d((\frac{1}{n\pi}, 0), (0, 0)) = \frac{1}{n\pi} < \epsilon$, we obtain

$$d(\gamma_1(x), \widetilde{S}) < \epsilon.$$

Next, if we define $\gamma : [\frac{1}{n\pi}, \infty) \to \widetilde{S}$ by

$$\gamma(t) = \begin{cases} \gamma_1(t), & \text{if } t \in \left[\dfrac{1}{n\pi}, x \right]; \\[2mm] \dfrac{1}{n\pi} \cdot \dfrac{2 - t}{2 - x}, & \text{if } t \in (x, 2]; \\[2mm] i(t - 2), & \text{if } t > 2, \end{cases} \tag{8.2}$$

then γ is a continuous curve on $[\frac{1}{n\pi}, \infty)$. Furthermore, if $t \in (x, 2]$, then $\gamma(t) < \frac{1}{n\pi}$ so that

$$d(\gamma(t), \widetilde{S}) < \epsilon.$$

Since $\gamma(\frac{1}{n\pi}) = \gamma_1(\frac{1}{n\pi}) = x + i \sin \frac{1}{x}$ and

$$\lim_{t \to \infty} \gamma(t) = \lim_{t \to \infty} i(t - 2) = \infty,$$

we see that our curve (8.2) satisfies the requirements of Definition 8.1 and hence D is simply connected, completing the proof of the problem. $\qquad\blacksquare$

---

[a] Recall that $(0, 0)$ is a limit point of the set $\{ x + i \sin \frac{1}{x} \mid 0 < x \le 1 \}$.

Problem 8.5

Bak and Newman Chapter 8 Exercise 5.[*]

Proof. Let $R = \{x + iy \mid a < x < b$ and $c < y < d\}$ be a rectangle containing z, $\Gamma = \partial R$ and γ be a continuous curve connecting z to ∞. We further set

$$E = \{t \in [0, \infty) \mid \gamma(t) \in R\} \quad \text{and} \quad t_0 = \sup E.$$

Since $\gamma(0) = z \in R$, E is not empty so that t_0 exists in $\mathbb{R}$. If $\gamma(t_0) \in \Gamma$, then there is nothing to prove. Therefore, without loss of generality, we may assume that $\gamma(t_0) \notin \Gamma$ in the following discussion.

Suppose that $t_0 \in E$, i.e., $\gamma(t_0) \in R$. Since R is open in $\mathbb{C}$, there exists a $\epsilon > 0$ such that $D(\gamma(t_0); \epsilon) \subseteq R$. Since γ is continuous on $[0, \infty)$, there exists a $\delta > 0$ such that $|t - t_0| < \delta$ implies

$$\gamma(t) \in D(\gamma(t_0); \epsilon).$$

Particularly, we have $\gamma(t_0 + \frac{\delta}{2}) \in R$ which means that $t_0 + \frac{\delta}{2} \in E$ but this contradicts the fact that $t_0 = \sup E$.

Now the observation in the previous paragraph forces that $t_0 \in \mathbb{C} \setminus E$. The assumption $\gamma(t_0) \notin \Gamma$ means that $\{\gamma(t_0)\} \cap \Gamma = \varnothing$. Since Γ is compact and $\{\gamma(t_0)\}$ is closed in $\mathbb{C}$, it follows from the well-known fact [22, Exercise 21, p. 101] that one can find a $\epsilon_1 > 0$ such that $d(\gamma(t_0), \Gamma) = 2\epsilon_1 > 0$. Consequently, we have

$$D(\gamma(t_0); \epsilon_1) \cap R = \varnothing. \tag{8.3}$$

Again, the continuity of γ implies that there is a $\delta_1 > 0$ such that $|t - t_0| < 2\delta_1$ must imply

$$\gamma(t) \in D(\gamma(t_0); \epsilon_1).$$

In particular, $\gamma(t_0 - \delta_1) \in D(\gamma(t_0); \epsilon_1)$ or equivalently, $\gamma(t_0 - \delta_1) \notin R$ by the result (8.3). Since $t_0 - \delta_1 < t_0$, the definition of t_0 shows that there is a $p \in E$ such that $t_0 - \delta_1 < p < t_0$. Obviously, we have $|p - t_0| < \delta_1 < 2\delta_1$, so $\gamma(p) \in D(\gamma(t_0); \epsilon_1)$ and then the set relation (8.3) ensures that $\gamma(p) \notin R$. However, $p \in E$ certainly implies that $\gamma(p) \in R$, a contradiction. Hence we conclude that $\gamma(t_0) \in \Gamma$ which completes the proof of the problem. $\blacksquare$

Remark 8.2

We note that only the continuity of γ is required in the proof of Problem 8.5.

Problem 8.6

Bak and Newman Chapter 8 Exercise 6.

Proof. Suppose that Γ is a simple closed polygonal path contained in a simply connected domain D. Recall from the proof of Lemma 8.3 that if Γ has only two levels, then it is the boundary of a single rectangle R. Thus it is natural to define the "inside" of Γ to be the points of the rectangle R. Assume that the "inside" of a simple closed polygonal path $\Gamma \subseteq D$ can be defined up to $k(\geq 2)$ levels and the "inside" of Γ is contained in D.

Let Γ' be a simple closed polygonal path of $(k + 1)$ levels, see Figure 8.2 below. Then we can write

$$\Gamma' = \Gamma \cup \partial R, \tag{8.4}$$

where Γ and ∂R are simple closed polygonal paths and boundaries of rectangles R respectively. Thus the "inside" of Γ' is just the union of the "inside" of Γ and the points of R.

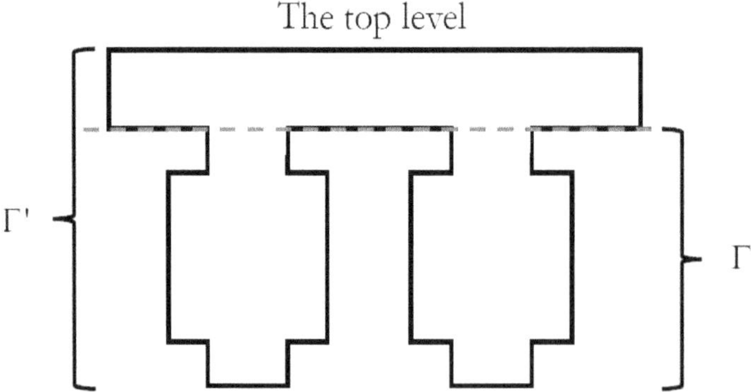

Figure 8.2: The "inside" of a simple closed polygonal path.

Therefore, if Γ' is contained in the simply connected domain D, then so are Γ and ∂R. By the assumption step, the "inside" of Γ is contained in D. Next, if $z \in R$, then we must have $z \in D$. Otherwise, we have $z \in D^c$. On the one hand, we know that D^c is closed in $\mathbb{C}$, ∂R is compact and $D^c \cap (\partial R) = \varnothing$, so there exists a $\delta > 0$ such that

$$d(\partial R, D^c) = \delta > 0. \tag{8.5}$$

On the other hand, we recall from Definition 8.1 that z is connected within ϵ to ∞, i.e., there exists a continuous curve $\gamma : [0, \infty) \to \infty$ such that $d(\gamma(t), D^c) < \delta$ for all $t \in [0, \infty)$. However, Problem 8.5 means that we can find a $t_0 \in [0, \infty)$ such that $\gamma(t_0) \in \partial R$ and

$$d(\gamma(t_0), D^c) < \delta,$$

contradicting the result (8.5). Therefore, R lies in D and we observe from the set relation (8.4) that the "inside" of Γ' must lie in D too. By induction, our desired result follows and we complete the proof of the problem. ◼

Problem 8.7

Bak and Newman Chapter 8 Exercise 7.

Proof. By Definition 4.13, let $z : [0, 1] \to \mathbb{C}$ be a non-simple closed polygonal path. Let p_{11}, p_{12} be a pair of points such that $0 < p_{11} < p_{12} < 1$ and $z(s) = z(t)$ for all $s, t \in [p_{11}, p_{12}]$, see Figure 8.3 for an example. In this example, we may define $z_1 : [0, 1] \to \mathbb{C}$ and $\gamma_1 : [0, 1] \to \mathbb{C}$ by

$$z_1(t) = \begin{cases} z(2p_{11}t), & \text{if } t \in [0, \tfrac{1}{2}]; \\[2mm] z(p_{12} + 2(1 - p_{12})(t - \tfrac{1}{2})), & \text{if } t \in [\tfrac{1}{2}, 1] \end{cases}$$

and

$$\gamma_1(t) = z(p_{11} + (p_{12} - p_{11})t).$$

Then it is easy to see that both z_1 and γ_1 are continuous on $[0, 1]$ so that we can write

$$z = z_1 \cup \gamma_1,$$

where z_1 is a simple closed polygonal path and γ_1 is a line segment traversed twice in opposite directions.[b]

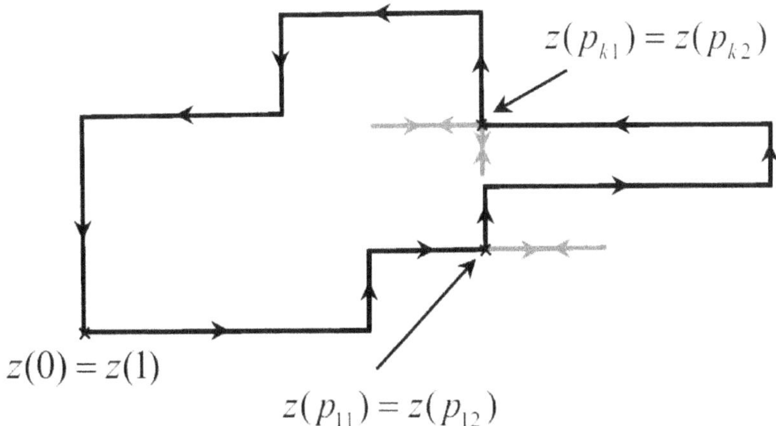

Figure 8.3: A decomposition of a closed polygonal path.

Generally speaking, if there are k pairs of points $\{p_{11}, p_{12}, p_{21}, p_{22}, \ldots, p_{k1}, p_{k2}\}$ such that $z(s_j) = z(t_j)$ for all $s_j, t_j \in [p_{j1}, p_{j2}]$ and $1 \le j \le k$. Then the above argument can be repeated to decompose z into

$$z = (z_1 \cup z_2 \cup \cdots \cup z_k) \cup (\gamma_1 \cup \gamma_2 \cup \cdots \cup \gamma_k),$$

where each pair of z_j and γ_j composes of simple closed polygonal paths and line segments traversed twice in opposite directions. Hence we have completed the proof of the problem. ∎

Problem 8.8

Bak and Newman Chapter 8 Exercise 8.

Proof. Let D be the whole plane minus the non-negative real axis. Then D is simply connected and $0 \notin D$. Since $-1 \in D$ and $\log(-1) = \log|-1| + i\operatorname{Arg}(-1) = \pi i$, Theorem 8.8 says that the function

$$f(z) = \pi i + \int_{-1}^{z} \frac{d\zeta}{\zeta} \tag{8.6}$$

is an analytic branch of $\log z$ in D.

Next, since $z \notin \mathbb{R}^+$, then we have $\operatorname{Arg} z \in (0, 2\pi)$. If z lies on the negative axis, then $z = -|z|$ and $\operatorname{Arg} z = \pi$ so that[c]

$$f(z) = \pi i + \int_{-1}^{-|z|} \frac{d\zeta}{\zeta} = \pi i + \log|\zeta|\Big|_{-1}^{-|z|} = \log|z| + \pi i. \tag{8.7}$$

Thus we have $\operatorname{Im}(\log z) = \operatorname{Im}(f(z)) = \operatorname{Arg} z = \pi$. If $z \in D \setminus \mathbb{R}^-$, then the line integral (8.6) goes from -1 to $-|z|$ and then from $-|z|$ to z to get

$$f(z) = \pi i + \int_{-1}^{-|z|} \frac{d\zeta}{\zeta} + \int_C \frac{d\zeta}{\zeta} = \log|z| + \pi i + \int_C \frac{d\zeta}{\zeta}, \tag{8.8}$$

---

[b]We remark that there are examples consisting of several simple closed polygonal paths (e.g. two rectangles intersecting at one of their vertices only) or line segments (see the horizontal and vertical line segments at $z(p_{k1}) = z(p_{k2})$ in Figure 8.3).

[c]The integral in the expression (8.7) is the usual Riemann integral.

where C is the circular arc from connecting $-|z|$ and z clockwise, see Figure 8.4. If C is

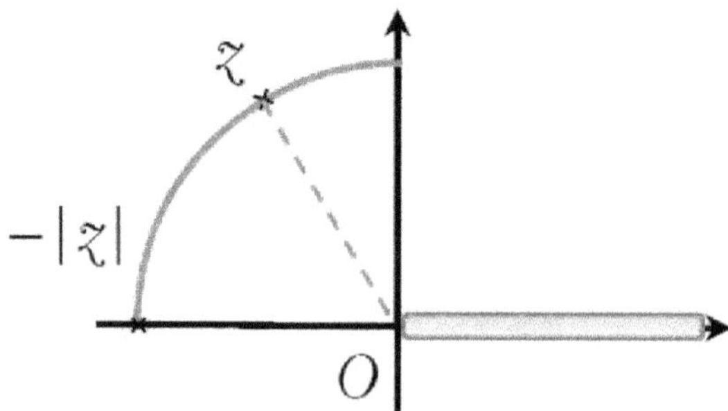

Figure 8.4: The circular arc C connecting $-|z|$ and z clockwise.

parameterized as $\gamma(t) = |z|e^{i(\pi-t)}$, where $0 \le t \le \pi - \operatorname{Arg} z$, then we deduce from the expression (8.8) that

$$
\begin{aligned}
f(z) &= \log|z| + \pi i + \int_0^{\pi-\operatorname{Arg} z} \frac{1}{|z|e^{i(\pi-t)}} \cdot \left[-i|z|e^{i(\pi-t)}\right] \mathrm{d}t \\
&= \log|z| + \pi i - i\int_0^{\pi-\operatorname{Arg} z} \mathrm{d}t \\
&= \log|z| + \pi i - i(\pi - \operatorname{Arg} z) \\
&= \log|z| + i\operatorname{Arg} z
\end{aligned}
$$

which implies that $\operatorname{Im}(\log z) = \operatorname{Arg} z \in (0, 2\pi)$. This completes the proof of the problem.

> **Problem 8.9**
>
> *Bak and Newman Chapter 8 Exercise 9.*

Proof. Let D be the plane minus the non-positive real axis. By the analysis on [4, p. 115], we can define an analytic branch of $\log z$ in D by

$$
\log z = \log|z| + i\operatorname{Arg} z, \tag{8.9}
$$

where $\operatorname{Arg} z \in (-\pi, \pi)$. Next, we define

$$
f(z) = e^{z \log z}. \tag{8.10}
$$

Since $\log z$ is analytic in D, f is also analytic in D. Besides, if $x > 0$, then $f(x) = e^{x \log x} = x^x$. In other words, the function f defined by the expression (8.10) satisfies our requirements.

Using the expressions (8.9) and (8.10) directly, we see that

$$
f(i) = e^{i \log i} = e^{i(\log|i| + i\operatorname{Arg}(i))} = e^{-\frac{\pi}{2}}
$$

and

$$
f(-i) = e^{-i \log -i} = e^{-i(\log|-i| + i\operatorname{Arg}(-i))} = e^{-\frac{\pi}{2}}.
$$

Finally, let $z = x + iy \in D$. If $y = 0$, then we have $f(\bar{x}) = \overline{f(x)}$, so we assume that $y \ne 0$ so that $\bar{z} = x - iy \in D$. In addition, we have

$$
\log \bar{z} = \log|\bar{z}| + i\operatorname{Arg} \bar{z} = \log|z| - i\operatorname{Arg} z = \overline{\log|z| + i\operatorname{Arg} z} = \overline{\log z}
$$

which gives
$$f(\overline{z}) = \exp(\overline{z} \cdot \log \overline{z}) = \exp(\overline{z \log z}) = \overline{\exp(z \log z)} = \overline{f(z)}.$$

This ends the proof of the problem.

CHAPTER **9**

Isolated Singularities of an Analytic Function

Proof. Assume that z_0 was a removable singularity of f. By Definition 9.2, there exists an analytic function g defined in a neighborhood of z_0 such that $f(z) = g(z)$ for all $z \neq z_0$. Now the continuity of g yields

$$\lim_{z \to z_0} |f(z)| = \lim_{z \to z_0} |g(z)| = |g(z_0)| < \infty$$

which contradicts our hypothesis. Consequently, z_0 is not a removable singularity of f.

Next, we assume that z_0 was an essential singularity of f. Since $f(z) \to \infty$ as $z \to z_0$, for every $\epsilon > 0$, there exists a $M > 0$ such that $|f(z)| > M$ for all $z \in D(z_0; \epsilon)$. Therefore, it implies that $f(D(z_0; \epsilon)) \cap D(0; M) = \varnothing$. In other words, $f(D(z_0; \epsilon))$ is *not* dense in $\mathbb{C}$ which contradicts Theorem 9.6 (The Casorati-Weierstrass Theorem). Hence z_0 cannot be an essential singularity of f and Definition 9.2 forces that f has a pole at z_0. This completes the analysis of the problem. ∎

Proof. The answer is negative. By the hypothesis, we get

$$\lim_{z \to 0} |f(z)| = \lim_{z \to 0} e^{\frac{1}{|z|}} = \infty,$$

so Problem 9.1 implies that f has a pole at 0 and then f has the form

$$f(z) = \sum_{n=-k}^{\infty} a_n z^n \tag{9.1}$$

for some $k \in \mathbb{N}$. Therefore, the form (9.1) implies that

$$|f(z)| \sim \frac{M}{|z|^k}$$

as $z \to 0$ for some positive constant M. This is certainly a contradiction and we complete the proof of the problem. ∎

> ### Problem 9.3
>
> *Bak and Newman Chapter 9 Exercise 3.**

Proof. Suppose that f is entire and injective. Then f is nonconstant. Consider the function $g : \mathbb{C} \setminus \{0\} \to \mathbb{C}$ defined by

$$g(z) = f(z^{-1})$$

which has a singularity at 0. Since f is entire, Theorem 5.5 (The Taylor Expansion of an Entire Function) says that

$$f(\omega) = \sum_{k=0}^{\infty} a_k \omega^k$$

for some $a_k \in \mathbb{C}$ and then

$$g(z) = \sum_{k=-\infty}^{0} a_k z^k.$$

If g has a removable singularity at 0, then $a_{-1} = a_{-2} = \cdots = 0$ which implies that $g(z) = a_0$. Consequently, f is also a constant which contradicts to our hypothesis. Therefore, g has either a pole or an essential singularity at 0.

Assume that the singularity was essential. By Theorem 9.6 (The Casorati-Weierstrass Theorem), the set $g(D(0;1) \setminus \{0\})$ is dense in $\mathbb{C}$. Since g is analytic in the region $D(2; \frac{1}{2})$, Problem 7.2 implies that $g(D(2, \frac{1}{2}))$ is also a region. In particular, $g(D(2, \frac{1}{2}))$ is open in $\mathbb{C}$. Thus we have

$$g(D(0;1) \setminus \{0\}) \cap g\left(D\left(2; \frac{1}{2}\right)\right) \neq \varnothing.$$

In other words, there exists $z_1 \in D(0;1) \setminus \{0\}$ and $z_2 \in D(2; \frac{1}{2})$ such that $z_1 \neq z_2$ but

$$f(z_1^{-1}) = g(z_1) = g(z_2) = f(z_2^{-1}).$$

This contradicts the fact that f is injective and hence the singularity is not essential.

The previous analysis ensures that the singularity of g at 0 is a pole, so f must be a polynomial. Suppose that $f(z) = a_0 + a_1 z + \cdots + a_n z^n$. By Theorem 5.12 (The Fundamental Theorem of Algebra), f has n roots counting multiplicity. Since f is injective, it cannot have any two *distinct* roots. In other words, f must have the form

$$f(z) = a(z - z_0)^n$$

for some $a, z_0 \in \mathbb{C}$ and $a \neq 0$. If $n \geq 2$, then the fact

$$f(z_0 + 1) = a = f(z_0 + e^{\frac{2\pi i}{n}})$$

shows that f is *not* injective, a contradiction. Now we have $n = 1$ and so $f(z) = a(z - z_0)$ which is a linear polynomial as desired. We end the proof of the problem.   ∎

> ### Problem 9.4
>
> *Bak and Newman Chapter 9 Exercise 4.*

Proof. If f is analytic at the origin, then f is entire. For $|z| > 1$, we have $\frac{1}{|z|} < |z|$ so that

$$|f(z)| < 2\sqrt{|z|} < 2|z|$$

there. Next, since $|f(z)|$ is continuous on the compact set $\overline{D(0;1)}$, $|f(z)|$ is bounded by a positive constant M there. These two facts imply that one can find a positive constant M' such that

$$|f(z)| \le M'|z|$$

holds for all $z \in \mathbb{C}$. Hence Theorem 5.11 (The Extended Liouville Theorem) guarantees that $f(z) = A + Bz$ for some $A, B \in \mathbb{C}$. Now the triangle inequality gives

$$2\sqrt{|z|} > |f(z)| = |Bz + A| \ge |B| \cdot |z| - |A|$$

for large z and this forces that $B = 0$. In other words, $f(z) = A$ in this case.

Suppose that f has an isolated singularity at 0. Then we consider its Laurent expansion at 0:

$$f(z) = \sum_{k=-\infty}^{\infty} a_k z^k.$$

Let $R > 0$. Now if $k \ne 0$, then we consider the circle $C = C(0; R)$ and we apply Theorem 4.10 (The M-L Formula) to obtain

$$|a_k| = \left| \frac{1}{2\pi i} \int_{C(0;R)} \frac{f(\omega)}{\omega^{k+1}} \, d\omega \right| \le \frac{1}{2\pi} \times 2\pi R \times \frac{1}{R^{k+1}}\left(\sqrt{R} + \frac{1}{\sqrt{R}}\right) = \frac{1}{R^k}\left(\sqrt{R} + \frac{1}{\sqrt{R}}\right). \quad (9.2)$$

If $k > 0$, then the inequality (9.2) shows that

$$\lim_{R \to \infty} |a_k| \le \lim_{R \to \infty} \frac{1}{R^k}\left(\sqrt{R} + \frac{1}{\sqrt{R}}\right) = 0$$

which means $a_1 = a_2 = \cdots = 0$. If $k < 0$, then the inequality (9.2) gives

$$\lim_{R \to 0} |a_k| \le \lim_{R \to 0} \frac{1}{R^k}\left(\sqrt{R} + \frac{1}{\sqrt{R}}\right) = 0.$$

In other words, $a_{-1} = a_{-2} = \cdots = 0$. Hence we have shown that $f(z) = a_0$ in this case, completing the proof of the problem. ∎

Problem 9.5

*Bak and Newman Chapter 9 Exercise 5.**

Proof. If $g \equiv 0$, then there is nothing to prove. Thus all the zeros of g are isolated. Furthermore, we have $\left|\frac{f(z)}{g(z)}\right| \le 1$ in a deleted neighborhood of a zero of g, so Corollary 9.4 asserts that all the singularities of $\frac{f}{g}$ are removable. As a result, $\frac{f}{g}$ can be extended to an entire function. By the continuity of the extended function, it is also true that

$$\left|\frac{f(z)}{g(z)}\right| \le 1$$

for all $z \in \mathbb{C}$. Now Theorem 5.10 (Liouville's Theorem) leads to us that $\frac{f}{g}$ is a constant which completes the proof of the problem. ∎

Problem 9.6

Bak and Newman Chapter 9 Exercise 6.

Proof. It is easy to see that $e^{\frac{1}{z}} \neq 0$ for all $z \in \mathbb{C}$. Then 0 is a missing value. Let $\omega \in \mathbb{C} \setminus \{0\}$. Set $\frac{1}{z} = \log \omega = \log |\omega| + i(\operatorname{Arg} \omega + 2\pi k)$, where $\operatorname{Arg} \omega \in (-\pi, \pi]$ and $k \in \mathbb{Z}$. Certainly, we have

$$e^{\frac{1}{z}} = e^{\log \omega} = \omega.$$

In addition, it is evident that

$$
\begin{aligned}
z &= \frac{1}{\log |\omega| + i(\operatorname{Arg} \omega + 2\pi k)} \\
&= \frac{1}{\log |\omega| + i(\operatorname{Arg} \omega + 2\pi k)} \cdot \frac{\log |\omega| - i(\operatorname{Arg} \omega + 2\pi k)}{\log |\omega| - i(\operatorname{Arg} \omega + 2\pi k)} \\
&= \frac{\log |\omega| - i(\operatorname{Arg} \omega + 2\pi k)}{(\log |\omega|)^2 + (\operatorname{Arg} \omega + 2\pi k)^2}
\end{aligned}
$$

which gives

$$|z| = \frac{1}{\sqrt{(\log |\omega|)^2 + (\operatorname{Arg} \omega + 2\pi k)^2}}. \tag{9.3}$$

Hence the modulus (9.3) will ensure that $0 < |z| < 1$ for all large enough k. This completes the proof of the problem. ∎

Problem 9.7

Bak and Newman Chapter 9 Exercise 7.

Proof. For simplicity, we assume that $z_0 = 0$ in the following discussion. Suppose that we have

$$f(z) = \sum_{k=-m}^{\infty} a_k z^k \quad \text{and} \quad g(z) = \sum_{p=-n}^{\infty} b_p z^p, \tag{9.4}$$

where $a_k, b_p \in \mathbb{C}$ for all $k = -m, -m+1, \ldots$ and $p = -n, -n+1, \ldots$. If $m \neq n$, then it is easy to see from the forms (9.4) that $f + g$ has a pole of order $\max\{m, n\}$ at 0. If $m = n$, then the order of the pole at 0 depends on the coefficients a_k and b_p. For example, if $a_n = -b_n$, then the order of pole of $f + g$ at 0 is less than n. Anyhow the order of pole of $f + g$ at 0 is less than or equal to n.

By Definition 9.2, let's write

$$f(z) = \frac{A(z)}{z^m B(z)} \quad \text{and} \quad g(z) = \frac{P(z)}{z^n Q(z)}, \tag{9.5}$$

where A, B, P and Q are analytic at 0, $A(0), B(0), P(0), Q(0) \neq 0$. It follows from the forms (9.5) that

$$f(z) \cdot g(z) = \frac{A(z)P(z)}{z^{m+n} B(z)Q(z)},$$

where $A(0)P(0) \neq 0$ and $B(0)Q(0) \neq 0$. Hence $f \cdot g$ has a pole of order $m+n$ at 0. Furthermore, the form (9.5) also implies that

$$\frac{f(z)}{g(z)} = \frac{A(z)}{z^m B(z)} \times \frac{z^n Q(z)}{P(z)} = \frac{A(z)Q(z)}{z^{m-n} B(z)P(z)} \tag{9.6}$$

which shows that $\frac{f}{g}$ has a zero of order $n - m$ at 0 if $n > m$ and similarly, it has a pole of order $m - n$ at 0 if $m > n$. Finally, if $m = n$, then the representation (9.6) reduces to

$$\frac{f(z)}{g(z)} = \frac{A(z)Q(z)}{B(z)P(z)}$$

so that

$$\lim_{z \to 0} z \cdot \frac{f(z)}{g(z)} = \lim_{z \to 0} z \cdot \frac{A(z)Q(z)}{B(z)P(z)} = 0.$$

By Theorem 9.3 (The Riemann's Principle of Removable Singularities), $\frac{f}{g}$ has a removable singularity at 0. This completes the proof of the problem. ∎

Problem 9.8

Bak and Newman Chapter 9 Exercise 8.[*]

Proof. Suppose that z_0 is an essential singularity of f. By Theorem 9.6 (The Casorati-Weierstrass Theorem), $f(D(z_0; \frac{1}{n}) \setminus \{z_0\})$ is dense in $\mathbb{C}$ for every $n \in \mathbb{N}$. In other words, it is true that

$$f\left(D\left(z_0; \frac{1}{n}\right) \setminus \{z_0\}\right) \cap D\left(0; \frac{1}{n}\right) \neq \varnothing \tag{9.7}$$

for every $n \in \mathbb{N}$. Equivalently, the set relation (9.7) asserts that there is an $\alpha_n \in D(z_0; \frac{1}{n}) \setminus \{z_0\}$ such that $f(\alpha_n) \in D(0; \frac{1}{n})$ for each $n \in \mathbb{N}$. Hence we obtain the result that both $\alpha_n \to z_0$ and $f(\alpha_n) \to 0$ hold. Similarly, we also have

$$f\left(D\left(z_0; \frac{1}{n}\right) \setminus \{z_0\}\right) \cap D\left(n; \frac{1}{n}\right) \neq \varnothing$$

for every $n \in \mathbb{N}$. Consequently, one can find a $\beta_n \in D(z_0; \frac{1}{n}) \setminus \{z_0\}$ such that $f(\beta_n) \in D(n, \frac{1}{n})$ for each $n \in \mathbb{N}$. Since $f(\beta_n) > n - \frac{1}{n}$ holds for all $n \in \mathbb{N}$, we conclude that $\beta_n \to z_0$ and $f(\beta_n) \to \infty$ as $n \to \infty$.

Conversely, we suppose that there exist sequences $\{\alpha_n\}$ and $\{\beta_n\}$ tending to z_0 such that $f(\alpha_n) \to 0$ and $f(\beta_n) \to \infty$. If z_0 was a removable singularity of f, then $\lim_{z \to z_0} f(z)$ exists by Definition 9.2 or by its Laurent expansion about z_0. However, this means that

$$\lim_{n \to \infty} f(\alpha_n) = \lim_{n \to \infty} f(\beta_n)$$

which is impossible. If it was a pole of f, then Definition 9.2 implies that

$$\lim_{z \to z_0} f(z) = \infty$$

which contradicts to our hypothesis. Hence they assert that z_0 is an essential singularity of f, completing the proof of the problem. ∎

Problem 9.9

Bak and Newman Chapter 9 Exercise 9.

Proof.

(a) We know that

$$\frac{1}{z^4 + z^2} = \frac{1}{z^2(z^2+1)} = \frac{1}{z^2(z+i)(z-i)}. \tag{9.8}$$

By Definition 9.2, it has a pole of order 2 at 0 and poles of order 1 at $\pm i$.

(b) We note that $\cot z = \frac{\cos z}{\sin z}$ and $\sin z$ is zero if and only if $z = n\pi$, where $n \in \mathbb{Z}$. Thus $\cot z$ is analytic in a deleted neighborhood of $n\pi$. By Lemma 7.2, we establish

$$\lim_{z \to n\pi} (z - n\pi) \cdot \frac{\cos z}{\sin z} = \lim_{\zeta \to 0} \zeta \cdot \frac{\cos(\zeta + n\pi)}{\sin(\zeta + n\pi)} = \lim_{\zeta \to 0} \zeta \cdot \frac{\cos \zeta}{\sin \zeta} = 1. \tag{9.9}$$

Furthermore, we have

$$\lim_{z \to n\pi} (z - n\pi)^2 \cdot \frac{\cos z}{\sin z} = \lim_{\zeta \to 0} \zeta \times \left(\zeta \cdot \frac{\cos \zeta}{\sin \zeta} \right) = 0.$$

Therefore, it follows from Theorem 9.5 that $\cot z$ has a pole of order 1 at $n\pi$, where $n \in \mathbb{Z}$.

(c) By similar analysis as part (b), $\csc z$ has a pole of order 1 at $n\pi$, where $n \in \mathbb{Z}$.

(d) The function

$$f(z) = \frac{\exp(\frac{1}{z^2})}{z - 1}$$

has singularities at 0 and 1. It is easily seen that f is analytic in a deleted neighborhood of 1. Since

$$\lim_{z \to 1} (z - 1) f(z) = \lim_{z \to 1} \exp\left(\frac{1}{z^2} \right) = e \quad \text{and} \quad \lim_{z \to 1} (z - 1)^2 f(z) = 0, \tag{9.10}$$

Theorem 9.5 ensures that f has a simple pole at 1.

For the singularities at 0, we consider the Laurent expansion of f at 0. In fact, we know that

$$\exp\left(\frac{1}{z^2} \right) = 1 + \frac{1}{z^2} + \frac{1}{z^4} + \cdots \quad \text{and} \quad \frac{1}{z-1} = -1 - z - z^2 - \cdots.$$

Thus we have

$$f(z) = -(1 + z + z^2 + \cdots) \times \left(1 + \frac{1}{z^2} + \frac{1}{z^4} + \cdots \right). \tag{9.11}$$

Since every term in the brackets of the expression (9.11) has positive coefficient, the resulting expansion of f will have infinitely many nonzero terms in its principal part. By the notes on [4, p. 125], we conclude that f has an essential singularity at 0.

Hence we end the proof of the problem. ∎

Proof. We note that $z^2 + 1 = (z - i)(z + i)$ so that

$$f(z) = \frac{1}{(z^2+1)^2} = \frac{1}{(z-i)^2} \cdot \frac{1}{(z+i)^2}. \tag{9.12}$$

Let $F(z) = \frac{1}{(z+i)^2}$. Then F is clearly analytic in a neighbourhood of $z = i$. By Theorem 6.5 (The Power Series Representation for Functions Analytic in a Disc), we know that

$$F(z) = \sum_{k=0}^{\infty} \frac{F^{(k)}(i)}{k!} (z - i)^k. \tag{9.13}$$

Combining the two expressions (9.12) and (9.13), we get

$$f(z) = \sum_{k=0}^{\infty} \frac{F^{(k)}(i)}{k!}(z-i)^{k-2} = \frac{F(i)}{(z-i)^2} + \frac{F'(i)}{z-i} + \sum_{k=0}^{\infty} \frac{F^{(k+2)}(i)}{(k+2)!}(z-i)^k.$$

Since $F(i) = \frac{1}{(2i)^2} = -\frac{1}{4}$ and $F'(i) = -\frac{i}{4}$, the principal part of the Laurent expansion of f is given by

$$-\frac{1}{4(z-i)^2} - \frac{i}{4(z-i)}.$$

This completes the proof of the problem.

> **Problem 9.11**
>
> *Bak and Newman Chapter 9 Exercise 11.*

Proof.

(a) We have

$$\frac{1}{z^4 + z^2} = \frac{1}{z^2(1+z^2)} = \frac{1}{z^2} \cdot (1 - z^2 + z^4 - \cdots) = \frac{1}{z^2} \sum_{k=0}^{\infty} (-1)^k z^{2k} = \sum_{k=-1}^{\infty} (-1)^{k+1} z^{2k}.$$

(b) Suppose that

$$\frac{\exp(\frac{1}{z^2})}{z-1} = \sum_{k=-\infty}^{\infty} a_k z^k.$$

Since

$$\exp\left(\frac{1}{z^2}\right) = 1 + \frac{1}{1!z^2} + \frac{1}{2!z^4} + \cdots \quad \text{and} \quad \frac{1}{z-1} = -(1 + z + z^2 + \cdots),$$

we have

$$-\left(1 + \frac{1}{1!z^2} + \frac{1}{2!z^4} + \cdots\right)(1 + z + z^2 + \cdots) = \sum_{k=-\infty}^{\infty} a_k z^k.$$

For $k \geq 0$, we obtain

$$a_k = -\left(1 + \frac{1}{1!} + \frac{1}{2!} + \cdots\right) = -e.$$

Let $n \in \mathbb{N}$. If $k = -2n$, then we get

$$a_{-2n} = -\left[\frac{1}{n!} + \frac{1}{(n+1)!} + \cdots\right] = -e + 1 + \frac{1}{2!} + \cdots + \frac{1}{(n-1)!}.$$

Similarly, if $k = -2n + 1$, then we conclude

$$a_{-2n+1} = -\left[\frac{1}{n!} + \frac{1}{(n+1)!} + \cdots\right] = -e + 1 + \frac{1}{2!} + \cdots + \frac{1}{(n-1)!}.$$

(c) We note that

$$\frac{1}{z^2 - 4} = \frac{1}{(z-2)(z+2)} = \frac{1}{z-2} \cdot \frac{1}{4(1 + \frac{z-2}{4})}$$

which implies that

$$\frac{1}{z^2 - 4} = \frac{1}{4(z-2)} \sum_{k=0}^{\infty} (-1)^k \left(\frac{z-2}{4}\right)^k = \sum_{k=-1}^{\infty} \frac{(-1)^{k+1}}{4^{k+2}}(z-2)^k.$$

We have completed the analysis of the problem.

> **Problem 9.12**
>
> *Bak and Newman Chapter 9 Exercise 12.**

Proof. We know that

$$f(z) = \frac{1}{z}\left(-\frac{1}{2-z} + \frac{1}{1-z}\right).$$

If $|z| < 1$, then we have

$$\frac{1}{1-z} = 1 + z + z^2 + \cdots . \tag{9.14}$$

If $|z| > 1$, then $\frac{1}{|z|} < 1$ and thus

$$\frac{1}{1-z} = -\frac{1}{z\left(1-\frac{1}{z}\right)} = -\frac{1}{z}\left(1 + \frac{1}{z} + \frac{1}{z^2} + \cdots\right) = -\frac{1}{z} - \frac{1}{z^2} - \cdots . \tag{9.15}$$

Combining the formulas (9.14) and (9.15), we see that

$$\frac{1}{1-z} = \begin{cases} \displaystyle\sum_{k=0}^{\infty} z^k, & \text{if } |z| < 1; \\[2em] \displaystyle -\sum_{k=-\infty}^{-1} z^k, & \text{if } |z| > 1. \end{cases} \tag{9.16}$$

Similarly, we have

$$\frac{1}{2-z} = \frac{1}{2} \times \frac{1}{1-\frac{z}{2}} = \begin{cases} \displaystyle\frac{1}{2}\sum_{k=0}^{\infty} \left(\frac{z}{2}\right)^k, & \text{if } |z| < 2; \\[2em] \displaystyle -\frac{1}{2}\sum_{k=-\infty}^{-1} \left(\frac{z}{2}\right)^k, & \text{if } |z| > 2. \end{cases} \tag{9.17}$$

(a) Now the formulas (9.16) and (9.17) imply that

$$f(z) = \frac{1}{z}\left[-\frac{1}{2}\sum_{k=0}^{\infty} \left(\frac{z}{2}\right)^k + \sum_{k=0}^{\infty} z^k\right] = \frac{1}{z}\sum_{k=0}^{\infty}\left(1 - \frac{1}{2^{k+1}}\right)z^k = \sum_{k=-1}^{\infty}\left(1 - \frac{1}{2^{k+2}}\right)z^k.$$

(b) In this case, the formulas (9.16) and (9.17) give

$$f(z) = \frac{1}{z}\left[-\frac{1}{2}\sum_{k=0}^{\infty} \left(\frac{z}{2}\right)^k - \sum_{k=-\infty}^{-1} z^k\right] = -\sum_{k=-\infty}^{-2} z^k - \sum_{k=-1}^{\infty}\frac{1}{2^{k+2}}z^k.$$

(c) Finally, we have

$$f(z) = \frac{1}{z}\left[\frac{1}{2}\sum_{k=-\infty}^{-1} \left(\frac{z}{2}\right)^k - \sum_{k=-\infty}^{-1} z^k\right] = \frac{1}{z}\sum_{k=-\infty}^{-1}\left(\frac{1}{2^{k+1}} - 1\right)z^k = \sum_{k=-\infty}^{-2}\left(\frac{1}{2^{k+2}} - 1\right)z^k.$$

We have completed the proof of the problem. ■

Bak and Newman Chapter 9 Exercise 13.

Proof. For $n = 1, 2, \ldots, k$, we notice that

$$z^{a_n} - 1 = (z-1)P_n(z),$$

where $P_n(z) = z^{a_n-1} + z^{a_n-2} + \cdots + z + 1$. Thus we have

$$R(z) = \frac{1}{(z-1)^k P_1(z) P_2(z) \cdots P_k(z)}.$$

Now the product $P(z) = P_1(z)P_2(z)\cdots P_k(z)$ is a polynomial in z, so Theorem 5.12 (The Fundamental Theorem of Algebra) ensures that there exists a $\delta > 0$ such that $C(1;\delta)$ *does not* contain any zero of $P(z)$. Thus we observe easily from Corollary 9.10 that if $C = C(1;\delta)$, then

$$c_{-k} = \frac{1}{2\pi i} \int_C \frac{R(z)}{(z-1)^{-k+1}}\, dz = \frac{1}{2\pi i} \int_C \frac{\frac{1}{P(z)}}{z-1}\, dz. \tag{9.18}$$

Since $\frac{1}{P(z)}$ is analytic inside and on C, we may apply Theorem 6.4 (The Cauchy Integral Formula) to the last integral (9.18) to obtain

$$c_{-k} = \frac{1}{P(1)} = \frac{1}{a_k a_{k-1} \cdots a_1}.$$

This ends the proof of the problem. $\blacksquare$

Bak and Newman Chapter 9 Exercise 14.

Proof. Suppose that

$$f(z) = \sum_{k=-\infty}^{\infty} a_k z^k. \tag{9.19}$$

Then we have

$$f(-z) = \sum_{k=-\infty}^{\infty} (-1)^k a_k z^k. \tag{9.20}$$

Since f is odd, we have $f(z) = \frac{f(z)-f(-z)}{2}$. Thus it follows from the two representations (9.19) and (9.20) that

$$\sum_{k=-\infty}^{\infty} a_k z^k = \sum_{k=-\infty}^{\infty} \frac{a_k + (-1)^{k+1}a_k}{2} z^k = \sum_{k=-\infty}^{\infty} a_{2k+1} z^{2k+1}.$$

Consequently, it means that $a_{2k} = 0$ for all $k \in \mathbb{Z}$. This completes the proof of the problem. $\blacksquare$

Bak and Newman Chapter 9 Exercise 15.

Proof.

(a) We observe from the expression (9.8) that

$$\frac{1}{z^4 + z^2} = \frac{A_1}{z} + \frac{A_2}{z^2} + \frac{B}{z-i} + \frac{C}{z+i} = \frac{A_1 z(z^2+1) + A_2(z^2+1) + Bz^2(z+i) + Cz^2(z-i)}{z^2(z-i)(z+i)}$$

which means that

$$A_1 z(z^2+1) + A_2(z^2+1) + Bz^2(z+i) + Cz^2(z-i) = 1$$

for all $z \in \mathbb{C}$. If $z = 0$, then $A_2 = 1$. If $z = i$, then $B = -\frac{1}{2i}$. If $z = -i$, then $C = \frac{1}{2i}$. Finally, if $z = 1$, then $A_1 = 0$. Consequently, we conclude that

$$\frac{1}{z^4 + z^2} = \frac{1}{z^2} - \frac{1}{2i(z-i)} + \frac{1}{2i(z+i)}.$$

(b) Similar as part (a), we have

$$\frac{1}{z^2 + 1} = \frac{1}{2i(z-i)} - \frac{1}{2i(z+i)}.$$

This completes the proof of the problem. ▨

Problem 9.16

Bak and Newman Chapter 9 Exercise 16.

Proof. We follow the given hint. Assume that there were $\omega \in \mathbb{C}$ and $\delta > 0$ such that $|f(z) - \omega| > \delta$ for all $z \in D$. Then we have

$$\left| \frac{1}{f(z) - \omega} \right| < \frac{1}{\delta}$$

throughout D. By Corollary 9.4, the function

$$g(z) = \frac{1}{f(z) - \omega}$$

has a removable singularity at z_0. Therefore, g is analytic at z_0. Now g is analytic in the region $D \cup \{z_0\}$ and $g(z_n) = 0$ for all $n \in \mathbb{N}$, where $z_n \to z_0$. It follows from Theorem 6.9 (The Uniqueness Theorem) that $g \equiv 0$ in $D \cup \{z_0\}$, i.e., $f(z) = \infty$ throughout D which is impossible. We have completed the analysis of the problem. ▨

Problem 9.17

Bak and Newman Chapter 9 Exercise 17.

Proof. Suppose that $D_0(0;1) = D(0;1) \setminus \{0\}$ and $g(z) = \sin \frac{1}{z}$.

(a) We know that

$$\sin \frac{1}{z} = \frac{1}{z} - \frac{1}{3!z^3} + \frac{1}{5!z^5} - \cdots , \tag{9.21}$$

so it has an essential singularity at 0 and Theorem 9.6 (The Casorati-Weierstrass Theorem) ensures that $g(D_0(0;1))$ is dense in $\mathbb{C}$. In other words, given $\omega \in \mathbb{C}$, there exists a sequence $\{z_n\} \subseteq D_0(0;1)$ such that $g(z_n) \to \frac{1}{\omega}$ as $n \to \infty$. Since $f = \frac{1}{g}$, we conclude that

$$\lim_{n\to\infty} f(z_n) = \lim_{n\to\infty} \frac{1}{g(z_n)} = \omega$$

which means that $f(D_0(0;1))$ is also dense in $\mathbb{C}$.

(b) Let $z_n = \frac{1}{n\pi}$, where $n \in \mathbb{Z}$. Clearly, we follow from Lemma 7.2 that

$$
\begin{aligned}
\lim_{z\to z_n} (z - z_n) \csc \frac{1}{z} &= \lim_{z\to z_n} \frac{z - z_n}{\sin \frac{1}{z}} \\
&= -\frac{1}{n\pi} \cdot \lim_{\zeta\to n\pi} \frac{\zeta - n\pi}{\sin \zeta} \\
&= -\frac{1}{n\pi} \cdot \lim_{\omega\to 0} \frac{\omega}{(-1)^n \sin \omega} \\
&= -\frac{(-1)^{n+1}}{n\pi} \\
&\neq 0
\end{aligned}
$$

and

$$\lim_{z\to z_n} (z - z_n)^2 \csc \frac{1}{z} = \lim_{z\to z_n} (z - z_n) \cdot \frac{z - z_n}{\sin \frac{1}{z}} = 0.$$

It follows from Theorem 9.5 that $\csc \frac{1}{z}$ has a simple pole at every z_n. Since $z_n \to 0$ as $n \to \pm\infty$. By Problem 9.16, we see that $f(D_0(0;1))$ is dense in $\mathbb{C}$.

Hence we complete the proof of the problem.

Problem 9.18

Bak and Newman Chapter 9 Exercise 18.

Proof. If f is a polynomial, then Theorem 5.12 (The Fundamental Theorem of Algebra) ensures that f takes every value in the plane $\mathbb{C}$. Suppose that f is not a polynomial and we consider $g(z) = f(\frac{1}{z})$. Similar to the proof of Problem 9.3, it can be shown that g has either a pole or an essential singularity at 0.

If 0 is a pole of g, then we have

$$g(z) = \sum_{n=-k}^{\infty} a_n z^n$$

for some $k \in \mathbb{N}$ so that

$$f(z) = \sum_{n=-\infty}^{k} a_n z^n. \tag{9.22}$$

Since f is analytic at 0, the principal part of the Laurent expansion (9.22) disappears, i.e., $a_{-1} = a_{-2} = \cdots = 0$. In other words, f reduces to a polynomial, a contradiction. Therefore, 0 is an essential singularity of g and Theorem 9.6 (The Casorati-Weierstrass Theorem) implies that $g(\mathbb{C} \setminus \{0\})$ is dense in $\mathbb{C}$. Since $g(\mathbb{C} \setminus \{0\}) \subseteq f(\mathbb{C})$, the set $f(\mathbb{C})$ is also dense in $\mathbb{C}$ and we complete the analysis of the problem.

CHAPTER **10**

The Residue Theorem

Problem 10.1

Bak and Newman Chapter 10 Exercise 1.

Proof.

(a) By Problem 9.9(a), $\frac{1}{z^4+z^2}$ has a pole of order 2 at 0 and poles of order 1 at $\pm i$. Furthermore, we apply [4, Eqn. (1), p. 129] to the expression (9.8) to get

$$\operatorname{Res}\left(\frac{1}{z^4+z^2};i\right) = \lim_{z\to i}(z-i)\times\frac{1}{z^4+z^2} = \lim_{z\to i}\frac{1}{z^2(z+i)} = -\frac{1}{2i} = \frac{i}{2}$$

and

$$\operatorname{Res}\left(\frac{1}{z^4+z^2};-i\right) = \lim_{z\to -i}(z+i)\times\frac{1}{z^4+z^2} = \lim_{z\to -i}\frac{1}{z^2(z-i)} = \frac{1}{2i} = -\frac{i}{2}.$$

Since we know

$$\frac{1}{z^4+z^2} = \frac{1}{z^2}\times\frac{1}{1+z^2} = \frac{1}{z^2}(1-z^2+z^4-\cdots) = \frac{1}{z^2}-1+z^2-\cdots,$$

Definition 10.1 implies that

$$\operatorname{Res}\left(\frac{1}{z^4+z^2};0\right) = 0.$$

(b) Recall from Problem 9.9(b) that $\cot z$ has a simple pole at $n\pi$ for every $n\in\mathbb{Z}$. Applying [4, Eqn. (1), p. 129] to the limit (9.9), we conclude that

$$\operatorname{Res}\left(\cot z; n\pi\right) = 1$$

for all $n\in\mathbb{Z}$.

(c) By Problem 9.9(c), $\csc z$ has a simple pole at $n\pi$ for every $n\in\mathbb{Z}$. Then we have

$$\lim_{z\to n\pi}(z-n\pi)\csc z = \lim_{z\to n\pi}\frac{z-n\pi}{\sin z-\sin n\pi} = \lim_{z\to n\pi}\frac{1}{\frac{\sin z-\sin n\pi}{z-n\pi}} = \frac{1}{\cos n\pi} = (-1)^n.$$

(d) By Problem 9.9(d), the function $\frac{\exp(\frac{1}{z^2})}{z-1}$ has a simple pole at 1 and an essential singularity at 0. By the limit (9.10), we conclude at once that

$$\operatorname{Res}\left(\frac{\exp(\frac{1}{z^2})}{z-1};1\right) = e.$$

Next, we get from Problem 9.11(b) (with $n = 1$) that

$$\mathrm{Res}\left(\frac{\exp(\frac{1}{z^2})}{z-1};0\right) = -e + 1.$$

(e) We note that

$$\frac{1}{z^2 + 3z + 2} = \frac{1}{(z+1)(z+2)},$$

so it has simple poles at -1 and -2. According to [4, Eqn. (1), p. 129], we see that

$$\mathrm{Res}\left(\frac{1}{z^2 + 3z + 2};-1\right) = \lim_{z \to -1}(z+1) \times \frac{1}{z^2 + 3z + 2} = \lim_{z \to -1}\frac{1}{z+2} = 1$$

and

$$\mathrm{Res}\left(\frac{1}{z^2 + 3z + 2};-2\right) = \lim_{z \to -2}(z+2) \times \frac{1}{z^2 + 3z + 2} = \lim_{z \to -2}\frac{1}{z+1} = -1.$$

(f) We know from the condition of Problem 9.17(a) that $\sin\frac{1}{z}$ has an essential singularity at 0. Thus we follow from the expression (9.21) that

$$\mathrm{Res}\left(\sin\frac{1}{z};0\right) = 1.$$

(g) It is clear that the function $ze^{\frac{3}{z}}$ has an essential singularity at 0. By [4, Example (iv), p. 124], we have

$$\mathrm{Res}\left(ze^{\frac{3}{z}};0\right) = \frac{1}{2 \cdot (\frac{1}{3})^2} = \frac{9}{2}.$$

(h) By the quadratic formula, the function

$$\frac{1}{az^2 + bz + c}$$

has simple poles at $z_\pm = \frac{-b \pm \sqrt{b^2 - 4ac}}{2a}$ if $b^2 - 4ac \neq 0$ and a pole of order 2 at $z_0 = -\frac{b}{2a}$ if $b^2 - 4ac = 0$. Therefore, we have

$$\begin{aligned}
\mathrm{Res}\left(\frac{1}{az^2 + bz + c};z_+\right) &= \lim_{z \to z_+}(z - z_+)\frac{1}{a(z - z_+)(z - z_-)} \\
&= \lim_{z \to z_+}\frac{1}{a(z - z_-)} \\
&= \frac{1}{a(z_+ - z_-)} \\
&= \frac{1}{\sqrt{b^2 - 4ac}}
\end{aligned}$$

and

$$\mathrm{Res}\left(\frac{1}{az^2 + bz + c};z_-\right) = \lim_{z \to z_-}(z - z_-)\frac{1}{a(z - z_+)(z - z_-)} = -\frac{1}{\sqrt{b^2 - 4ac}}.$$

Finally, we get

$$\mathrm{Res}\left(\frac{1}{az^2 + bz + c};z_0\right) = \frac{d}{dz}\left[(z - z_0)^2 \times \frac{1}{a(z - z_0)^2}\right]\Bigg|_{z=z_0} = \frac{d}{dz}\left(\frac{1}{a}\right)\Bigg|_{z=z_0} = 0.$$

Hence we complete the proof of the problem.

Bak and Newman Chapter 10 Exercise 2.

Proof.

(a) Obviously, $\cot z$ is analytic in the simply connected domain $D(0; 1 + \delta)$ except 0 for some small $\delta > 0$ and $C(0; 1) \subseteq D(0; 1 + \delta)$ is a closed curve such that 0 lies inside $C(0; 1)$, it follows from Theorem 10.5 (The Cauchy's Residue Theorem) and Problem 10.1(b) that

$$\int_{|z|=1} \cot z \, \mathrm{d}z = 2\pi i \mathrm{Res}\left(\cot z; 0\right) = 2\pi i.$$

(b) The function $f(z) = \frac{1}{(z-4)(z^3-1)}$ has simple poles at 4 and $z_k = \mathrm{e}^{\frac{2\pi k i}{3}}$, where $k = 1, 2, 3$. Now f is analytic in the simply connected domain $D(0; 2 + \delta)$ except z_1, z_2, z_3 for some small $\delta > 0$ and $C(0; 2) \subseteq D(0; 2 + \delta)$ is a closed curve such that z_k lie inside $C(0; 2)$ and 4 lies outside $C(0; 2)$. If $B(z) = (z - 4)(z^3 - 1)$, then $B'(z) = (z^3 - 1) + 3z^2(z - 4)$. Thus we obtain from [4, Eqn. (1), p. 129] that

$$\mathrm{Res}\left(f(z); 4\right) = \frac{1}{4^3 - 1} = \frac{1}{63} \quad \text{and} \quad \mathrm{Res}\left(f(z); z_k\right) = \frac{1}{3z_k^2(z_k - 4)},$$

where $k = 1, 2, 3$. Since $z_k^3 = 1$, we have $z_k^2 = \frac{1}{z_k}$ so that

$$\mathrm{Res}\left(f(z); z_k\right) = \frac{z_k}{3(z_k - 4)},$$

where $k = 1, 2, 3$. Next, recall that $z_1 + z_2 + z_3 = z_1 z_2 + z_1 z_3 + z_2 z_3 = 0$ and $z_1 z_2 z_3 = 1$, so direct computation gives

$$
\begin{aligned}
&\frac{z_1}{z_1 - 4} + \frac{z_2}{z_2 - 4} + \frac{z_3}{z_3 - 4} \\
&= \frac{z_1(z_2 - 4)(z_3 - 4) + z_2(z_1 - 4)(z_3 - 4) + z_3(z_1 - 4)(z_2 - 4)}{(z_1 - 4)(z_2 - 4)(z_3 - 4)} \\
&= \frac{3z_1 z_2 z_3 - 8(z_1 z_2 + z_1 z_3 + z_2 z_3) + 16(z_1 + z_2 + z_3)}{z_1 z_2 z_3 - 4(z_1 z_2 + z_1 z_3 + z_2 z_3) + 16(z_1 + z_2 + z_3) - 64} \\
&= -\frac{1}{21}.
\end{aligned}
$$

Hence Theorem 10.5 (The Cauchy's Residue Theorem) leads to us that

$$
\begin{aligned}
\int_{|z|=2} \frac{\mathrm{d}z}{(z - 4)(z^3 - 1)} &= 2\pi i\left(0 \times \frac{1}{63} + 1 \times \frac{z_1}{3(z_1 - 4)} + 1 \times \frac{z_2}{3(z_2 - 4)} + 1 \times \frac{z_3}{3(z_3 - 4)}\right) \\
&= -\frac{2\pi i}{63}.
\end{aligned}
$$

(c) Now $\sin \frac{1}{z}$ has an essential singularity at 0 and it is analytic in the simply connected domain $D(0; 1 + \delta)$ except 0 for some small $\delta > 0$ and $C(0; 1) \subseteq D(0; 1 + \delta)$ is a closed curve such that 0 lies inside $C(0; 1)$. Combining Problem 10.1(f) and Theorem 10.5 (The Cauchy's Residue Theorem), we establish

$$\int_{|z|=1} \sin \frac{1}{z} \, \mathrm{d}z = 2\pi i \mathrm{Res}\left(\sin \frac{1}{z}; 0\right) = 2\pi i.$$

(d) By using Problem 10.1(g) and Theorem 10.5 (The Cauchy's Residue Theorem) directly, we see that

$$\int_{|z|=2} ze^{\frac{3}{z}}\,dz = 2\pi i \mathrm{Res}\left(ze^{\frac{3}{z}};0\right) = 9\pi i.$$

Hence we end the proof of the problem. $\blacksquare$

Problem 10.3

Bak and Newman Chapter 10 Exercise 3.

Proof. It is clear that the function $f(z) = \frac{1}{[1-\exp(-z)]^n}$ is analytic everywhere except $z_k = 2\pi ki$, where $k \in \mathbb{Z}$. If C is a regular closed curve surrounding only the singularity $z_0 = 0$, then it deduces from Theorem 10.5 (The Cauchy's Residue Theorem) that

$$\mathrm{Res}\left(\frac{1}{(1-e^{-z})^n};0\right) = \frac{1}{2\pi i}\int_C \frac{dz}{(1-e^{-z})^n}. \tag{10.1}$$

By the substitution $\omega = \omega(z) = 1 - e^{-z}$, the integral in the formula (10.1) becomes

$$\int_C \frac{dz}{(1-e^{-z})^n} = \int_{C'} \frac{d\omega}{\omega^n(1-\omega)}, \tag{10.2}$$

where $C' = \omega(C)$.

The function $\frac{1}{\omega^n(1-\omega)}$ has (simple) poles at 0 and 1. Now we need to construct a suitable C so that the curve C' surrounds *only* the pole 0 but not 1. To this end, we consider the rectangle

$$R = \{z = x + iy \,|\, -1 \le x \le 1 \text{ and } -\tfrac{\pi}{2} \le y \le \tfrac{\pi}{2}\}.$$

Now its boundary $C = \partial R$ is a regular closed curve. Obviously, we have

$$\omega = \omega(z) = 1 - e^{-x-iy} = (1 - e^{-x}\cos y) + ie^{-x}\sin y$$

which implies that

$$|\omega(z) - 1| = e^{-x},$$

where $-1 \le x \le 1$, so we have actually

$$\frac{1}{e} \le |\omega(z) - 1| \le e$$

for all $z \in R$. Furthermore, $\mathrm{Re}\,\omega(z) = 1 - e^{-x}\cos y$ implies that $\mathrm{Re}\,\omega(z) > 1$ if and only if $\cos y < 0$ if and only if $\frac{\pi}{2} < y < \frac{3\pi}{2}$. In other words, we have $\mathrm{Re}\,\omega(z) \le 1$ for all $z \in R$ which assures us that $\omega(R)$ is a subset of the left half of the annulus A centred at 1 with inner and outer radii e^{-1} and e respectively. Finally, it is easy to see from the definition that the curves

$$\gamma_1 = \left\{\omega(1+iy) = (1 - e^{-1}\cos y) + ie^{-1}\sin y \,\middle|\, -\frac{\pi}{2} \le y \le \frac{\pi}{2}\right\}$$

and

$$\gamma_2 = \left\{\omega(-1+iy) = (1 - e\cos y) + ie\sin y \,\middle|\, -\frac{\pi}{2} \le y \le \frac{\pi}{2}\right\}$$

are exactly the inner arc and the outer arc of A respectively. Similarly, the curves

$$\gamma_3 = \left\{\omega\left(x + i\frac{\pi}{2}\right) = 1 + ie^{-x} \,\middle|\, -1 \le x \le 1\right\}$$

and

$$\gamma_4 = \left\{ \omega\left(x - i\frac{\pi}{2}\right) = 1 - ie^{-x} \,\middle|\, -1 \le x \le 1 \right\}$$

are the remaining boundary of the annulus A. Consequently, we obtain the desired result that the regular closed curve $C' = \omega(C)$ surrounds *only* the origin.

Hence it deduces from Theorem 10.5 (The Cauchy's Residue Theorem) that

$$\int_{C'} \frac{d\omega}{\omega^n(1-\omega)} = \mathrm{Res}\left(\frac{1}{\omega^n(1-\omega)}; 0\right) = 1. \tag{10.3}$$

Now our desired result follows immediately from the comparison of the expressions (10.1), (10.2) and (10.3), completing the proof of the problem. ∎

> **Problem 10.4**
>
> *Bak and Newman Chapter 10 Exercise 4.*

Proof. We know that $f(z) = (z + \frac{1}{z})^{2m+1}$ has a pole of order $2m+1$ at 0 for every $m = 0, 1, 2, \ldots$. By Theorem 10.5 (The Cauchy's Residue Theorem), we establish

$$\int_{|z|=1} \left(z + \frac{1}{z}\right)^{2m+1} dz = 2\pi i \mathrm{Res}\left(\left(z + \frac{1}{z}\right)^{2m+1}; 0\right).$$

By the binomial theorem, we have

$$\left(z + \frac{1}{z}\right)^{2m+1} = \sum_{k=0}^{2m+1} C_k^{2m+1} z^{2m+1-k} z^{-k} = \sum_{k=0}^{2m+1} C_k^{2m+1} z^{2m+1-2k}.$$

If $k = m + 1$, then C_{m+1}^{2m+1} is the coefficient of z^{-1} so that

$$\int_{|z|=1} \left(z + \frac{1}{z}\right)^{2m+1} dz = 2\pi i C_{m+1}^{2m+1} = 2\pi i C_m^{2m+1},$$

completing the proof of the problem. ∎

> **Problem 10.5**
>
> *Bak and Newman Chapter 10 Exercise 5.*

Proof. Suppose that $f(\omega)$ is analytic in the region D, D contains the regular closed curve C and

$$F(\omega) = \frac{f(\omega)}{p(\omega)} \cdot \frac{p(\omega) - p(z)}{\omega - z},$$

where $\omega \in D \backslash \{\omega_1, \omega_2, \ldots, \omega_n\}$ and $z \in \mathbb{C}$. Intuitively, one may apply Corollary 10.6 and Theorem 10.11 (The Generalized Cauchy Integral Formula) directly but we can't do that because they require the domain of the analyticity of f should be *simply connected*. Thus we have to search another versions of the results that fit our case. In fact, such variations can be found in [3, Theorem 3.8.6, p. 213; Theorem 5.1.2, p. 294].

Since C encloses the distinct points $\omega_1, \omega_2, \ldots, \omega_n$, it follows from [3, Theorem 5.1.2, p. 294] that

$$\int_C F(\omega)\, d\omega = 2\pi i \sum_{k=1}^{n} \mathrm{Res}\left(F(\omega); \omega_k\right) = 2\pi i \sum_{k=1}^{n} \mathrm{Res}\left(\frac{f(\omega)}{p(\omega)} \cdot \frac{p(\omega) - p(z)}{\omega - z}; \omega_k\right). \tag{10.4}$$

Since p is a polynomial of degree n, direct computation shows that the quotient $\frac{p(\omega)-p(z)}{\omega-z}$ is a polynomial of degree $n-1$. Thus if we may let

$$\frac{p(\omega)-p(z)}{\omega-z} = \sum_{m=0}^{n-1} Q_m(\omega) z^m,$$

where $Q_k(\omega)$ is a polynomial in ω, then we get from [4, Eqn. (1), p. 129] that

$$
\begin{aligned}
\operatorname{Res}\left(\frac{f(\omega)}{p(\omega)} \cdot \frac{p(\omega)-p(z)}{\omega-z}; \omega_k\right) &= \operatorname{Res}\left(\frac{f(\omega)\cdot \frac{p(\omega)-p(z)}{\omega-z}}{(\omega-\omega_1)(\omega-\omega_2)\cdots(\omega-\omega_n)}; \omega_k\right)\\
&= \frac{f(\omega_k)\cdot \frac{p(\omega_k)-p(z)}{\omega_k-z}}{\displaystyle\prod_{j\neq k}(\omega-\omega_j)}\\
&= \sum_{m=0}^{n-1}\frac{f(\omega_k)\cdot Q_m(\omega_k)}{\displaystyle\prod_{j\neq k}(\omega-\omega_j)} z^m.
\end{aligned}
$$

(10.5)

Using the expressions (10.4) and (10.5), we conclude that

$$P(z) = \frac{1}{2\pi i}\int_C F(\omega)\,d\omega = \sum_{k=1}^{n}\sum_{m=0}^{n-1}\frac{f(\omega_k)\cdot Q_m(\omega_k)}{\displaystyle\prod_{j\neq k}(\omega-\omega_j)} z^m$$

which is a polynomial in z of degree $n-1$.

Besides, we follow from the definition and [3, Theorem 3.8.6, p. 213] that

$$P(\omega_k) = \frac{1}{2\pi i}\int_C \frac{f(\omega)}{p(\omega)}\cdot\frac{p(\omega)}{\omega-\omega_k}\,d\omega = \frac{1}{2\pi i}\int_C \frac{f(\omega)}{\omega-\omega_k}\,d\omega = f(\omega_k)$$

for every $k=1,2,\ldots,n$. We complete the analysis of the problem. ▪

Problem 10.6

Bak and Newman Chapter 10 Exercise 6.

Proof. For small h, we observe that

$$\frac{f(z+h)-f(z)}{h} = \int_\gamma \frac{\phi(\omega)}{h}\left(\frac{1}{\omega-z-h}-\frac{1}{\omega-z}\right)d\omega = \int_\gamma \frac{\phi(\omega)}{(\omega-z-h)(\omega-z)}\,d\omega.$$

Then we have

$$f'(z) = \lim_{h\to 0}\frac{f(z+h)-f(z)}{h} = \lim_{h\to 0}\int_\gamma \frac{\phi(\omega)}{(\omega-z-h)(\omega-z)}\,d\omega.$$

(10.6)

If $\gamma: \omega = \omega(t)$ with $a\leq t\leq b$, then it follows from Definition 4.3 that

$$G(h) = \int_\gamma \frac{\phi(\omega)}{(\omega-z-h)(\omega-z)}\,d\omega = \int_a^b F(t,h)\,dt,$$

(10.7)

where

$$F(t,h) = \frac{\phi(\omega(t))}{[\omega(t)-z-h][\omega(t)-z]}\dot\omega(t).$$

We claim that G is a continuous function of h. To this end, since $z \notin \gamma$, we obtain from [22, Exercise 20, p. 101] that $d(z, \gamma) = \inf_{\omega \in \gamma} |\omega - z| > 0$. Furthermore, we know that $d(z, \gamma)$ is a continuous function in z, so we also have $d(z + h, \gamma) > 0$ for sufficiently small $h > 0$. These facts imply that

$$\frac{1}{[\omega(t) - z - h][\omega(t) - z]}$$

is a continuous function of t and h. Finally, the continuity of ϕ on γ and the piecewise continuity of $\dot{\omega}(t)$ on $[a, b]$ ensure that $F(t, h)$ is a piecewise continuous function of t and h. By the definition (10.7), we have

$$|G(h) - G(h')| \le \int_a^b |F(t, h) - F(t, h')| \, dt,$$

so G is a continuous function of h, as desired.

Consequently, it is legitimate by the above claim that we can interchange the limit and the integral in the expression (10.6) and get

$$f'(z) = \int_\gamma \lim_{h \to 0} \frac{\phi(\omega)}{(\omega - z - h)(\omega - z)} \, d\omega = \int_\gamma \frac{\phi(\omega)}{(\omega - z)^2} \, d\omega$$

which proves the first assertion.

For the second assertion, suppose that f is analytic in a simply connected domain D and γ is a regular closed curve contained in D. By Definition 4.3, γ is compact. Since f is obviously continuous on γ, the first assertion can be applied here to conclude

$$f'(z) = \frac{1}{2\pi i} \int_\gamma \frac{f(\omega)}{(\omega - z)^2} \, d\omega,$$

where z lies inside γ. Suppose that we have

$$f^{(k)}(z) = \frac{k!}{2\pi i} \int_\gamma \frac{f(\omega)}{(\omega - z)^{k+1}} \, d\omega.$$

Using similar attack as proving the first assertion,

$$\frac{f^{(k)}(z + h) - f^{(k)}(z)}{h} = \frac{k!}{2\pi i} \int_\gamma f(\omega) \cdot \frac{1}{h} \left[\frac{1}{(\omega - z - h)^{k+1}} - \frac{1}{(\omega - z)^{k+1}} \right] d\omega.$$

Recall that

$$a^{k+1} - b^{k+1} = (a - b)(a^k + a^{k-1}b + \cdots + ab^{k-1} + b^k),$$

so we see that

$$\frac{1}{(\omega - z - h)^{k+1}} - \frac{1}{(\omega - z)^{k+1}} = \frac{h}{(\omega - z - h)(\omega - z)} \times$$

$$\left[\frac{1}{(\omega - z - h)^k} + \frac{1}{(\omega - z - h)^{k-1}(\omega - z)} + \cdots + \frac{1}{(\omega - z)^k} \right]$$

and thus

$$f^{(k+1)}(z) = \lim_{h \to 0} \frac{f^{(k)}(z + h) - f^{(k)}(z)}{h}$$

$$= \lim_{h \to 0} \frac{k!}{2\pi i} \int_\gamma f(\omega) \cdot \frac{1}{h} \left[\frac{1}{(\omega - z - h)^{k+1}} - \frac{1}{(\omega - z)^{k+1}} \right] d\omega$$

$$= \frac{k!}{2\pi i} \int_\gamma f(\omega) \cdot \lim_{h \to 0} \frac{1}{h} \left[\frac{1}{(\omega - z - h)^{k+1}} - \frac{1}{(\omega - z)^{k+1}} \right] d\omega$$

$$= \frac{k!}{2\pi i} \int_\gamma f(\omega) \cdot \lim_{h \to 0} \frac{1}{(\omega - z - h)(\omega - z)} \times$$

$$\left[\frac{1}{(\omega - z - h)^k} + \frac{1}{(\omega - z - h)^{k-1}(\omega - z)} + \cdots + \frac{1}{(\omega - z)^k} \right] d\omega$$

$$= \frac{k!}{2\pi i} \int_\gamma f(\omega) \cdot \frac{k + 1}{(\omega - z)^{k+2}} \, d\omega$$

$$= \frac{(k + 1)!}{2\pi i} \int_\gamma \frac{f(\omega)}{(\omega - z)^{k+2}} \, d\omega.$$

Hence Theorem 10.11 (The Generalized Cauchy Integral Formula) follows immediately from the induction. It completes the proof of the problem.

> **Problem 10.7**
>
> *Bak and Newman Chapter 10 Exercise 7.*

Proof. It is impossible that $f(z) \equiv 0$ on $\mathbb{C}$ because it contradicts our hypothesis. Assume that f had more than one zero, say z_1 and z_2. Let r be so large that z_1 and z_2 lie inside $C(0; r)$. By the hypothesis again, we know that both z_1 and z_2 must be real. Suppose that $z_1 > z_2$ which gives

$$r > z_1 > z_2 > -r.$$

Furthermore, we know from the hypothesis that $f(\pm r)$ are the *only* real values on $C(0; r)$. Thus if $f(r) = 0$, then we can select another $r > r_1$ so that $D(0; r_1)$ contains z_1, z_2, i.e.,

$$r > r_1 > z_1 > z_2 > -r_1 > -r.$$

This process will stop in a finite number of steps, otherwise, one can find a decreasing sequence $\{r_n\}$ such that

$$r > r_1 > r_2 > \cdots > z_1 > z_2 > \cdots > -r_2 > -r_1 > -r.$$

Since the sequence $\{r_n\}$ is obviously bounded, it follows from [22, Theorem 3.14, p. 55] that it converges. Since $f(r_n) = 0$ for all $n \in \mathbb{N}$, Theorem 6.9 (The Uniqueness Theorem) ensures that $f \equiv 0$ in $\mathbb{C}$, a contradiction. Hence we may suppose that our chosen r satisfies $f(r) \neq 0$. Similarly, we also assume that $f(-r) \neq 0$.

Next, we define $g : (0, \pi) \to \mathbb{R}$ by

$$g(\theta) = f(re^{i\theta}).$$

Since f is entire, g is a continuous function of θ and the previous paragraph shows that its imaginary part is always positive or negative, i.e., $\{f(re^{i\theta}) \,|\, 0 < \theta < \pi\}$ lies entirely in the upper half-plane or the lower half-plane. Similarly, the set $\{f(re^{i\theta}) \,|\, \pi < \theta < 2\pi\}$ lies entirely in the upper half-plane or the lower half-plane. Since f is analytic in $\overline{D(0; r)}$ and $f(z) \neq 0$ on $C(0; r)$, we obtain from Corollary 10.9 (The Argument Principle) that

$$\mathbb{Z}(f) = \frac{1}{2\pi} \Delta \mathrm{Arg}\, f(z) \tag{10.8}$$

as z travels around the circle $C(0; r)$ from $re^{0 \cdot i}$ to $re^{2\pi i}$. Since $\Delta \mathrm{Arg}\, f(z) \leq \pi$ in the upper half-plane or the lower half-plane, we establish from the formula (10.8) that

$$\mathbb{Z}(f) \leq 1,$$

completing the proof of the problem.

Problem 10.8

*Bak and Newman Chapter 10 Exercise 8.**

Proof.

(a) If z is a zero of f, then $|f(z)| \geq |g(z)|$ implies that $g(z) = 0$ and so $f(z) + g(z) = 0$, a contradiction. In other words, f has no zero on γ so that $1 + \frac{g}{f}$ is well-defined on γ. Now we can follow the proof of Theorem 10.10 (Rouché's Theorem) to get

$$\mathbb{Z}(f + g) = \mathbb{Z}(f) + \frac{1}{2\pi i} \int_\gamma \frac{\mathrm{d}(1 + \frac{g}{f})}{1 + \frac{g}{f}} = \mathbb{Z}(f) + n\left(\left(1 + \frac{g}{f}\right)(\gamma); 0\right).$$

By the hypothesis, we know that $|(1 + \frac{g}{f}) - 1| \leq 1$ on γ, so $(1 + \frac{g}{f})(\gamma)$ remains *within* or *on* the disc of radius 1 centred at 1. In addition, we observe that $1 + \frac{g(z)}{f(z)} = 0$ if and only if $f(z) + g(z) = 0$, so the analytic function $1 + \frac{g}{f}$ will *not* touch the origin. Finally, we note that γ is a closed curve. Hence these observations ensure that $(1 + \frac{g}{f})(\gamma)$ is also a closed curve lying entirely in the right half-plane without surrounding 0. As a consequence, we have

$$n\left(\left(1 + \frac{g}{f}\right)(\gamma); 0\right) = 0,$$

so Rouché's Theorem holds in this case.

(b) Let $f(z) = 2z^4$ and $g(z) = z^5 + 1$. If $e^{5i\theta} + 2e^{4i\theta} + 1 = 0$ for some $\theta \in [0, 2\pi]$, then $|e^{i\theta} + 2| = 1$ or

$$(2 + \cos\theta)^2 + \sin^2\theta = 1. \tag{10.9}$$

After simplification, the equation (10.9) reduces to $\cos\theta = -2$ which is impossible. Hence $f(z) + g(z) = z^5 + 2z^4 + 1 \neq 0$ on $C(0; 1)$. Since $|g(z)| \leq 2 = |f(z)|$, part (a) implies that $\mathbb{Z}(z^5 + 2z^4 + 1) = \mathbb{Z}(2z^4) = 4$ in the unit disc.

We complete the proof of the problem. $\blacksquare$

Problem 10.9

Bak and Newman Chapter 10 Exercise 9.

Proof.

(a) Now the functions $f(z) = 3e^z$ and $g(z) = -z$ which are analytic inside and on the regular closed curve $C(0; 1)$. Since $|f(z)| = 3|e^z| \geq 3e^{-1} > |z|$ for all $z \in C(0; 1)$, Theorem 10.10 (Rouché's Theorem) implies that

$$\mathbb{Z}(f_1) = \mathbb{Z}(f) = \mathbb{Z}(3e^z) = 0$$

inside $C(0; 1)$.

(b) The functions $f(z) = -z$ and $g(z) = \frac{1}{3}e^z$ are analytic inside and on the regular closed curve $C(0; 1)$. Since $|f(z)| = |z| = 1 > \frac{e}{3} \geq |g(z)|$ on $C(0; 1)$, it follows from Theorem 10.10 (Rouché's Theorem) that

$$\mathbb{Z}(f_2) = \mathbb{Z}(f) = \mathbb{Z}(-z) = 1$$

inside $C(0; 1)$.

(c) The functions $f(z) = z^4$ and $g(z) = -5z + 1$ are analytic inside and on the regular closed curve $C(0; 2)$. Since $|f(z)| = |z^4| = 32 > |5z - 1| = |g(z)|$ on $C(0; 2)$, it follows from Theorem 10.10 (Rouché's Theorem) that

$$\mathbb{Z}(f_3) = \mathbb{Z}(f) = \mathbb{Z}(z^4) = 4 \tag{10.10}$$

inside $C(0; 2)$. Similarly, the functions $F(z) = -5z$ and $G(z) = z^4 + 1$ are analytic inside and on the regular closed curve $C(0; 1)$. Since $|F(z)| = |5z| = 5 > |z^4 + 1| = |G(z)|$, it follows from Theorem 10.10 (Rouché's Theorem) that

$$\mathbb{Z}(f_3) = \mathbb{Z}(F) = \mathbb{Z}(5z) = 1 \tag{10.11}$$

inside $C(0; 1)$. Combining the results (10.10) and (10.11), $\mathbb{Z}(f_3) = 4 - 1 = 3$ inside $1 \le |z| \le 2$.

(d) Suppose that $f(z) = -5z^4 + 3z^2$ and $g(z) = z^6 - 1$. Both f and g are analytic inside and on the regular closed curve $C(0; 1)$. The triangle inequality implies that

$$|f(z)| = |-z^4 + 3z^2| \ge |3z^2| - |z^4| = 2 \quad \text{and} \quad |g(z)| \le 2$$

on $|z| = 1$. Put $z = e^{i\theta}$ into $|f(z)| = 2$ to get $|-5e^{2i\theta} + 3| = 2$. Then we derive from this that $\cos 2\theta = 1$ so that $\theta = 0, \pi$. Consequently, $|f(z)| = 2$ if and only if $z = \pm 1$. Since $|g(\pm 1)| = 0 < 2$, it means that we always have $|f(z)| > |g(z)|$ on $C(0; 1)$. By Theorem 10.10 (Rouché's Theorem), we see that

$$\mathbb{Z}(f_4) = \mathbb{Z}(f) \tag{10.12}$$

inside $C(0; 1)$. Notice that $f(z) = 0$ if and only if $z = 0$ (repeated roots) and $z = \pm\sqrt{\frac{3}{5}}$, so all the zeros of f lie inside $C(0; 1)$ and hence the formula (10.12) gives $\mathbb{Z}(f_3) = 4$ inside $C(0; 1)$.

We complete the analysis of the problem. ■

> **Problem 10.10**
>
> *Bak and Newman Chapter 10 Exercise 10.*

Proof. We take $f(z) = \lambda - z$ and $g(z) = -e^{-z}$. It is clear that $z = \lambda$ is the *only* zero of $f(z)$ and it lies in the right half-plane. Let $R > 2\lambda$ and we consider the rectangle in the right half-plane

$$S_R = \{z = x + iy \,|\, 0 \le x \le R \text{ and } -R \le y \le R\}.$$

Now the boundary $\gamma_R = \partial S_R$ is easily seen to be a regular closed curve and S_R becomes the right half-plane as $R \to \infty$. Furthermore, our functions f and g are entire.

Let $z \in \gamma_R$. If $x = 0$, then $z = iy$ and so we observe that

$$|\lambda - iy| = \sqrt{\lambda^2 + y^2} \ge \lambda > 1 = |e^{-iy}|. \tag{10.13}$$

Next, we suppose that $x > 0$. Since $z = x + iR$ for $0 < x \le R$, we have $|z| = \sqrt{x^2 + R^2} > R$ so that the triangle inequality gives

$$|\lambda - z| \ge |z| - \lambda > \lambda > 1 \ge |e^{-z}|. \tag{10.14}$$

Finally, if $z = R + iy$, then we also have $|z| = \sqrt{R^2 + y^2} \geq R$ and thus the inequality (10.14) remains valid in this case. Consequently, the inequalities (10.13) and (10.14) together imply that

$$|f(z)| > |g(z)|$$

for all $z \in \gamma_R$. Hence it yields from Theorem 10.10 (Rouché's Theorem) that

$$\mathbb{Z}(\lambda - z - e^{-z}) = \mathbb{Z}(f + g) = \mathbb{Z}(f) = 1$$

inside γ_R and the desired result follows if we let $R \to \infty$. This completes the proof of the problem. ■

Problem 10.11

Bak and Newman Chapter 10 Exercise 11.

Proof. If f is a nonzero constant, then there is nothing to prove. Without loss of generality, we may assume that f is nonconstant. Let $F(z) = z^m \frac{f'(z)}{f(z)}$. Since $f \neq 0$ on γ, F is analytic inside and on γ except for isolated singularities inside γ. According to [3, Theorem 5.1.2, p. 294], we see immediately that

$$\frac{1}{2\pi i} \int_\gamma z^m \frac{f'(z)}{f(z)}\, \mathrm{d}z = \frac{1}{2\pi i} \int_\gamma F(z)\, \mathrm{d}z = \sum_k \mathrm{Res}\,(F; z_k),$$

where the z_k denote the isolated singularities of F inside γ which are exactly the zeros of f inside γ. By Lemma 7.1, there exists a function g analytic inside and on the regular closed curve γ such that $f(z) = (z - z_k)^{p_k} g(z)$ and $g(z_k) \neq 0$. Therefore, we have

$$F(z) = z^m \frac{f'(z)}{f(z)} = \frac{p_k z^m}{z - z_k} + \frac{z^m g'(z)}{g(z)}. \tag{10.15}$$

If $z_k = 0$, then F is actually analytic at 0 so that $\mathrm{Res}\,(F; z_k) = 0$. Otherwise, F has a simple pole at each zero of f. Thus it yields from [4, Eqn. (1), p. 129] and the expression (10.15) that

$$\mathrm{Res}\,(F; z_k) = \lim_{z \to z_k} (z - z_k) F(z) = p_k z_k^m.$$

Consequently, we obtain

$$\frac{1}{2\pi i} \int_\gamma z^m \frac{f'(z)}{f(z)}\, \mathrm{d}z = \frac{1}{2\pi i} \int_\gamma F(z)\, \mathrm{d}z = \sum_k p_k z_k^m,$$

where the sum takes over all zeros of f inside γ. This completes the proof of the problem. ■

Problem 10.12

Bak and Newman Chapter 10 Exercise 12.

Proof. On the one hand, we have $e^z \neq 0$ for all $z \in \overline{D(0; R)}$ with $R > 0$, so there exists a $\epsilon_R > 0$ such that

$$|e^z| > \epsilon_R \tag{10.16}$$

for all $z \in \overline{D(0; R)}$. On the other hand, since $|\frac{z^k}{k!}| \leq \frac{R^k}{k!}$ and the series

$$\sum_{k=0}^{\infty} \frac{z^k}{k!}$$

is obviously convergent, we deduce from Theorem 1.9 (The Weierstrass M-Test) that $P_n(z) \to \mathrm{e}^z$ uniformly in $\overline{D(0; R)}$. In other words, there exists an $N \in \mathbb{N}$ such that $n \geq N$ implies

$$|P_n(z) - \mathrm{e}^z| < \epsilon_R \tag{10.17}$$

for all $z \in \overline{D(0; R)}$. By combining the inequalities (10.16) and (10.17), it is easy to see that

$$|P_n(z) - \mathrm{e}^z| < |\mathrm{e}^z|$$

holds for all $n \geq N$ and $z \in \overline{D(0; R)}$. As both $P_n(z) - \mathrm{e}^z$ and e^z are analytic in $\overline{D(0; R)}$, Theorem 10.10 (Rouché's Theorem) asserts that

$$\mathbb{Z}(P_n(z)) = \mathbb{Z}(P_n(z) - \mathrm{e}^z + \mathrm{e}^z) = \mathbb{Z}(\mathrm{e}^z) = 0$$

inside $C(0; R)$, completing the proof of the problem. ∎

> **Problem 10.13**
>
> *Bak and Newman Chapter 10 Exercise 13.**

Proof.

(a) Direct computation gives

$$(1 - z)P(z) = a_0 + a_1 z + a_2 z^2 + \cdots + a_n z^n - a_0 z - a_1 z^2 - a_2 z^3 - \cdots - a_n z^{n+1}$$

$$= a_0 + \sum_{k=1}^{n} (a_k - a_{k-1}) z^k - a_n z^{n+1}. \tag{10.18}$$

Let $R > 1$. If $|z| = R$, then we have $|z|^{n+1} > |z|^k$ for all $k = 0, 1, 2, \ldots, n$. Therefore, it follows from the expression (10.18) and the fact $a_n \geq a_{n-1} \geq \cdots \geq a_0 \geq 0$ that

$$
\begin{aligned}
|(1 - z)P(z) - (-a_n z^{n+1})| &= \left| a_0 + \sum_{k=1}^{n} (a_k - a_{k-1}) z^k \right| \\
&< \left[a_0 + \sum_{k=1}^{n} (a_k - a_{k-1}) \right] \cdot |z^{n+1}| \\
&= |-a_n z^{n+1}|.
\end{aligned}
$$

By Theorem 10.10 (Rouché's Theorem), we conclude that

$$\mathbb{Z}((1 - z)P(z)) = \mathbb{Z}(-a_n z^{n+1}) = n + 1$$

inside $C(0; R)$. By letting $R \to 1$, all zeros of $(1 - z)P(z)$ (except $z = 1$) and then the zeros of $P(z)$ lie in $C(0; 1)$.

(b) We consider

$$P(z) = \frac{1}{(1-z)^2} \tag{10.19}$$

which is analytic in $D(0;1)$. Since $P^{(k)}(z) = (k+1)!(1-z)^{-k-2}$ for every $k = 0, 1, 2, \ldots$, we observe from Theorem 6.5 (The Power Series Representation for Functions Analytic in a Disc) that

$$C_k = \frac{P^{(k)}(0)}{k!} = k + 1$$

and so

$$P(z) = \sum_{k=1}^{\infty} (k+1) z^k \tag{10.20}$$

in $D(0;1)$.

Now if $|z| < 1$, then $|1 - z| < 2$ and thus it follows from the definition (10.19) that

$$|P(z)| > \frac{1}{4}. \tag{10.21}$$

Next, since the power series (10.20) has radius of convergence 1, we recall from [4, p. 27] that the series $P_n(z) \to P(z)$ uniformly in the smallest disc $\overline{D(0;\rho)}$. Consequently, there exists an $N \in \mathbb{N}$ such that $n \geq N$ implies

$$|P_n(z) - P(z)| \leq \frac{1}{4} \tag{10.22}$$

for all $z \in \overline{D(0;\rho)}$. Combining the inequalities (10.21) and (10.22), we know that

$$|P_n(z) - P(z)| < |P(z)|$$

on $C(0;\rho)$. Hence we apply Theorem 10.10 (Rouché's Theorem) that if $n \geq N$, then we have

$$\mathbb{Z}(P_n) = \mathbb{Z}(P)$$

inside $C(0;\rho)$. Since P has no zeros in $D(0;1)$, we establish that $\mathbb{Z}(P_n) = 0$ inside $C(0;\rho)$ for all $n \geq N$.

This completes the proof of the problem. $\blacksquare$

Problem 10.14

Bak and Newman Chapter 10 Exercise 14.

Proof. Let $P(z) = z^n + a_{n-1} z^{n-1} + \cdots + a_1 z + a_0$, where $a_0, a_1, \ldots, a_{n-1} \in \mathbb{C}$. Let $f(z) = z^n$ and $g(z) = a_{n-1} z^{n-1} + \cdots + a_1 z + a_0$. On the regular closed curve $C(0;R)$ for large enough R, we know that

$$|g(z)| \leq |a_{n-1}| \cdot |z|^{n-1} + |a_{n-2}| \cdot |z|^{n-2} + \cdots + |a_0| \leq MR^{n-1} < R^n = |f(z)|,$$

where M is a positive constant. Then Theorem 10.10 (Rouché's Theorem) applies in this case to conclude that

$$\mathbb{Z}(P) = \mathbb{Z}(f + g) = \mathbb{Z}(f) = n$$

inside $C(0;R)$. It completes the proof of the problem. $\blacksquare$

Problem 10.15

Bak and Newman Chapter 10 Exercise 15.

Proof. Since $|f| > |g|$ on γ, we get from the triangle inequality that

$$|f(z) + \lambda g(z)| \geq |f(z)| - \lambda|g(z)| \geq |f(z)| - |g(z)| > 0 \tag{10.23}$$

for all $z \in \gamma$ and all $\lambda \in [0,1]$. Thus $J(\lambda)$ is well-defined on $[0,1]$. If $s, t \in [0,1]$, then we have

$$|J(s) - J(t)| = \frac{1}{2\pi} \cdot \left| \int_\gamma \left(\frac{f' + sg'}{f + sg} - \frac{f' + tg'}{f + tg} \right) dz \right| = \frac{|s - t|}{2\pi} \cdot \left| \int_\gamma \frac{fg' - f'g}{(f + tg)(f + sg)} \, dz \right|. \tag{10.24}$$

Using the inequality (10.23), we know that

$$|(f + tg)(f + sg)| = |f + tg| \cdot |f - sg| \geq (|f| - |g|)^2$$

on γ and then

$$\left| \frac{fg' - f'g}{(f + tg)(f + sg)} \right| \leq \frac{|fg' - f'g|}{(|f| - |g|)^2} \tag{10.25}$$

on γ.

By Definition 4.2, $\gamma : [a, b] \to \mathbb{C}$ is continuous, so $\gamma([a, b])$ is compact and hence it is of finite length L. Since f, g, f' and g' are analytic on γ, they are continuous functions on $\gamma([a, b])$ so that all the sets $f(\gamma([a, b]))$, $g(\gamma([a, b]))$, $f'(\gamma([a, b]))$ and $g'(\gamma([a, b]))$ are also compact. Thus there exists a $M > 0$ such that

$$f, g, f', g' \ll M$$

throughout γ which shows that

$$|fg' - f'g| \leq 2M^2 \tag{10.26}$$

throughout γ. By the inequality (10.23) again, there exists a $\epsilon > 0$ such that

$$|f(z)| - |g(z)| \geq \sqrt{\epsilon} > 0 \tag{10.27}$$

on γ. Substutiting the inequalities (10.26) and (10.27) into the inequality (10.25), we conclude that

$$\left| \frac{fg' - f'g}{(f + tg)(f + sg)} \right| \leq \frac{2M^2}{\epsilon}$$

on γ. Now we apply Theorem 4.10 (The M-L Formula) to the inequality (10.24) to obtain

$$|J(s) - J(t)| \leq \frac{|s - t|}{2\pi} \times \frac{2M^2}{\epsilon} \times L = A|s - t|$$

for some constant A. Hence $J(\lambda)$ is a continuous function of λ.

Next, direct computation shows

$$\frac{(f + \lambda g)'}{f + \lambda g} = \frac{f' + \lambda g'}{f + \lambda g} = \frac{f'}{f} + \frac{\frac{f(\lambda g)' - f'(\lambda g)}{f^2}}{1 + \frac{\lambda g}{f}} = \frac{f'}{f} + \frac{(1 + \frac{\lambda g}{f})'}{1 + \frac{\lambda g}{f}}$$

and then

$$J(\lambda) = \frac{1}{2\pi i} \int_\gamma \frac{(f + \lambda g)'}{f + \lambda g} \, dz = \frac{1}{2\pi i} \int_\gamma \frac{f'}{f} \, dz + \frac{1}{2\pi i} \int_\gamma \frac{(1 + \frac{\lambda g}{f})'}{1 + \frac{\lambda g}{f}} \, dz. \tag{10.28}$$

Our hypothesis definitely implies that $|f(z)| > 0$ on γ, so we are able to employ Theorem 10.8 to conclude that the second integral in the expression (10.28) is an integer N. Next, the inequality (10.23) implies that $|1 + \frac{\lambda g}{f}| > 0$ on γ, so the third integral in the expression (10.28) is also an integer N_λ by Theorem 10.8. In other words, $J(\lambda) \in \mathbb{Z}$ for every $\lambda \in [0,1]$. Recall from the previous paragraph that $J(\lambda)$ is a continuous function of λ, so it must be a constant. In particular, $J(0) = J(1)$ and Theorem 10.8 establishes that

$$\mathbb{Z}(f) = \frac{1}{2\pi i} \int_\gamma \frac{f'}{f}\,\mathrm{d}z = \frac{1}{2\pi i} \int_\gamma \frac{(f+g)'}{f+g}\,\mathrm{d}z = \mathbb{Z}(f+g)$$

as required. We end the proof of the problem.　■

Problem 10.16

Bak and Newman Chapter 10 Exercise 16.

Proof. By the discussion on [4, p. 115], we know that the function

$$F(z) = \sqrt{z^2 - 1} = \exp\left(\frac{1}{2}\log(z^2 - 1)\right) = \exp\left(\frac{1}{2}\int_{\sqrt{2}}^{z} \frac{2\zeta}{\zeta^2 - 1}\,\mathrm{d}\zeta\right) \tag{10.29}$$

is analytic in $\mathbb{C} \setminus (-\infty, 1]$ because $f(z) = z^2 - 1$ is analytic there and $f(\sqrt{2}) \neq 0$. Suppose that $D = \mathbb{C} \setminus [-1, 1]$. Then D is open in $\mathbb{C}$.

- **Step 1: F is continuous on D.** If $\operatorname{Im} z > 0$, then the path of integration in the integral (10.29) is taken to be in the upper half-plane. Similarly, if $\operatorname{Im} z < 0$, then the path of integration in the integral (10.29) is taken to be in the lower half-plane. Since F is analytic in $\mathbb{C} \setminus (-\infty, 1]$, F is clearly continuous there. Therefore, it remains to show that F is continuous on $(-\infty, -1)$, i.e., for every $x_0 \in (-\infty, -1)$, we have

$$F(x_0) = \lim_{\substack{z \to x_0 \\ z \in D}} F(z). \tag{10.30}$$

To this end, suppose that $\operatorname{Im} z > 0$. Let γ_1 and γ_2 be the paths from z to x_0 and $\sqrt{2}$ in the upper half-plane respectively. Let γ_3 be the path from $\sqrt{2}$ to x_0 in the lower half-plane. See Figure 10.1 for an illustration.

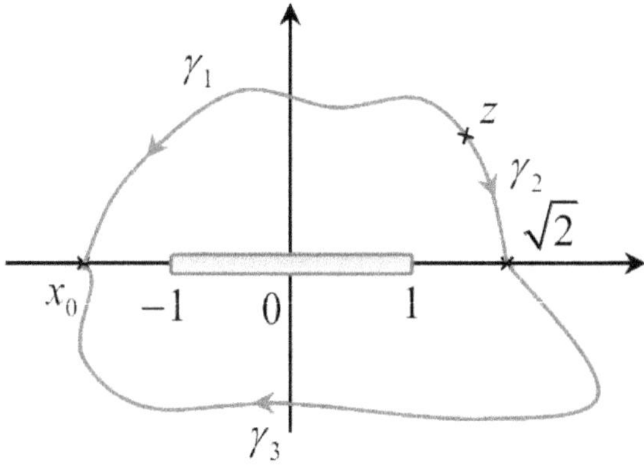

Figure 10.1: The regular closed curve C surrounding ± 1.

Let $C = \gamma_1 - \gamma_2 - \gamma_3$ be the regular closed curve connecting $\sqrt{2}$ and x_0 in D. Then C must surround ± 1. Since the function $g(z) = z^2 - 1$ is analytic and nonzero on C, it observes from Corollary 10.9 (The Argument Principle) that

$$\int_C \frac{2\zeta}{\zeta^2 - 1}\, d\zeta = 4\pi i$$

and then

$$\frac{1}{2}\int_{\gamma_1} \frac{2\zeta}{\zeta^2 - 1}\, d\zeta - \frac{1}{2}\int_{\gamma_2+\gamma_3} \frac{2\zeta}{\zeta^2 - 1}\, d\zeta = 2\pi i$$

which implies that

$$\exp\left(\frac{1}{2}\int_{\gamma_1} \frac{2\zeta}{\zeta^2 - 1}\, d\zeta\right) = \exp\left(2\pi i + \frac{1}{2}\int_{\gamma_2+\gamma_3} \frac{2\zeta}{\zeta^2 - 1}\, d\zeta\right)$$
$$= \exp\left(\frac{1}{2}\int_{\gamma_2+\gamma_3} \frac{2\zeta}{\zeta^2 - 1}\, d\zeta\right).$$

Hence we conclude that

$$\lim_{\substack{z \to x_0 \\ z \in \gamma_1}} F(z) = \lim_{\substack{z \to x_0 \\ z \in \gamma_2+\gamma_3}} F(z).$$

In other words, this proves the validity of (10.30) in the case $\operatorname{Im} z > 0$. Now the case $\operatorname{Im} z < 0$ can be verified similarly and so we have established the desired result (10.30).

- **Step 2: F is analytic in D.** Let N be a large positive integer. By **Step 1**, we see that F is continuous in the open set $D_N = D(0; N) \setminus [-1, 1] \subseteq D$ and analytic there except on the line segment $(-N, -1)$, so Theorem 7.7 ensures that F is analytic throughout D_N. Since N is arbitrary and $D_N \to D$ as $N \to \infty$, we conclude that F is analytic in D.

This completes the proof of the problem. ■

Problem 10.17

Bak and Newman Chapter 10 Exercise 17.

Proof. Similar to Problem 10.16, we define

$$\begin{aligned}
F(z) &= \sqrt[3]{(z-1)(z-2)(z-3)} \\
&= \exp\left(\frac{1}{3}\log[(z-1)(z-2)(z-3)]\right) \\
&= \exp\left(\frac{1}{3}\int_4^z \frac{[(\zeta-1)(\zeta-2)(\zeta-3)]'}{(\zeta-1)(\zeta-2)(\zeta-3)}\, d\zeta + \frac{1}{3}\log 6\right) \\
&= \sqrt[3]{6}\exp\left(\frac{1}{3}\int_4^z \frac{[(\zeta-1)(\zeta-2)(\zeta-3)]'}{(\zeta-1)(\zeta-2)(\zeta-3)}\, d\zeta\right)
\end{aligned} \tag{10.31}$$

is analytic in $\mathbb{C} \setminus (-\infty, 3]$ because $f(z) = (z-1)(z-2)(z-3)$ is analytic there and $f(4) \neq 0$. If we define $D = \mathbb{C} \setminus [1, 3]$, then D is open in $\mathbb{C}$.

- **Step 1: F is continuous on D.** If $\operatorname{Im} z > 0$, then the path of integration in the integral (10.31) is supposed to be in the upper half-plane. Similarly, if $\operatorname{Im} z < 0$, then the path of integration in the integral (10.31) is assumed to be in the lower half-plane. Since F is analytic in $\mathbb{C} \setminus (-\infty, 3]$, F is clearly continuous there. Therefore, it remains to show that F is continuous on $(-\infty, 1)$, i.e., for every $x_0 \in (-\infty, 1)$, we have

$$F(x_0) = \lim_{\substack{z \to x_0 \\ z \in D}} F(z). \tag{10.32}$$

To this end, suppose that $\operatorname{Im} z > 0$. Similar to Figure 10.1, we let γ_1 and γ_2 be the paths from z to x_0 and 4 in the upper half-plane respectively. Furthermore, let γ_3 be the path from 4 to x_0 in the lower half-plane.

Let $C = \gamma_1 - \gamma_2 - \gamma_3$ be the regular closed curve connecting 4 and x_0 in D. Then C must surround $1, 2$ and 3. Since the function $g(z) = (z-1)(z-2)(z-3)$ is analytic and nonzero on C, it observes from Corollary 10.9 (The Argument Principle) that

$$\int_C \frac{[(\zeta-1)(\zeta-2)(\zeta-3)]'}{(\zeta-1)(\zeta-2)(\zeta-3)} \, d\zeta = 6\pi i$$

and then

$$\frac{1}{3} \int_{\gamma_1} \frac{[(\zeta-1)(\zeta-2)(\zeta-3)]'}{(\zeta-1)(\zeta-2)(\zeta-3)} \, d\zeta - \frac{1}{3} \int_{\gamma_2+\gamma_3} \frac{[(\zeta-1)(\zeta-2)(\zeta-3)]'}{(\zeta-1)(\zeta-2)(\zeta-3)} \, d\zeta = 2\pi i$$

which implies that

$$\exp\left(\frac{1}{3} \int_{\gamma_1} \frac{[(\zeta-1)(\zeta-2)(\zeta-3)]'}{(\zeta-1)(\zeta-2)(\zeta-3)} \, d\zeta\right) = \exp\left(\frac{1}{3} \int_{\gamma_2+\gamma_3} \frac{[(\zeta-1)(\zeta-2)(\zeta-3)]'}{(\zeta-1)(\zeta-2)(\zeta-3)} \, d\zeta\right).$$

Hence we conclude that

$$\lim_{\substack{z \to x_0 \\ z \in \gamma_1}} F(z) = \lim_{\substack{z \to x_0 \\ z \in \gamma_2+\gamma_3}} F(z),$$

i.e., the limit (10.32) exists in the case $\operatorname{Im} z > 0$. Now the case $\operatorname{Im} z < 0$ can be verified similarly and so we have proved that the limit (10.32) exists.

- **Step 2: F is analytic in D.** Let N be a large positive integer. Now our F is continuous in the open set $D_N = D(0; N) \setminus [1, 3] \subseteq D$ and analytic there except on the line segment $(-N, 1)$. Hence Theorem 7.7 guarantees that F is analytic throughout D_N and then throughout D as desired if we take $N \to \infty$.

We complete the proof of the problem.

CHAPTER **11**

Applications of the Residue Theorem to the Evaluation of Integrals and Sums

Proof.

(a) It is a **Type I** integral, so we have

$$\int_{-\infty}^{\infty} \frac{x^2 \, dx}{(1 + x^2)^2} = 2\pi i \operatorname{Res}\left(\frac{z^2}{(1 + z^2)^2}; i\right).$$
(11.1)

Since i is a pole of order 2 of the function

$$f(z) = \frac{z^2}{(z + i)^2 (z - i)^2},$$

we see that

$$\operatorname{Res}\left(\frac{z^2}{(1 + z^2)^2}; i\right) = \frac{d}{dz}\left[(z - i)^2 f(z)\right]\Big|_{z=i} = \frac{d}{dz}\left[\frac{z^2}{(z + i)^2}\right]\Big|_{z=i} = -\frac{i}{4}.$$
(11.2)

Combining the expressions (11.1) and (11.2), we conclude immediately that

$$\int_{-\infty}^{\infty} \frac{x^2 \, dx}{(1 + x^2)^2} = \frac{\pi}{2}.$$

(b) Since $\frac{x^2}{(x^2+4)^2(x^2+9)}$ is an even function, we have

$$\int_{0}^{\infty} \frac{x^2 \, dx}{(x^2 + 4)^2 (x^2 + 9)} = \frac{1}{2} \int_{-\infty}^{\infty} \frac{x^2 \, dx}{(x^2 + 4)^2 (x^2 + 9)}$$

which is a **Type I** integral, so we have

$$\int_{0}^{\infty} \frac{x^2 \, dx}{(x^2 + 4)^2 (x^2 + 9)} = \frac{1}{2} \cdot 2\pi i \left[\operatorname{Res}\left(\frac{z^2}{(z^2 + 4)^2 (z^2 + 9)}; 3i\right)\right.$$
$$\left. + \operatorname{Res}\left(\frac{z^2}{(z^2 + 4)^2 (z^2 + 9)}; 2i\right)\right].$$
(11.3)

Here $3i$ and $2i$ are a simple pole and a pole of order 2 of the function $f(z) = \frac{z^2}{(z^2+4)^2(z^2+9)}$ respectively.

Since $(z - 2i)^2 f(z) = \frac{z^2}{(z+2i)^2(z^2+9)}$, we have

$$\text{Res}\left(\frac{z^2}{(z^2+4)^2(z^2+9)}; 2i\right) = \frac{d}{dz}\left[\frac{z^2}{(z+2i)^2(z^2+9)}\right]\Big|_{z=2i} = -\frac{13i}{200}. \qquad (11.4)$$

Using [4, Eqn. (1), p. 129], we see that

$$\text{Res}\left(\frac{z^2}{(z^2+4)^2(z^2+9)}; 3i\right) = \frac{z^2}{2(z^2+4)\cdot(2z)(z^2+9) + (z^2+4)^2\cdot(2z)}\Big|_{z=3i}$$

$$= -\frac{3}{50i}. \qquad (11.5)$$

Therefore, if we put the residues (11.4) and (11.5) back into the expression (11.3), then we obtain

$$\int_0^\infty \frac{x^2\,dx}{(x^2+4)^2(x^2+9)} \frac{13\pi}{200} - \frac{3\pi}{50} = \frac{\pi}{200}.$$

(c) Since $\frac{1}{x^4+x^2+1}$ is an even function, we have

$$\int_0^\infty \frac{dx}{x^4 + x^2 + 1} = \frac{1}{2}\int_{-\infty}^\infty \frac{dx}{x^4 + x^2 + 1}$$

which is a **Type I** integral, so we have

$$\int_0^\infty \frac{dx}{x^4 + x^2 + 1} = \frac{1}{2}\cdot 2\pi i\left[\text{Res}\left(\frac{1}{z^4 + z^2 + 1}; z_1\right) + \text{Res}\left(\frac{1}{z^4 + z^2 + 1}; z_2\right)\right], \qquad (11.6)$$

where $z_1 = \text{cis}\,\frac{\pi}{3}$ and $z_2 = \text{cis}\,\frac{2\pi}{3}$. Since both z_1 and z_2 are simple poles of $f(z) = \frac{1}{z^4+z^2+1}$, [4, Eqn. (1), p. 129] gives

$$\text{Res}\left(\frac{1}{z^4 + z^2 + 1}; z_1\right) = \frac{1}{4z_1^3 + 2z_1} = \frac{1}{-3 + i\sqrt{3}} \qquad (11.7)$$

and

$$\text{Res}\left(\frac{1}{z^4 + z^2 + 1}; z_2\right) = \frac{1}{4z_2^3 + 2z_2} = \frac{1}{3 + i\sqrt{3}}. \qquad (11.8)$$

Hence, by substituting the residues (11.7) and (11.8) into the expression (11.6), we obtain

$$\int_0^\infty \frac{dx}{x^4 + x^2 + 1} = \pi i\left(\frac{1}{-3 + i\sqrt{3}} + \frac{1}{3 + i\sqrt{3}}\right) = \frac{\sqrt{3}\pi}{6}.$$

(d) It is clear that it is a **Type II** integral. Since the function $\frac{\sin x}{x(1+x^2)}$ is even, we have

$$\int_0^\infty \frac{\sin x\,dx}{x(1+x^2)} = \frac{1}{2}\int_{-\infty}^\infty \frac{\sin x\,dx}{x(1+x^2)} = \frac{1}{2}\text{Im}\int_{-\infty}^\infty \frac{e^{ix}}{x}\,dx.$$

Since $\frac{e^{iz}}{z}$ has a simple pole at 0, we follow the idea of the example on [4, p. 146], it is true that

$$\int_0^\infty \frac{\sin x\,dx}{x(1+x^2)} = \frac{1}{2}\text{Im}\int_{-\infty}^\infty \frac{e^{ix} - 1}{x(1+x^2)}\,dx = \frac{1}{2}\text{Im}\left[2\pi i\text{Res}\left(\frac{e^{iz} - 1}{z(1 + z^2)}; i\right)\right].$$

Using [4, Eqn. (1), p. 129], we know that

$$\mathrm{Res}\left(\frac{e^{iz}-1}{z(1+z^2)};i\right)=\left.\frac{e^{iz}-1}{(1+z^2)+2z^2}\right|_{z=i}=-\frac{e^{-1}-1}{2}$$

and hence

$$\int_0^\infty \frac{\sin x\,\mathrm{d}x}{x(1+x^2)}=\frac{1}{2}\mathrm{Im}\left((1-e^{-1})\pi i\right)=\frac{(1-e^{-1})\pi}{2}.$$

(e) Again, it is a **Type II** integral. Since the integrand is even, we can write

$$\int_0^\infty \frac{\cos x\,\mathrm{d}x}{1+x^2}=\frac{1}{2}\int_{-\infty}^\infty \frac{\cos x\,\mathrm{d}x}{1+x^2}=\frac{1}{2}\mathrm{Re}\left[2\pi i\,\mathrm{Res}\left(\frac{e^{iz}}{1+z^2};i\right)\right].$$

By [4, Eqn. (1), p. 129], we have

$$\mathrm{Res}\left(\frac{e^{iz}}{1+z^2};i\right)=\left.\frac{e^{iz}}{2z}\right|_{z=i}=\frac{e^{-1}}{2i}$$

which implies

$$\int_0^\infty \frac{\cos x\,\mathrm{d}x}{1+x^2}=\frac{\pi}{2e}.$$

(f) It is a **Type III (A)** integral, so

$$\int_0^\infty \frac{\mathrm{d}x}{x^3+8}=-\sum_{k=0}^{2}\mathrm{Res}\left(\frac{\log z}{z^3+8};z_k\right),$$

where $z_k=2\exp\left[\frac{(2k+1)\pi i}{3}\right]$ for $k=0,1,2$. Since they are simple poles of $\frac{1}{z^3+8}$, it follows from [4, Eqn. (1), p. 129] that

$$\begin{aligned}
\mathrm{Res}\left(\frac{\log z}{z^3+8};z_k\right)&=\left.\frac{\log z}{3z^2}\right|_{z=z_k}\\
&=-\frac{z_k\log z_k}{24}\\
&=-\frac{z_k(\log|z_k|+i\mathrm{Arg}\,z_k)}{24}\\
&=-\frac{(2+i\frac{(2k+1)\pi}{3})\exp\frac{(2k+1)\pi i}{3}}{12}.
\end{aligned}$$

Hence we conclude that

$$\begin{aligned}
\int_0^\infty \frac{\mathrm{d}x}{x^3+8}&=\frac{(2+i\frac{\pi}{3})\exp\frac{\pi i}{3}}{12}+\frac{(2+i\pi)\exp(\pi i)}{12}+\frac{(2+i\frac{5\pi}{3})\exp\frac{5\pi i}{3}}{12}\\
&=\frac{1}{12}\left[\left(2+\frac{\pi i}{3}\right)\left(\frac{1}{2}+\frac{\sqrt{3}i}{2}\right)-(2+i\pi)+\left(2+\frac{5\pi i}{3}\right)\left(\frac{1}{2}-\frac{\sqrt{3}i}{2}\right)\right]\\
&=\frac{1}{12}\times\frac{4\sqrt{3}\pi}{6}\\
&=\frac{\sqrt{3}\pi}{18}.
\end{aligned}$$

(g) It is a **Type III (C)** integral. Thus

$$[1-e^{2\pi i(\alpha-1)}]\int_0^\infty \frac{x^{\alpha-1}}{1+x}\,\mathrm{d}x=2\pi i\,\mathrm{Res}\left(\frac{z^{\alpha-1}}{1+z};-1\right)=2\pi i(-1)^{\alpha-1}=2\pi i e^{(\alpha-1)\pi i}$$

which implies that

$$\int_0^\infty \frac{x^{\alpha-1}}{1+x}\,\mathrm{d}x=\frac{2\pi i e^{(\alpha-1)\pi i}}{1-e^{2\pi i(\alpha-1)}}=\frac{-2\pi i e^{\alpha\pi i}}{1-e^{2\alpha\pi i}}=\frac{-2\pi i}{e^{-\alpha\pi i}-e^{\alpha\pi i}}=\frac{\pi}{\sin(\alpha\pi)}.$$

(h) The integral is of **Type IV**, so [4, Eqn. (5), p. 150] gives

$$\int_0^{2\pi} \frac{\mathrm{d}x}{(2+\cos x)^2} = \int_{|z|=1} \frac{1}{(2 + \frac{z+z^{-1}}{2})^2} \times \frac{\mathrm{d}z}{iz}$$

$$= \frac{4}{i} \int_{|z|=1} \frac{z\,\mathrm{d}z}{(z^2 + 4z + 1)^2}$$

$$= 8\pi \operatorname{Res}\left(\frac{z}{(z^2+4z+1)^2}; \sqrt{3} - 2 \right). \qquad (11.9)$$

If $f(z) = \frac{z}{(z^2+4z+1)^2} = \frac{z}{(z+2-\sqrt{3})^2(z+2+\sqrt{3})^2}$, then we know that

$$\operatorname{Res}\left(\frac{z}{(z^2+4z+1)^2}; \sqrt{3} - 2 \right) = \frac{\mathrm{d}}{\mathrm{d}z}\left[(z+2-\sqrt{3})^2 f(z) \right]\Big|_{z=-2+\sqrt{3}}$$

$$= \frac{\mathrm{d}}{\mathrm{d}z}\left[\frac{z}{(z+2+\sqrt{3})^2} \right]\Big|_{z=-2+\sqrt{3}}$$

$$= \frac{\sqrt{3}}{18}. \qquad (11.10)$$

Now it yields from the expressions (11.9) and (11.10) that

$$\int_0^{2\pi} \frac{\mathrm{d}x}{(2+\cos x)^2} = 8\pi \times \frac{\sqrt{3}}{18} = \frac{4\sqrt{3}\pi}{9}.$$

(i) It is a **Type IV** integral, so [4, Eqn. (5), p. 150] implies

$$\int_0^{2\pi} \frac{\sin^2 x\,\mathrm{d}x}{5+3\cos x} = \int_{|z|=1} \frac{-\frac{1}{4}(z-\frac{1}{z})^2}{5 + \frac{3}{2}(z+\frac{1}{z})} \times \frac{\mathrm{d}z}{iz}$$

$$= \frac{i}{2} \int_{|z|=1} \frac{(z^2-1)^2\,\mathrm{d}z}{z^2(3z^2+10z+3)}$$

$$= \frac{i}{2} \times 2\pi i \sum_k \operatorname{Res}\left(\frac{(z^2-1)^2}{z^2(3z^2+10z+3)}; z_k \right), \qquad (11.11)$$

where z_k are the zeros of $z^2(3z^2+10z+3)$ inside the unit circle $|z|=1$. Therefore, the formula (11.11) reduces to

$$\int_0^{2\pi} \frac{\sin^{2x}\,\mathrm{d}x}{5+3\cos x} = -\pi\left[\operatorname{Res}\left(\frac{(z^2-1)^2}{z^2(3z^2+10z+3)}; 0 \right) + \operatorname{Res}\left(\frac{(z^2-1)^2}{z^2(3z^2+10z+3)}; -\frac{1}{3} \right) \right].$$

Notice that

$$\operatorname{Res}\left(\frac{(z^2-1)^2}{z^2(3z^2+10z+3)}; 0 \right) = \frac{\mathrm{d}}{\mathrm{d}z}\left[\frac{(z^2-1)^2}{3z^2+10z+3} \right]\Big|_{z=0} = -\frac{10}{9}$$

and

$$\operatorname{Res}\left(\frac{(z^2-1)^2}{z^2(3z^2+10z+3)}; -\frac{1}{3} \right) = \frac{(z^2-1)^2}{2z(3z^2+10z+3)+z^2(6z+10)}\Big|_{z=-\frac{1}{3}} = \frac{8}{9}.$$

Thus we get

$$\int_0^{2\pi} \frac{\sin^2 x\,\mathrm{d}x}{5+3\cos x} = -\pi\left(-\frac{10}{9} + \frac{8}{9} \right) = \frac{2\pi}{9}.$$

(j) The integral is of **Type IV** and then

$$\int_0^{2\pi} \frac{\mathrm{d}x}{a + \cos x} = \int_{|z|=1} \frac{1}{a + \frac{1}{2}\left(z + \frac{1}{z}\right)} \times \frac{\mathrm{d}z}{iz}$$

$$= \frac{2}{i} \int_{|z|=1} \frac{\mathrm{d}z}{z^2 + 2az + 1}$$

$$= 4\pi \sum_k \mathrm{Res}\left(\frac{1}{z^2 + 2az + 1}; z_k\right), \tag{11.12}$$

where z_k are zeros of $z^2 + 2az + 1$ lying inside the unit circle. We note that $z^2 + 2az + 1 = 0$ if and only if

$$z = -a \pm \sqrt{a^2 - 1}.$$

If $a > 1$, then $|-a + \sqrt{a^2 - 1}| < 1$ but $|-a - \sqrt{a^2 - 1}| > 1$. Similarly, if $a < -1$, then $|-a - \sqrt{a^2 - 1}| < 1$ but $|-a + \sqrt{a^2 - 1}| > 1$. Therefore, we get from the expression (11.12) that

$$\int_0^{2\pi} \frac{\mathrm{d}x}{a + \cos x} = \begin{cases} 4\pi\mathrm{Res}\left(\dfrac{1}{z^2 + 2az + 1}; -a + \sqrt{a^2 - 1}\right), & \text{if } a > 1; \\[2ex] 4\pi\mathrm{Res}\left(\dfrac{1}{z^2 + 2az + 1}; -a - \sqrt{a^2 - 1}\right), & \text{if } a < -1 \end{cases}$$

$$= \begin{cases} 4\pi \times \dfrac{1}{2z + 2a}\Big|_{z=-a+\sqrt{a^2-1}}, & \text{if } a > 1; \\[2ex] 4\pi \times \dfrac{1}{2z + 2a}\Big|_{z=-a-\sqrt{a^2-1}}, & \text{if } a < -1 \end{cases}$$

$$= \begin{cases} \dfrac{2\pi}{\sqrt{a^2 - 1}}, & \text{if } a > 1; \\[2ex] -\dfrac{2\pi}{\sqrt{a^2 - 1}}, & \text{if } a < -1. \end{cases}$$

We have completed the proof of the problem. $\blacksquare$

Problem 11.2

Bak and Newman Chapter 11 Exercise 2.

Proof. By the power series expansion of e^{2iz} around 0, we have

$$\frac{e^{2iz} - 1 - 2iz}{z^2} = -2 - \frac{4i}{3}z + \cdots.$$

Since $\frac{e^{2iz}-1-2iz}{z^2}$ has a removable singularity at 0, it is entire. Let C_R be the closed contour used on [4, p. 143], we follow from Theorem 6.3 (The Closed Curve Theorem) that

$$\int_{C_R} \frac{e^{2iz} - 1 - 2iz}{z^2} \,\mathrm{d}z = 0.$$

Recall that

$$\int_{C_R} = \int_{\Gamma_R} + \int_{-R}^{R}.$$

Therefore, we get

$$\int_{-R}^{R} \frac{e^{2ix} - 1 - 2ix}{x^2}\, dx + \int_{\Gamma_R} \frac{e^{2iz} - 1 - 2iz}{z^2}\, dz = 0. \tag{11.13}$$

On the one hand, if $z \in \Gamma_R$, then we have

$$\left| \frac{e^{2iz} - 1}{z^2} \right| \le \frac{e^{-2\operatorname{Im} z} + 1}{R^2} \le \frac{2}{R^2}$$

and Theorem 4.10 (The M-L Formula) implies that

$$\left| \int_{\Gamma_R} \frac{e^{2iz} - 1}{z^2}\, dz \right| \le \frac{2}{R^2} \times \pi R = \frac{2\pi}{R}.$$

Consequently, we obtain

$$\lim_{R \to \infty} \int_{\Gamma_R} \frac{e^{2iz} - 1}{z^2}\, dz = 0. \tag{11.14}$$

By Definition 4.3, we know that

$$\int_{\Gamma_R} \frac{dz}{z} = \int_0^\pi \frac{iRe^{i\theta}}{Re^{i\theta}}\, d\theta = \int_0^\pi i\, d\theta = i\pi \tag{11.15}$$

for every $R > 0$. Now it follows from the results (11.14) and (11.15) that

$$\lim_{R \to \infty} \int_{\Gamma_R} \frac{e^{2iz} - 1 - 2iz}{z^2}\, dz = -2i \cdot i\pi = 2\pi. \tag{11.16}$$

On the other hand, since

$$\int_{-R}^{0} \frac{e^{2ix} - 1 - 2ix}{x^2}\, dx = \int_0^R \frac{e^{-2ix} - 1 + 2ix}{x^2}\, dx,$$

the first integral in the expression (11.13) can be written as

$$\begin{aligned}
\int_{-R}^{R} \frac{e^{2ix} - 1 - 2ix}{x^2}\, dx &= \int_{-R}^{0} \frac{e^{2ix} - 1 - 2ix}{x^2}\, dx + \int_0^R \frac{e^{2ix} - 1 - 2ix}{x^2}\, dx \\
&= \int_0^R \frac{e^{-2ix} - 1 + 2ix}{x^2}\, dx + \int_0^R \frac{e^{2ix} - 1 - 2ix}{x^2}\, dx \\
&= \int_0^R \frac{e^{2ix} + e^{-2ix} - 2}{x^2}\, dx \\
&= \int_0^R \frac{-4\sin^2 x}{x^2}\, dx. \tag{11.17}
\end{aligned}$$

Hence, by putting the results (11.16) and (11.17) into the expression (11.13), we conclude that

$$-4 \int_0^\infty \frac{\sin^2 x}{x^2}\, dx + 2\pi = 0$$

so that

$$\int_0^\infty \frac{\sin^2 x}{x^2}\, dx = \frac{\pi}{2}.$$

This completes the proof of the problem.

> **Problem 11.3**
>
> *Bak and Newman Chapter 11 Exercise 3.*

Proof. Let C_R be the given contour of large enough radius R. Furthermore, let Γ_R, η_R and η'_R be the circular arc, the ray on the real axis and the ray with angle $\frac{2\pi}{n}$ with the direction indicated in the exercise respectively. Note that if R is large enough, then the point $\exp(\frac{\pi i}{n})$ lies inside C_R.

On the one hand, since $\exp(\frac{\pi i}{n})$ is a simple zero of the equation $z^n + 1 = 0$, Theorem 10.5 (The Cauchy's Residue Theorem)[a] implies that

$$\int_{C_R} \frac{\mathrm{d}z}{1 + z^n} = 2\pi i \mathrm{Res}\left(\frac{1}{1 + z^n}; \exp\left(\frac{\pi i}{n}\right)\right)$$

$$= 2\pi i \times \left.\frac{1}{n z^{n-1}}\right|_{z=\exp(\frac{\pi i}{n})}$$

$$= -2\pi i \times \left.\frac{z}{n}\right|_{z=\exp(\frac{\pi i}{n})}$$

$$= \frac{-2\pi i}{n} \exp\left(\frac{\pi i}{n}\right). \tag{11.18}$$

On the other hand, if $z \in \Gamma_R$, then $z = Re^{i\theta}$ with $0 \le \theta \le \frac{2\pi}{n}$. Since $|z^n + 1| \ge |z|^n - 1 = R^n - 1$ on Γ_R for large R, Theorem 4.10 (The *M-L* Formula) gives

$$\left|\int_{\Gamma_R} \frac{\mathrm{d}z}{1 + z^n}\right| \le \frac{2\pi R}{n} \times \frac{1}{R^n - 1}. \tag{11.19}$$

Since $n \ge 2$, the inequality (11.19) shows that

$$\lim_{R \to \infty} \int_{\Gamma_R} \frac{\mathrm{d}z}{1 + z^n} = 0. \tag{11.20}$$

Next, it is obvious that

$$\int_0^\infty \frac{\mathrm{d}x}{1 + x^n} = \lim_{R \to \infty} \int_{\eta_R} \frac{\mathrm{d}x}{1 + x^n} = \lim_{R \to \infty} \int_{\eta_R} \frac{\mathrm{d}z}{1 + z^n}. \tag{11.21}$$

Finally, on η'_R, we have $z = (R - x)\exp(\frac{2\pi i}{n})$ with $0 \le x \le R$ so that

$$\int_{\eta'_R} \frac{\mathrm{d}z}{1 + z^n} = -\int_0^R \frac{\exp(\frac{2\pi i}{n})\,\mathrm{d}x}{1 + (R - x)^n} = \int_R^0 \frac{\exp(\frac{2\pi i}{n})\,\mathrm{d}y}{1 + y^n} = -\exp\left(\frac{2\pi i}{n}\right)\int_0^R \frac{\mathrm{d}x}{1 + x^n}. \tag{11.22}$$

Recall that $C_R = \Gamma_R + \eta_R + \eta'_R$, we substitute the results (11.20), (11.21) and (11.22) into the expression (11.18) to get

$$\lim_{R \to \infty} \int_{\eta_R} \frac{\mathrm{d}z}{1 + z^n} + \lim_{R \to \infty} \int_{\eta'_R} \frac{\mathrm{d}z}{1 + z^n} + \lim_{R \to \infty} \int_{\Gamma_R} \frac{\mathrm{d}z}{1 + z^n} = -\frac{2\pi i}{n} \exp\left(\frac{\pi i}{n}\right)$$

$$\int_0^\infty \frac{\mathrm{d}x}{1 + x^n} - \exp\left(\frac{2\pi i}{n}\right)\int_0^\infty \frac{\mathrm{d}x}{1 + x^n} = -\frac{2\pi i}{n}\exp\left(\frac{\pi i}{n}\right)$$

$$\int_0^\infty \frac{\mathrm{d}x}{1 + x^n} = -\frac{2\pi i}{n} \cdot \frac{\exp(\frac{\pi i}{n})}{1 - \exp(\frac{2\pi i}{n})}$$

$$= \frac{\pi}{n \sin \frac{\pi}{n}}.$$

This ends the proof of the problem.

---

[a]Of course, we apply the version [3, Theorem 5.1.2, p. 294] of Asmar and Grafakos here.

Bak and Newman Chapter 11 Exercise 4.

Proof.

(a) By the substitution $y = ax$, we have

$$\int_0^\infty \frac{\cos ax}{(1+x^2)^2}\,\mathrm{d}x = a^3 \int_0^\infty \frac{\cos y\,\mathrm{d}y}{(a^2+y^2)^2} = \frac{a^3}{2}\int_{-\infty}^\infty \frac{\cos y\,\mathrm{d}y}{(a^2+y^2)^2}$$

which is a **Type II** integral. Since

$$\int_{-\infty}^\infty \frac{\cos y\,\mathrm{d}y}{(a^2+y^2)^2} = \operatorname{Re}\left[2\pi i \operatorname{Res}\left(\frac{e^{iz}}{(a^2+z^2)^2}; ai\right)\right]$$

$$= \operatorname{Re}\left\{2\pi i \frac{\mathrm{d}}{\mathrm{d}z}\left[\frac{e^{iz}}{(z+ai)^2}\right]\Big|_{z=ai}\right\}$$

$$= \frac{(a+1)e^{-a}\pi}{2a^3},$$

we have

$$\int_0^\infty \frac{\cos ax}{(1+x^2)^2}\,\mathrm{d}x = \frac{(a+1)e^{-a}\pi}{4}.$$

(b) Using the same strategy of Problem 11.3, we see that

$$\int_{C_R} \frac{z^2\,\mathrm{d}z}{1+z^{10}} = 2\pi i \operatorname{Res}\left(\frac{z^2}{1+z^{10}}; \exp\left(\frac{\pi i}{10}\right)\right) = -\frac{\pi i}{5}\exp\left(\frac{3\pi i}{10}\right). \tag{11.23}$$

On Γ_R, similar to the inequality (11.19), we have

$$\left|\int_{\Gamma_R} \frac{z^2\,\mathrm{d}z}{1+z^{10}}\right| \le \frac{\pi R}{5} \times \frac{R^2}{R^{10}-1}$$

so that

$$\lim_{R\to\infty}\int_{\Gamma_R} \frac{z^2\,\mathrm{d}z}{1+z^{10}} = 0. \tag{11.24}$$

On η_R, we have

$$\int_0^\infty \frac{x^2\,\mathrm{d}x}{1+x^{10}} = \lim_{R\to\infty}\int_{\eta_R}\frac{x^2\,\mathrm{d}x}{1+x^{10}} = \lim_{R\to\infty}\int_{\eta_R}\frac{z^2\,\mathrm{d}z}{1+z^{10}}. \tag{11.25}$$

On η_R', we have

$$\int_{\eta_R'} \frac{z^2\,\mathrm{d}z}{1+z^{10}} = -\int_0^R \frac{(R-x)^2\exp(\frac{3\pi i}{5})}{1+(R-x)^2}\,\mathrm{d}x = -\exp\left(\frac{3\pi i}{5}\right)\int_0^R \frac{x^2\,\mathrm{d}x}{1+x^{10}}. \tag{11.26}$$

After putting the results (11.24), (11.25) and (11.26) into the expression (11.23), we establish that

$$\left[1 - \exp\left(\frac{3\pi i}{5}\right)\right]\int_0^\infty \frac{x^2\,\mathrm{d}x}{1+x^{10}} = -\frac{\pi i}{5}\exp\left(\frac{3\pi i}{10}\right)$$

$$\int_0^\infty \frac{x^2\,\mathrm{d}x}{1+x^{10}} = -\frac{\pi i}{5}\times\frac{1}{\exp(-\frac{3\pi i}{10})-\exp(\frac{3\pi i}{10})}$$

$$= -\frac{\pi i}{5}\times\frac{1}{-2i\sin\frac{3\pi}{10}}$$

$$= \frac{\pi}{10\sin\frac{3\pi}{10}}.$$

(c) We note that

$$\int_0^{2\pi} \exp(e^{i\theta})\,\mathrm{d}\theta = \int_{|z|=1} e^z \frac{\mathrm{d}z}{iz} = -i\int_{|z|=1} \frac{e^z}{z}\,\mathrm{d}z = 2\pi\mathrm{Res}\left(\frac{e^z}{z};0\right) = 2\pi.$$

Hence we have completed the analysis of the problem.

> **Problem 11.5**
>
> *Bak and Newman Chapter 11 Exercise 5.*

Proof. It is clear that

$$\int_0^{2\pi}(\cos x)^{2m}\,\mathrm{d}x = \int_{|z|=1}\frac{1}{4^m}\times\left(z+\frac{1}{z}\right)^{2m}\frac{\mathrm{d}z}{iz}$$
$$= \frac{1}{i4^m}\int_{|z|=1}\frac{(z^2+1)^{2m}}{z^{2m+1}}\,\mathrm{d}z$$
$$= \frac{2\pi}{4^m}\mathrm{Res}\left(\frac{(z^2+1)^{2m}}{z^{2m+1}};0\right).$$

Since $(z^2+1)^{2m} = \displaystyle\sum_{k=0}^{2m}C_k^{2m}z^{4m-2k}$, we have

$$\frac{(z^2+1)^{2m}}{z^{2m+1}} = \sum_{k=0}^{2m}C_k^{2m}z^{2m-2k-1} = \frac{C_m^{2m}}{z} + \sum_{\substack{k=0\\k\neq m}}^{2m}C_k^{2m}z^{2m-2k-1}. \tag{11.27}$$

Hence we immediately conclude from the expansion (11.27) that

$$\int_0^{2\pi}(\cos x)^{2m}\,\mathrm{d}x = \frac{2\pi}{4^m}C_m^{2m},$$

completing the proof of the problem.

> **Problem 11.6**
>
> *Bak and Newman Chapter 11 Exercise 6.*

Proof. This is a **Type I** integral so that

$$\int_{-\infty}^{\infty}\frac{\mathrm{d}x}{(1+x^2)^{n+1}} = 2\pi i\mathrm{Res}\left(\frac{1}{(1+z^2)^{n+1}};i\right). \tag{11.28}$$

Since $f(z) = (z+i)^{-n-1}(z-i)^{-n-1}$ has a pole of order $n+1$ at i, we know from the formula on [4, p. 130] that

$$\mathrm{Res}\left(\frac{1}{(1+z^2)^{n+1}};i\right) = \frac{1}{n!}\cdot\frac{\mathrm{d}^n}{\mathrm{d}z^n}(z+i)^{-n-1}\Big|_{z=i}$$
$$= \frac{1}{n!}\cdot(-n-1)(-n-2)\cdots(-2n)(z+i)^{-2n-1}\Big|_{z=i}$$
$$= \frac{1}{n!}\times\frac{(-1)^n(n+1)(n+2)\cdots(2n)}{(2i)^{2n+1}}$$

$$= -i \cdot \frac{(2n)!}{(n!)^2 \cdot 2^{2n+1}}$$

$$= -i \cdot \frac{1 \cdot 3 \cdot 5 \cdots (2n-1)}{2 \cdot 4 \cdot 6 \cdots (2n) \cdot 2}. \tag{11.29}$$

Hence our desired result follows immediately if we combine the expressions (11.28) and (11.29). This ends the proof of the problem.

Problem 11.7

Bak and Newman Chapter 11 Exercise 7.

Proof. Consider the contour $C_R = \Gamma_R + \eta_R + \eta_R'$ as described in Problem 11.3 with the central angle $\frac{\pi}{4}$, see Figure 11.1 below.

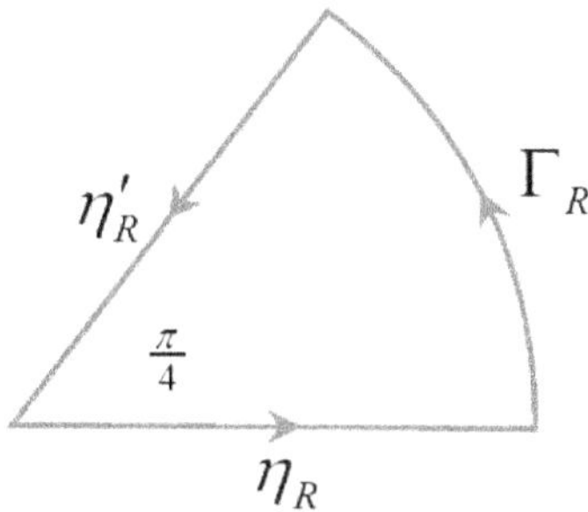

Figure 11.1: The contour C_R with central angle $\frac{\pi}{4}$.

Since $f(z) = e^{iz^2}$ is entire and C_R is a closed curve, Theorem 4.16 (The Closed Curve Theorem) ensures that

$$\int_{\Gamma_R} e^{iz^2}\, dz + \int_{\eta_R} e^{iz^2}\, dz + \int_{\eta_R'} e^{iz^2}\, dz = \int_{C_R} e^{iz^2}\, dz = 0. \tag{11.30}$$

(a) On Γ_R, if $z = R\operatorname{cis}\theta$, where $0 \le \theta \le \frac{\pi}{4}$, then

$$\left|e^{iz^2}\right| = e^{-R^2 \sin 2\theta}. \tag{11.31}$$

Recall from [22, Exercise 7, p. 197] that $\frac{2x}{\pi} \le \sin x \le x \le \pi$ for all $x \in [0, \frac{\pi}{2}]$, so if we put $x = 2\theta$, then we have

$$\frac{4\theta}{\pi} \le \sin 2\theta \le \pi \tag{11.32}$$

for every $\theta \in [0, \frac{\pi}{4}]$. By applying the result (11.32) to the expression (11.31) we conclude that

$$\left|e^{iz^2}\right| \le \exp\left(-R^2 \cdot \frac{4\theta}{\pi}\right)$$

on Γ_R and thus Lemma 4.9 gives

$$\left|\int_{\Gamma_R} e^{iz^2}\, dz\right| = \left|\int_0^{\frac{\pi}{4}} \underbrace{e^{i[z(\theta)]^2} \cdot R(-\sin\theta + i\cos\theta)}_{\text{This is } G(\theta).}\, d\theta\right|$$

$$\leq \int_0^{\frac{\pi}{4}} \underbrace{R \exp\left(-R^2 \cdot \frac{4\theta}{\pi}\right)}_{\text{This is } |G(\theta)|.} d\theta$$

$$= -R \cdot \frac{\pi}{4R^2} \exp\left(-R^2 \cdot \frac{4\theta}{\pi}\right)\Big|_0^{\frac{\pi}{4}}$$

$$= \frac{\pi}{4R}(1 - e^{-R^2})$$

which obviously tends to 0 as $R \to \infty$.

(b) On η_R, we know that

$$\int_{\eta_R} e^{iz^2}\, dz = \int_0^R e^{ix^2}\, dx = \int_0^R \cos x^2\, dx + i \int_0^R \sin x^2\, dx. \tag{11.33}$$

Next, on η_R', we have $z = (R - x)e^{i\frac{\pi}{4}}$ with $0 \leq x \leq R$ so that

$$\int_{-\eta_R'} e^{iz^2}\, dz = -e^{i\frac{\pi}{4}} \int_0^R \exp[-(R-x)^2]\, dx = -e^{i\frac{\pi}{4}} \int_0^R e^{-y^2}\, dy. \tag{11.34}$$

Hence we put the result of part (a), the expressions (11.33) and (11.34) into the equation (11.30) to obtain

$$\lim_{R \to \infty} \left[\int_{\eta_R} e^{iz^2}\, dz + \int_{\eta_R'} e^{iz^2}\, dz \right] = 0$$

$$\int_0^\infty \cos x^2\, dx + i \int_0^\infty \sin x^2\, dx = e^{i\frac{\pi}{4}} \int_0^\infty e^{-x^2}\, dx. \tag{11.35}$$

Recall the fact [11, §3.321, Eqn. (3), p. 336] that

$$\int_0^\infty e^{-x^2}\, dx = \frac{\sqrt{\pi}}{2},$$

so the equation (11.35) becomes

$$\int_0^\infty \cos x^2\, dx = \int_0^\infty \sin x^2\, dx = \frac{\sqrt{\pi}}{2} \cdot \cos\frac{\pi}{4} = \frac{\sqrt{2\pi}}{4}.$$

This completes the analysis of the problem. ∎

Remark 11.1

The integrals considered in Problem 11.7(b) are called **Fresnel integrals**.

Problem 11.8

Bak and Newman Chapter 11 Exercise 8.

Proof. Let $z_1, z_2, \ldots, z_m$ be the poles of f (i.e., the zeros of Q). Let R be large enough so that all poles of f lie inside the circle $C_R = C(0; R)$. Then we follow from Theorem 10.5 (The Cauchy's Residue Theorem) that

$$\int_{C_R} \frac{P(z)}{Q(z)}\, dz = 2\pi i \sum_{k=1}^m \mathrm{Res}\left(\frac{P(z)}{Q(z)}; z_k\right). \tag{11.36}$$

Let $P(z) = a_n z^n + a_{n-1} z^{n-1} + \cdots + a_0$ and $Q(z) = b_m z^m + b_{m-1} z^{m-1} + \cdots + b_0$, where $m - n \geq 2$ and $a_n b_m \neq 0$. If $M = \max\{|a_0|, |a_1|, \ldots, |a_n|\}$, then it is clear that $M > 0$ and

$$|P(z)| \leq M |z|^n \tag{11.37}$$

for large $|z|$. Furthermore, we recall from Problem 1.26 that

$$|Q(z)| \geq \frac{|b_m|}{2} |z|^m \tag{11.38}$$

for large enough $|z|$. Hence if $|z| = R$ is large enough, then we yield from the inequalities (11.37) and (11.38) that

$$\left| \frac{P(z)}{Q(z)} \right| \leq \frac{2M}{|b_m| R^{m-n}}$$

on $C(0; R)$. Thus we deduce from Theorem 4.10 (The M-L Formula) that

$$\left| \int_{C_R} \frac{P(z)}{Q(z)} \, dz \right| \leq \frac{2M}{|b_m| R^{m-n}} \times 2\pi R = \frac{4M\pi}{|b_m| R^{m-n-1}} \to 0$$

as $R \to \infty$ because $m - n - 1 \geq 1$, and hence the equation (11.36) becomes

$$\sum_{k=1}^{m} \operatorname{Res}\left(\frac{P(z)}{Q(z)}; z_k \right) = \frac{1}{2\pi i} \lim_{R \to \infty} \int_{C_R} \frac{P(z)}{Q(z)} \, dz = 0.$$

This ends the proof of the problem.   ▨

Problem 11.9

Bak and Newman Chapter 11 Exercise 9.

Proof.

(a) This is a **Type I** sum. Let $f(z) = \frac{1}{1+z^2}$. Then we have

$$\sum_{\substack{n=-\infty \\ n \neq 0}}^{\infty} f(n) = -\left[\operatorname{Res}\left(\frac{\pi \cot \pi z}{1+z^2}; 0 \right) + \operatorname{Res}\left(\frac{\pi \cot \pi z}{1+z^2}; i \right) + \operatorname{Res}\left(\frac{\pi \cot \pi z}{1+z^2}; -i \right) \right]. \tag{11.39}$$

Since 0 and $\pm i$ are simple poles of $\frac{\pi \cot \pi z}{1+z^2}$, we establish from [4, Eqn. (1), p. 129] that

$$\operatorname{Res}\left(\frac{\pi \cot \pi z}{1+z^2}; 0 \right) = \frac{\pi}{2z \tan \pi z + \pi(1+z^2) \sec^2 \pi z} \Big|_{z=0} = 1,$$

$$\operatorname{Res}\left(\frac{\pi \cot \pi z}{1+z^2}; i \right) = \operatorname{Res}\left(\frac{\pi \cot \pi z}{1+z^2}; -i \right)$$

$$= \frac{\pi \cot \pi i}{2i}.$$

By the identities of $\sin z$ and $\cos z$ on [4, p. 41], we have

$$\cot \pi i = \frac{i(e^{-\pi} + e^{\pi})}{e^{-\pi} - e^{\pi}},$$

so we get from the formula (11.39) that

$$2\sum_{n=1}^{\infty} \frac{1}{n^2 + 1} = -\left(1 + \frac{\pi \cot \pi i}{i} \right) = -1 + \frac{\pi(e^{-\pi} + e^{\pi})}{e^{\pi} - e^{-\pi}} = -1 + \frac{\pi(e^{2\pi} + 1)}{e^{2\pi} - 1}.$$

As a consequence, we have

$$\sum_{n=1}^{\infty} \frac{1}{n^2 + 1} = \frac{1}{2}\left[-1 + \frac{\pi(e^{2\pi} + 1)}{e^{2\pi} - 1} \right].$$

(b) Again it is a **Type I** sum, so if $f(z) = \frac{1}{z^4}$ which has a pole of order 4 at 0, then we have

$$\sum_{\substack{n=-\infty \\ n\neq 0}}^{\infty} f(n) = -\mathrm{Res}\left(\frac{\pi \cot \pi z}{z^4}; 0 \right).$$

Since the Laurent series of $\cot \pi z$ around 0 is given by[b]

$$\cot \pi z = \frac{1}{\pi} - \frac{\pi z}{3} - \frac{\pi^3 z^3}{45} - \frac{2}{945}\pi^5 z^5 + \cdots,$$

we obtain at once that

$$\sum_{n=1}^{\infty} \frac{1}{n^4} = \frac{\pi^4}{90}.$$

(c) This is a **Type II** sum. Suppose that $f(z) = \frac{1}{1+z^2}$ which has simple poles at $\pm i$. By [4, Eqn. (1), p. 129], we have

$$\sum_{\substack{n=-\infty \\ n\neq 0}}^{\infty} \frac{(-1)^n}{1 + n^2} = -\mathrm{Res}\left(\frac{\pi \csc \pi z}{1 + z^2}; i \right) - \mathrm{Res}\left(\frac{\pi \csc \pi z}{1 + z^2}; -i \right)$$

$$= -\frac{\pi \csc \pi z}{2z}\bigg|_{z=i} - \frac{\pi \csc \pi z}{2z}\bigg|_{z=-i}$$

$$= -\frac{\pi}{i \sin \pi i}$$

$$= \frac{2\pi}{e^\pi - e^{-\pi}}$$

which implies that

$$\sum_{n=1}^{\infty} \frac{(-1)^n}{1 + n^2} = \frac{e^{2\pi} + 2\pi e^\pi - 1}{2(e^{2\pi} - 1)}.$$

Hence we have completed the analysis of the problem. $\blacksquare$

> **Problem 11.10**
>
> *Bak and Newman Chapter 11 Exercise 10.*[*]

Proof.

(a) Figure 11.2 shows the square $C_N = \gamma_1 + \gamma_2 + \gamma_3 + \gamma_4$ considered in the problem. Let $A = N + \frac{1}{2}$. On γ_1, we have $z = A + it$ with $t \in [-A, A]$. Therefore, we obtain from Definition 4.3 that

$$\int_{\gamma_1} \frac{dz}{z \sin \pi z} = (-1)^N \int_{-A}^{A} \frac{i\,dt}{(A + it)\cos it\pi}.$$

---

[b]See [16, Exercise A(5)(a), p. 730].

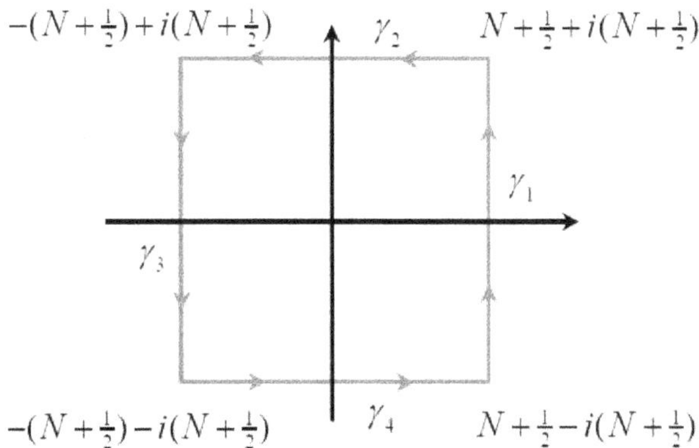

Figure 11.2: The square C_N with vertices $\pm(N + \frac{1}{2}) \pm i(N + \frac{1}{2})$.

Similarly, on $-\gamma_3$, we have $z = -A + it$ with $t \in [-A, A]$ so that

$$\int_{\gamma_3} \frac{dz}{z \sin \pi z} = -\int_{-\gamma_3} \frac{dz}{z \sin \pi z} = -(-1)^N \int_{-A}^{A} \frac{i\,dt}{(A - it)\cos it\pi}. \tag{11.40}$$

Using the substitution $t = -u$ to the last integral (11.40), we see that

$$\int_{\gamma_3} \frac{dz}{z \sin \pi z} = -(-1)^N \int_{-A}^{A} \frac{i\,du}{(A + iu)\cos iu\pi}.$$

Consequently, we have

$$\int_{\gamma_1} \frac{dz}{z \sin \pi z} + \int_{\gamma_3} \frac{dz}{z \sin \pi z} = 0$$

and then

$$\int_{C_N} \frac{dz}{z \sin \pi z} = \int_{\gamma_2} \frac{dz}{z \sin \pi z} + \int_{\gamma_4} \frac{dz}{z \sin \pi z}. \tag{11.41}$$

On γ_2 or γ_4, we have $z = t \pm iA$, where $t \in [-A, A]$, so

$$\left| \frac{1}{z \sin \pi z} \right| \le \frac{2}{\sqrt{A^2 + t^2} \cdot |e^{-A\pi} - e^{A\pi}|} \le \frac{2}{A(e^{A\pi} - e^{-A\pi})}$$

for large enough A and so for large N. By Theorem 4.10 (The M-L Formula), we conclude that

$$\left| \int_{\gamma_2} \frac{dz}{z \sin \pi z} \right| \le \frac{4}{e^{A\pi} - e^{-A\pi}} \quad \text{and} \quad \left| \int_{\gamma_4} \frac{dz}{z \sin \pi z} \right| \le \frac{4}{e^{A\pi} - e^{-A\pi}}$$

which tend to 0 as $A \to \infty$ or equivalently as $N \to \infty$. By the relation (11.41), we see at once that

$$\lim_{N \to \infty} \int_{C_N} \frac{dz}{z \sin \pi z} = 0.$$

(b) Similar to **Type II** series, if $f(z) = \frac{1}{2z-1}$ which has a simple pole at $\frac{1}{2}$, then we have

$$\int_{C_N} \frac{\pi\,dz}{(2z - 1)\sin \pi z} = 2\pi i \left[\sum_{n=-(N-1)}^{N} \frac{(-1)^n}{2n - 1} + \operatorname{Res}\left(\frac{\pi}{(2z - 1)\sin \pi z}; \frac{1}{2} \right) \right]$$

$$= 2\pi i \left[\sum_{n=-N}^{N} \frac{(-1)^n}{2n - 1} + \operatorname{Res}\left(\frac{\pi}{(2z - 1)\sin \pi z}; \frac{1}{2} \right) - \frac{(-1)^N}{2N - 1} \right]$$

$$= 2\pi i \left[2 \sum_{n=1}^{N} \frac{(-1)^n}{2n-1} + \operatorname{Res}\left(\frac{\pi}{(2z-1)\sin \pi z}; \frac{1}{2} \right) - \frac{(-1)^N}{2N-1} \right], \quad (11.42)$$

where C_N is the rectangle with vertices $(N+\frac{1}{2}) \pm i(N+\frac{1}{2})$ and $-(N-\frac{1}{2}) \pm i(N+\frac{1}{2})$ for large positive integer N, see Figure 11.3.

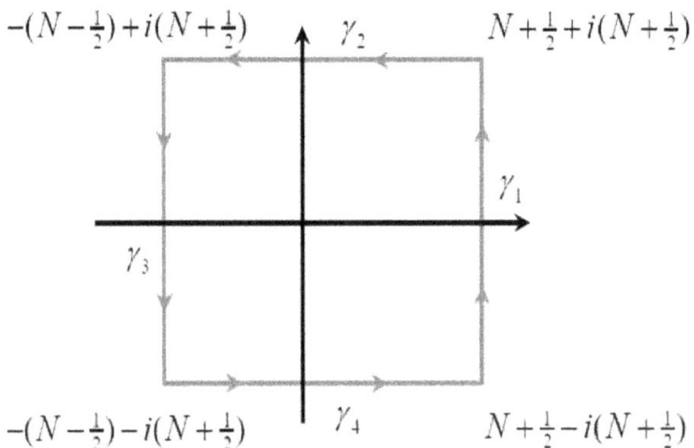

Figure 11.3: The rectangle C_N with vertices $(N+\frac{1}{2}) \pm i(N+\frac{1}{2})$ and $-(N-\frac{1}{2}) \pm i(N+\frac{1}{2})$.

Let $A = N + \frac{1}{2}$, if $z = A + it$ with $|t| \leq A$, then we get

$$\int_{\gamma_1} \frac{\pi \, dz}{(2z-1)\sin \pi z} = \frac{(-1)^N i}{2} \int_{-A}^{A} \frac{dt}{(N+it)\cos i\pi t}$$

and if $z = -(N - \frac{1}{2}) + it$ with $|t| \leq A$, then we have

$$\int_{\gamma_3} \frac{\pi \, dz}{(2z-1)\sin \pi z} = -\frac{(-1)^N i}{2} \int_{-A}^{A} \frac{dt}{(N+it)\cos i\pi t}.$$

Hence, by similar strategy applied in part (a), we can show that

$$\lim_{N \to \infty} \int_{C_N} \frac{\pi \, dz}{(2z-1)\sin \pi z} = 0,$$

so the equation (11.42) reduces to

$$2 \sum_{n=1}^{\infty} \frac{(-1)^{n+1}}{2n-1} = \operatorname{Res}\left(\frac{\pi}{(2z-1)\sin \pi z}; \frac{1}{2} \right)$$

$$= \frac{\pi}{2\sin \pi z + \pi(2z-1)\cos \pi z} \Big|_{z=\frac{1}{2}}$$

$$= \frac{\pi}{2}$$

which definitely implies that

$$1 - \frac{1}{3} + \frac{1}{5} - \frac{1}{7} + \cdots = \sum_{n=1}^{\infty} \frac{(-1)^{n+1}}{2n-1} = \frac{\pi}{4}.$$

We end the proof of the problem.

Problem 11.11

Bak and Newman Chapter 11 Exercise 11.[*]

Proof. Suppose that $f(z) = \frac{e^{kz}}{1+e^z}$ and consider the rectangle Γ_R with vertices at $\pm R$ and $\pm R + 2\pi i$ as shown in Figure 11.4. It is clear that $1+e^z$ vanishes inside the rectangle if and only if $z = \pi i$.

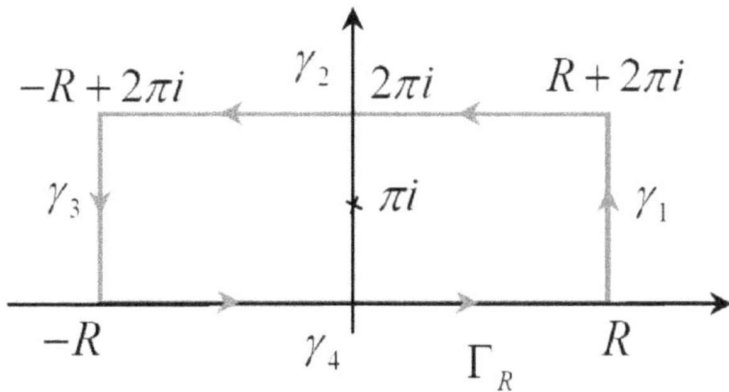

Figure 11.4: The rectangle Γ_R with vertices at $\pm R$ and $\pm R + 2\pi i$.

Thus we represent

$$(z - \pi i)f(z) = e^{kz} \cdot \frac{z - \pi i}{1 + e^z} = e^{kz} \cdot \frac{z - \pi i}{e^z - e^{\pi i}} = e^{kz} \cdot \frac{1}{\frac{e^z - e^{\pi i}}{z - \pi i}}$$

which implies

$$\lim_{z \to \pi i}(z - \pi i)f(z) = \lim_{z \to \pi i} e^{kz} \cdot \frac{1}{\frac{e^z - e^{\pi i}}{z - \pi i}} = -e^{k\pi i}.$$

By Theorem 10.5 (The Cauchy's Residue Theorem), we see that

$$\int_{\Gamma_R} \frac{e^{kz}}{1 + e^z}\, dz = 2\pi i \operatorname{Res}\left(\frac{e^{kz}}{1 + e^z}; \pi i\right) = -2\pi i e^{k\pi i}. \tag{11.43}$$

Now it is obvious that, on γ_4, we have

$$\int_{\gamma_4} \frac{e^{kz}}{1 + e^z}\, dz = \int_{-R}^{R} \frac{e^{kx}}{1 + e^x}\, dx. \tag{11.44}$$

Similarly, we know that points on $-\gamma_2$ have the form $z = x + 2\pi i$ with $-R \le x \le R$, so it follows from Proposition 4.7 that

$$\int_{\gamma_2} \frac{e^{kz}}{1 + e^z}\, dz = -\int_{-\gamma_2} \frac{e^{kz}}{1 + e^z}\, dz = -\int_{-R}^{R} \frac{e^{k(x+2\pi i)}}{1 + e^{x+2\pi i}}\, dx = -e^{2k\pi i} \int_{-R}^{R} \frac{e^{kx}}{1 + e^x}\, dx. \tag{11.45}$$

Next, suppose that $z = R + it \in \gamma_1$ with $t \in [0, 2\pi]$. For large enough R, the triangle inequality implies that

$$|1 + e^{R+it}| \ge |e^{R+it}| - 1 \ge \frac{e^R}{2}$$

so that

$$\left|\frac{e^{kz}}{1 + e^z}\right| = \left|\frac{e^{k(R+it)}}{1 + e^{R+it}}\right| \le 2e^{(k-1)R}.$$

Hence Theorem 4.10 (The M-L Formula) gives

$$\left| \int_{\gamma_1} \frac{e^{kz}}{1 + e^z} \, dz \right| \leq 4\pi e^{(k-1)R}$$

which ensures that the integral along γ_1 tends to 0 as $R \to \infty$ because $k < 1$. By similar analysis, the integral along γ_3 tends to 0 as $R \to \infty$ because $k > 0$. Combining these facts and the expressions (11.44) and (11.45), this assures us that

$$(1 - e^{2k\pi i}) \int_{-\infty}^{\infty} \frac{e^{kx}}{1 + e^x} \, dx = -2\pi i e^{k\pi i}$$

$$\int_{-\infty}^{\infty} \frac{e^{kx}}{1 + e^x} \, dx = -2\pi i \cdot \frac{e^{k\pi i}}{1 - e^{2k\pi i}}$$

$$= 2\pi i \cdot \frac{1}{e^{k\pi i} - e^{-k\pi i}}$$

$$= \frac{\pi}{\sin k\pi}.$$

Hence we complete the proof of the problem.

Problem 11.12

Bak and Newman Chapter 11 Exercise 12.

Proof. We consider the keyhole contour $K_{\epsilon,M}$ used on [4, p. 147], but this time we let $M \to 1$. By Theorem 10.5 (The Cauchy's Residue Theorem), we have

$$\int_{K_{\epsilon,M}} f(z) \log z \, dz = 2\pi i \sum_{k} \text{Res} \left(f(z) \log z; z_k \right),$$

where the sum is taken over all the poles of f inside $K_{\epsilon,M}$. Since f is analytic for $|z| \leq 1$, f has no poles so that

$$\int_{K_{\epsilon,M}} f(z) \log z \, dz = 0 \tag{11.46}$$

for any $0 < \epsilon < M < 1$.

By Theorem 4.10 (The M-L Formula), we have

$$\left| \int_{C_\epsilon} f(z) \log z \, dz \right| \leq \pi \epsilon \max_{C_\epsilon} |f(z) \log z|. \tag{11.47}$$

Since f is continuous at 0 and $|\log z| < |\log |z|| + 2\pi = |\log \epsilon| + 2\pi$, we obtain from the inequality (11.47) that

$$\left| \int_{C_\epsilon} f(z) \log z \, dz \right| \leq A\pi \epsilon (|\log \epsilon| + 2\pi) \to 0 \tag{11.48}$$

as $\epsilon \to 0$, where A is a positive constant. Next, we recall that C_M is the circular arc of radius M traced counterclockwise from $\sqrt{M^2 - \epsilon^2} + i\epsilon$ to $\sqrt{M^2 - \epsilon^2} - i\epsilon$, so it is true that

$$\lim_{\substack{\epsilon \to 0 \\ M \to 1}} \int_{C_M} f(z) \log z \, dz = \int_{|z|=1} f(z) \log z \, dz. \tag{11.49}$$

Finally, since

$$\lim_{\substack{\epsilon \to 0 \\ M \to 1}} \int_{I_1} f(z) \log z \, dz = \int_0^1 f(x) \log x \, dx \tag{11.50}$$

and similarly,

$$\lim_{\substack{\epsilon \to 0 \\ M \to 1}} \int_{I_2} f(z) \log z \, \mathrm{d}z = - \int_0^1 f(x)(\log x + 2\pi i) \, \mathrm{d}x. \tag{11.51}$$

Since $K_{\epsilon,M} = C_M + I_2 + C_\epsilon + I_1$, we put all the results (11.48), (11.49), (11.50) and (11.51) into the equation (11.46) to conclude that

$$\lim_{\substack{\epsilon \to 0 \\ M \to 1}} \int_{K_{\epsilon,M}} f(z) \log z \, \mathrm{d}z = 0$$

$$\lim_{\substack{\epsilon \to 0 \\ M \to 1}} \int_{C_M} f(z) \log z \, \mathrm{d}z + \lim_{\substack{\epsilon \to 0 \\ M \to 1}} \int_{I_1} f(z) \log z \, \mathrm{d}z + \lim_{\substack{\epsilon \to 0 \\ M \to 1}} \int_{I_1} f(z) \log z \, \mathrm{d}z = 0$$

$$\int_{|z|=1} f(z) \log z \, \mathrm{d}z + \int_0^1 f(x) \log x \, \mathrm{d}x - \int_0^1 f(x)(\log x + 2\pi i) \, \mathrm{d}x = 0$$

$$\frac{1}{2\pi i} \int_{|z|=1} f(z) \log z \, \mathrm{d}z = \int_0^1 f(x) \, \mathrm{d}x.$$

We end the proof of the problem.   ▨

Problem 11.13

Bak and Newman Chapter 11 Exercise 13.

Proof. We note that C_n^{3n} is the coefficient of z^n in $(1+z)^{3n}$. Therefore, we have

$$C_n^{3n} = \frac{1}{2\pi i} \int_C \frac{(1+z)^{3n}}{z^{n+1}} \, \mathrm{d}z$$

where C is a simple closed contour surrounding the origin. Similar to the analysis of [4, Example 1, p. 155], we choose C to be the circle $C(0; \frac{1}{2})$ because

$$\left| \frac{(1+z)^3}{8z} \right| \le \frac{(1+|z|)^3}{8|z|} = \frac{27}{32} < 1$$

throughout $C(0; \frac{1}{2})$ and the convergence of the series

$$\sum_{n=0}^{\infty} \frac{(1+z)^{3n}}{(8z)^n}$$

is uniform there. Hence we have

$$\sum_{n=0}^{\infty} C_n^{3n} \cdot \frac{1}{8^n} = \frac{1}{2\pi i} \sum_{n=0}^{\infty} \int_{C(0;\frac{1}{2})} \frac{(1+z)^{3n}}{(8z)^n} \cdot \frac{\mathrm{d}z}{z}$$

$$= \frac{1}{2\pi i} \int_{C(0;\frac{1}{2})} \left[\sum_{n=0}^{\infty} \frac{(1+z)^{3n}}{(8z)^n} \right] \cdot \frac{\mathrm{d}z}{z}$$

$$= \frac{1}{2\pi i} \int_{C(0;\frac{1}{2})} \frac{1}{1 - \frac{(1+z)^3}{8z}} \cdot \frac{\mathrm{d}z}{z}$$

$$= \frac{1}{2\pi i} \int_{C(0;\frac{1}{2})} \frac{8 \, \mathrm{d}z}{8z - (1+z)^3}$$

$$= \frac{4i}{\pi} \int_{C(0;\frac{1}{2})} \frac{\mathrm{d}z}{z^3 + 3z^2 - 5z + 1}$$

$$= \frac{4i}{\pi} \int_{C(0;\frac{1}{2})} \frac{\mathrm{d}z}{(z-1)(z^2+4z-1)}. \tag{11.52}$$

Since $z^2 + 4z - 1$ has a simple zero at $-2 \pm \sqrt{5}$, the application of Theorem 10.5 (The Cauchy's Residue Theorem) shows that the expression (11.52) further reduces to

$$\sum_{n=0}^{\infty} C_n^{3n} \cdot \frac{1}{8^n} = \frac{4i}{\pi} \cdot 2\pi i \mathrm{Res}\left(\frac{1}{(z-1)(z^2+4z-1)}; -2+\sqrt{5}\right)$$

$$= -8 \cdot \frac{1}{z^2+4z-1+(z-1)(2z+4)}\Big|_{z=-2+\sqrt{5}}$$

$$= \frac{-8}{2\sqrt{5}(-3+\sqrt{5})}$$

$$= \frac{4}{\sqrt{5}(3-\sqrt{5})}$$

$$= \frac{5+3\sqrt{5}}{5}.$$

This completes the proof of the problem.

Problem 11.14

Bak and Newman Chapter 11 Exercise 14.

Proof. The result is trivial if $x = 0$, so we assume that $x \neq 0$. Similar to [4, Example 1, p. 155], we have

$$\sum_{n=0}^{\infty} C_n^{2n} x^n = \frac{1}{2\pi i} \sum_{n=0}^{\infty} \int_C \left[\frac{(1+z)^2 x}{z}\right]^n \cdot \frac{\mathrm{d}z}{z}, \tag{11.53}$$

where C is a simple contour surrounding the origin. Particularly, we choose $C = C(0;1)$. It follows from the hypothesis $|x| < \frac{1}{4}$ that

$$\left|\frac{(1+z)^2 x}{z}\right| \leq \frac{|1+z|^2}{|z|} |x| \leq 4|x| < 1$$

throughout $C(0;1)$. Therefore, the convergence is uniform and we are able to interchange the order of summation and integration in the expression (11.53) and get

$$\sum_{n=0}^{\infty} C_n^{2n} x^n = \frac{1}{2\pi i} \int_{C(0;1)} \sum_{n=0}^{\infty} \left[\frac{(1+z)^2 x}{z}\right]^n \cdot \frac{\mathrm{d}z}{z} = -\frac{1}{2\pi i} \int_{C(0;1)} \frac{\mathrm{d}z}{(1+z)^2 x - z},$$

Since $(1+z)^2 x - z = 0$ is the same as

$$xz^2 + (2x-1)z + x = 0$$

whose solutions are given by

$$z_{\pm} = \frac{1-2x \pm \sqrt{1-4x}}{2x}.$$

Now the hypothesis $|x| < \frac{1}{4}$ again implies that $0 < 1 - 4x < 2$ so that

$$\frac{1-2x}{2x} < z_+ < \frac{1+\sqrt{2}-2x}{2x} \quad \text{and} \quad \frac{1-\sqrt{2}-2x}{2x} < z_- < \frac{1-2x}{2x}$$

if $x > 0$ as well as

$$\frac{1 + \sqrt{2} - 2x}{2x} < z_+ < \frac{1 - 2x}{2x} \quad \text{and} \quad \frac{1 - 2x}{2x} < z_- < \frac{1 - \sqrt{2} - 2x}{2x}$$

if $x < 0$. Direct computation shows that $z_+ > 1$ if $x > 0$ and $z_+ < -1$ if $x < 0$. Hence it follows from Theorem 10.5 (The Cauchy's Residue Theorem) that

$$\sum_{n=0}^{\infty} C_n^{2n} x^n = -\mathrm{Res}\left(\frac{1}{xz^2 + (2x - 1)z + x}; z_-\right)$$

$$= -\frac{1}{2xz + (2x - 1)}\Big|_{z=z_-}$$

$$= \frac{1}{\sqrt{1 - 4x}}$$

which completes the proof of the problem.

Proof. To find the maximum of $a^2 b$, where $a^2 + b^2 = 4$ and $0 \le a, b \le 2$. We define the function $f(b) = (4 - b^2)b = 4b - b^3$. Then we have

$$\max_{0 \le a,b \le 2} a^2 b = \max_{0 \le b \le 2} f(b).$$

Note that $f'(b) = 4 - 3b^2 = 0$ if and only if $b = \frac{2\sqrt{3}}{3}$ and $f''(\frac{2\sqrt{3}}{3}) = -6 \cdot \frac{2\sqrt{3}}{3} = -4\sqrt{3} < 0$, so f attains its maximum at $b = \frac{2\sqrt{3}}{3}$ and its maximum value is given by

$$\max_{0 \le b \le 2} f(b) = f\left(\frac{2\sqrt{3}}{3}\right) = \frac{16\sqrt{3}}{9}. \tag{11.54}$$

If $|z| = 1$, then we see from the figure provided in the problem that $|z - 1|^2 + |z + 1|^2 = 4$. Thus we may let $a = |z - 1|$ and $b = |z + 1|$ so that our a and b satisfy the hypotheses in the previous paragraph. Consequently, we conclude from the value (11.54) that

$$|(z - 1)^2(z + 1)| \le \frac{16\sqrt{3}}{9},$$

completing the analysis of the problem.

Proof. Since part (b) considers points on the circle of radius $R > 0$, we do this general case in part (a) and then take $R = \sqrt{2}$ to obtain the desired result. Let $z \in C(0; R)$, $a = |z + 1|$ and $b = |z - 1|$. Now the points $z, -z, 1$ and -1 form a parallelogram in the plane $\mathbb{C}$.

(a) It is well-known that the sum of the squares of the diagonals of parallelogram is equal to the sum of the squares of its sides.[c] In other words, we have $(2R)^2 + 2^2 = 2(a^2 + b^2)$ which reduces to

$$a^2 + b^2 = 2(1 + R^2), \tag{11.55}$$

where $0 \le a, b \le \sqrt{2(1 + R^2)}$. If $R = \sqrt{2}$, then we have $a^2 = 6 - b^2$ and $0 \le a, b \le \sqrt{6}$. To find the maximum of $\frac{a^2 b}{2}$, we define

$$f(b) = \frac{6b - b^3}{2},$$

where $b \in [0, \sqrt{6}]$. Then their maximums are the same, i.e.,

$$\max_{a, b \in [0, \sqrt{6}]} \frac{a^2 b}{2} = \max_{b \in [0, \sqrt{6}]} f(b).$$

Now $f'(b) = \frac{6 - 3b^2}{2} = 0$ if and only if $b = \sqrt{2}$ and $f''(b) = -3b$ so that $f''(\sqrt{2}) < 0$. Consequently, f attains its maximum at $b = \sqrt{2}$ and the maximum value is

$$\max_{0 \le b \le \sqrt{6}} f(b) = f(\sqrt{2}) = \frac{\sqrt{2}(6 - 2)}{2} = 2\sqrt{2}$$

which gives the desired result.

(b) In the general case, we note that $R > 0$ and

$$f(b) = \frac{[2(1 + R^2) - b^2]b}{R^2},$$

where $0 \le b \le \sqrt{2(1 + R^2)}$. As usual, we have

$$f'(b) = \frac{2(1 + R^2) - 3b^2}{R^2} = 0$$

if and only if

$$b = \sqrt{\frac{2(1 + R^2)}{3}}.$$

Since $f''(b) = -\frac{6b}{R^2}$, we must have

$$f''\left(\sqrt{\frac{2(1 + R^2)}{3}}\right) = -\frac{6}{R^2} \cdot \sqrt{\frac{2(1 + R^2)}{3}} < 0.$$

Hence f attains its maximum at $b = \sqrt{\frac{2(1 + R^2)}{3}}$ and the maximum value is given by

$$\max_{0 \le b \le \sqrt{2(1 + R^2)}} f(b) = f\left(\sqrt{\frac{2(1 + R^2)}{3}}\right) = \frac{4\sqrt{6}}{9} \cdot \frac{(1 + R^2)^{\frac{3}{2}}}{R^2}. \tag{11.56}$$

To find the minimum of the expression (11.56), we define $F : (0, \infty) \to (0, \infty)$ by

$$F(R) = \frac{(1 + R^2)^{\frac{3}{2}}}{R^2}.$$

____

[c]Read [22, Exercise 17, p. 23].

Direct computation gives

$$F'(R) = \frac{(R^2 - 2)\sqrt{1 + R^2}}{R^3},$$

and then $F'(R) = 0$ if and only if $R = \sqrt{2}$. It can be easily seen from the First Derivative Test that F attains its minimum at $R = \sqrt{2}$ and actually,

$$\min_{0 < R < \infty} F(R) = \frac{(1 + 2)^{\frac{3}{2}}}{2} = \frac{3\sqrt{3}}{2}. \tag{11.57}$$

Finally, by putting the value (11.57) back into the expression (11.56), we see that

$$\max_{|z|=R} \left| \frac{(z-1)^2(z+1)}{z^2} \right| = \max_{0 \le b \le \sqrt{2(1+R^2)}} f(b) \ge \frac{4\sqrt{6}}{9} \cdot \frac{3\sqrt{3}}{2} = 2\sqrt{2}.$$

Hence we have completed the analysis of the problem.   ◾

Problem 11.17

*Bak and Newman Chapter 11 Exercise 17.**

Proof.

(a) Note that C_k^n and $(-1)^k C_k^{3n}$ are the coefficients of z^k and z^{-k} in $(1+z)^n$ and $(1 - z^{-1})^{3n}$ respectively. Therefore, the sum

$$\sum_{k=0}^{n} (-1)^k C_k^{3n} C_k^n \tag{11.58}$$

is the constant term in the product $(1+z)^n (1 - z^{-1})^{3n}$. Using [4, Eqn. (4), p. 123], it is given by

$$\frac{1}{2\pi i} \int_C \frac{(1+z)^n (1 - z^{-1})^{3n}}{z} \, dz = \frac{1}{2\pi i} \int_C \frac{[(z-1)^3(1+z)]^n}{z^{3n+1}} \, dz, \tag{11.59}$$

where $C = C(0; R)$ for $R > 0$. Combining the expression (11.58) and the integral (11.59), we get

$$\sum_{k=0}^{n} (-1)^k C_k^{3n} C_k^n = \frac{1}{2\pi i} \int_C \frac{[(z-1)^3(1+z)]^n}{z^{3n+1}} \, dz. \tag{11.60}$$

(b) Similar to Problem 11.16, we have to find the value

$$\max_{|z|=R} \left| \frac{(z-1)^3(z+1)}{z^3} \right|.$$

Let $a = |z - 1|$ and $b = |z + 1|$. To find the maximum of $\frac{a^3 b}{R^3}$, where a and b satisfy the condition (11.55), we define

$$f(b) = \left(\frac{a^3 b}{R^3} \right)^2 = \frac{[2(1 + R^2) - b^2]^3 b^2}{R^6}, \tag{11.61}$$

where $0 \le b \le \sqrt{2(1 + R^2)}$. Particularly, we put $R = \sqrt{3}$ into the formula (11.61) to get

$$f(b) = -\frac{b^2(b^2 - 8)^3}{27},$$

where $0 \le b \le \sqrt{8}$. Consider the function $F(u) = -u(u-8)^3$ with domain $[0,8]$. Elementary calculus shows that

$$F'(u) = -4(u-8)^2(u-2),$$

so $F'(u) = 0$ if and only if $u = 2, 8$. Now the First Derivative Test ensures that F takes the maximum at $u = 2$ and its maximum value is given by

$$\max_{0 \le u \le 8} F(u) = F(2) = 432$$

so that

$$\max_{|z|=R} \left| \frac{(z-1)^3(z+1)}{z^3} \right| = \max_{0 \le a,b \le \sqrt{8}} \frac{a^3 b}{3^{\frac{3}{2}}} = \sqrt{\max_{0 \le b \le \sqrt{8}} f(b)} = \sqrt{\frac{432}{27}} = 4. \tag{11.62}$$

Combining this fact (11.62) and Theorem 4.10 (The M-L Formula), we conclude that

$$\left| \frac{1}{2\pi i} \int_{C(0;\sqrt{3})} \frac{[(z-1)^3(1+z)]^n}{z^{3n+1}} \, dz \right| \le \frac{1}{2\pi} \cdot \left(\max_{|z|=R} \left| \frac{(z-1)^3(z+1)}{z^3} \right| \right)^n \cdot \frac{1}{\sqrt{3}} \cdot 2\pi\sqrt{3} = 4^n.$$

Finally, the formula (11.60) guarantees that

$$\left| \sum_{k=0}^{n} (-1)^k C_k^{3n} C_k^n \right| \le 4^n.$$

Now we complete the proof of the problem.

CHAPTER **12**

Further Contour Integral Techniques

Proof.

(a) Let $z = \omega - 1$. Then I becomes the line I' given by $\omega(t) = 1 + it$, where $-\infty < t < \infty$. Furthermore, we have

$$\int_I \frac{e^z}{(z+1)^3}\, dz = e^{-1} \int_{I'} \frac{e^\omega}{\omega^4}\, d\omega. \tag{12.1}$$

Using similar argument as [4, Example 1, pp. 161, 162], we see that

$$\int_{I'} \frac{e^\omega}{\omega^4}\, d\omega = 2\pi i \operatorname{Res}\left(\frac{e^\omega}{\omega^4};0\right) = \frac{2\pi i}{5!}. \tag{12.2}$$

Combining the expressions (12.1) and (12.2), we assert that

$$\int_I \frac{e^z}{(z+1)^3}\, dz = \frac{\pi i}{60 e}.$$

(b) We use the same contour considered in [4, Example 1, pp. 161, 162] to write

$$\int_{1-iR}^{1+iR} \frac{a^z}{z^2}\, dz + \int_{C_R} \frac{a^z}{z^2}\, dz = 2\pi i \operatorname{Res}\left(\frac{a^z}{z^2};0\right), \tag{12.3}$$

where $R > 3$. Since $a^z = e^{z \log a}$, if $z = x + iy$ for $x \leq 1$, then we have

$$|a^z| = |e^{z \log a}| = |e^{x \log a}| = e^{x \log a} \leq \max(1, e^{\log a}) = \max(1, a)$$

in the left half-plane as $R \to \infty$. As a result, Theorem 4.10 (The M-L Formula) implies that

$$\lim_{R \to \infty} \int_{C_R} \frac{a^z}{z^2}\, dz = 0$$

and hence the equation (12.3) reduces to

$$\int_I \frac{a^z}{z^2}\, dz = 2\pi i \operatorname{Res}\left(\frac{a^z}{z^2};0\right) = 2\pi i \operatorname{Res}\left(\frac{e^{z \log a}}{z^2};0\right) = 2\pi i \log a$$

because

$$e^{z \log a} = 1 + \frac{z \log a}{1!} + \frac{(z \log a)^2}{2!} + \cdots .$$

We end the proof of the problem. $\blacksquare$

163

Remark 12.1

There is a general formula for the integral

$$\int_{x_0-i\infty}^{x_0+i\infty} a^z F(z)\,\mathrm{d}z$$

in [16, Eqn. (4.13.4), p. 952] for the case $a \geq 1$ and $x_0 \in \mathbb{R}$.

Problem 12.2

*Bak and Newman Chapter 12 Exercise 2.**

Proof. The function $4z^2 - 8z + 3$ has zeros at $z = \frac{1}{2}$ and $z = \frac{3}{2}$, so $\sqrt{4z^2 - 8z + 3}$ is analytic in the plane minus $[\frac{1}{2}, \frac{3}{2}]$.[a] Since $\sqrt{4z^2 - 8z + 3} \sim 2z$ for large z, it follows that

$$\lim_{R\to\infty} \int_{|z|=R} \frac{\mathrm{d}z}{\sqrt{4z^2 - 8z + 3}} = \frac{1}{2} \lim_{R\to\infty} \int_{|z|=R} \frac{\mathrm{d}z}{z} = \pi i. \tag{12.4}$$

Now Figure 12.1 shows the formations of the contours $C(0; R)$ and $C(0; 2)$.

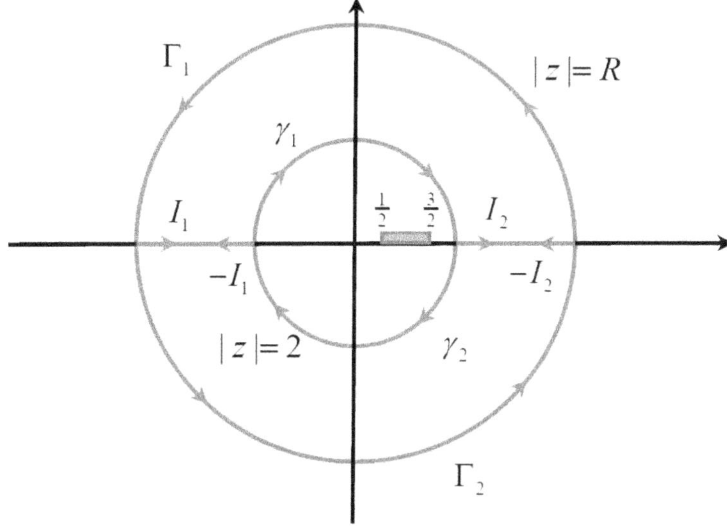

Figure 12.1: The formations of the contours $C(0; R)$ and $C(0; 2)$.

Note that

$$C(0; R) - C(0; 2) = \Gamma_1 + \Gamma_2 + (\gamma_1 + \gamma_2) = (\Gamma_1 + I_1 + \gamma_1 + I_2) + (\Gamma_2 - I_2 + \gamma_2 - I_1).$$

Since $(4z^2 - 8z + 3)^{-\frac{1}{2}}$ is analytic in a simply connected domain containing the smooth closed curve $\Gamma_1 + I_1 + \gamma_1 + I_2$, Theorem 8.6 (The General Closed Curve Theorem) ensures that

$$\int_{\Gamma_1+I_1+\gamma_1+I_2} \frac{\mathrm{d}z}{\sqrt{4z^2 - 8z + 3}} = 0. \tag{12.5}$$

---

[a] Refer to Problem 10.16.

Similarly, we also have

$$\int_{\Gamma_2 - I_2 + \gamma_2 - I_1} \frac{dz}{\sqrt{4z^2 - 8z + 3}} = 0. \qquad (12.6)$$

Therefore, we follow from the results (12.5) and (12.6) that

$$\int_{C(0;2)} \frac{dz}{\sqrt{4z^2 - 8z + 3}} = \int_{C(0;R)} \frac{dz}{\sqrt{4z^2 - 8z + 3}}$$

and hence the limit (12.4) implies

$$\int_{C(0;2)} \frac{dz}{\sqrt{4z^2 - 8z + 3}} = \pi i.$$

This completes the proof of the problem.

> **Problem 12.3**
>
> *Bak and Newman Chapter 12 Exercise 3.*

Proof. Let $R > 0$ be large and $\epsilon > 0$ be small. We consider the closed contour Γ_R formed by the following curves and lines:

- $\gamma_1 : [-R, R] \to \mathbb{C}$ is the parabola defined by

$$\gamma_1(t) = 1 - t^2 + it;$$

- the straight line γ_2 connecting $\gamma_1(R) = 1 - R^2 + iR$ and the point $\epsilon \exp(i\theta_R)$, where $\theta_R = \operatorname{Arg} \gamma_1(R)$;

- $\gamma_3 : [\theta_R, \theta_{-R}] \to \mathbb{C}$ defined by

$$\gamma_3(t) = \epsilon e^{it},$$

  where $\theta_{-R} = \operatorname{Arg} \gamma_1(-R)$;

- the straight line γ_4 connecting $\epsilon \exp(i\theta_{-R})$ to $\gamma_1(-R) = 1 - R^2 - iR$.

This contour is shown in Figure 12.2 below:

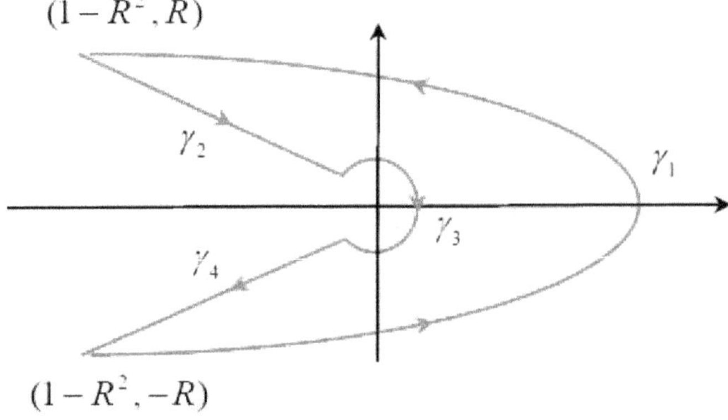

Figure 12.2: The closed contour formed by $\gamma_1, \gamma_2, \gamma_3$ and γ_4.

It is clear that $f(z) = e^z \log z$ is analytic inside and on the closed contour Γ_R and f has no singularities inside Γ_R, Theorem 10.5 (The Cauchy's Residue Theorem), we find

$$\int_{\Gamma_R} e^z \log z \, \mathrm{d}z = 0$$

and thus

$$\int_{\gamma_1} e^z \log z \, \mathrm{d}z = -\int_{\gamma_2} e^z \log z \, \mathrm{d}z - \int_{\gamma_3} e^z \log z \, \mathrm{d}z - \int_{\gamma_4} e^z \log z \, \mathrm{d}z. \tag{12.7}$$

Notice that both $-\gamma_2$ and γ_4 have the same form $\gamma : [\epsilon, \sqrt{R^4 - R^2 + 1}] \to \mathbb{C}$ given by

$$\gamma(t) = t e^{i\theta},$$

where θ satisfies either $\tan\theta = \frac{R}{1-R^2}$ or $\tan\theta = -\frac{R}{1-R^2}$. When $R \to \infty$ and $\epsilon \to 0$, in the case of $-\gamma_2$, we have $\theta \to -\pi$ and for γ_4, we find $\theta \to \pi$. By Proposition 4.7 and then Definition 4.3, we have

$$\int_{\gamma_2} f(z) \, \mathrm{d}z = -\int_{-\gamma_2} f(z) \, \mathrm{d}z$$

$$= -\int_{\epsilon}^{\sqrt{R^4 - R^2 + 1}} f(-\gamma_2(t))(-\gamma_2)'(t) \, \mathrm{d}t$$

$$= -\int_{\epsilon}^{\sqrt{R^4 - R^2 + 1}} e^{i\theta} f(t e^{i\theta}) \, \mathrm{d}t$$

$$= -e^{i\theta} \int_{\epsilon}^{\sqrt{R^4 - R^2 + 1}} \exp(t e^{i\theta}) \log(t e^{i\theta}) \, \mathrm{d}t$$

$$= -e^{i\theta} \int_{\epsilon}^{\sqrt{R^4 - R^2 + 1}} \exp(t e^{i\theta})(\log t + i\theta) \, \mathrm{d}t$$

so that

$$\lim_{\substack{R\to\infty \\ \epsilon\to 0}} \int_{\gamma_2} f(z) \, \mathrm{d}z = \int_0^\infty e^{-t}(\log t - i\pi) \, \mathrm{d}t. \tag{12.8}$$

Similarly, we also have

$$\int_{\gamma_4} f(z) \, \mathrm{d}z = \int_{\epsilon}^{\sqrt{R^4 - R^2 + 1}} f(\gamma_4(t))\gamma_4'(t) \, \mathrm{d}t = e^{i\theta} \int_{\epsilon}^{\sqrt{R^4 - R^2 + 1}} \exp(t e^{i\theta})(\log t + i\theta) \, \mathrm{d}t$$

which gives

$$\lim_{\substack{R\to\infty \\ \epsilon\to 0}} \int_{\gamma_4} f(z) \, \mathrm{d}z = -\int_0^\infty e^{-t}(\log t + i\pi) \, \mathrm{d}t. \tag{12.9}$$

By [11, §4.331, Eqn. (1), p. 571], we know that

$$\int_0^\infty e^{-t} \log t \, \mathrm{d}t = -C,$$

where C is the Euler's constant, so the integrals (12.8) and (12.9) can be split into two convergent integrals and then

$$\lim_{\substack{R\to\infty \\ \epsilon\to 0}} \int_{\gamma_2} f(z) \, \mathrm{d}z + \lim_{\substack{R\to\infty \\ \epsilon\to 0}} \int_{\gamma_4} f(z) \, \mathrm{d}z = 2\pi i \int_0^\infty e^{-t} \, \mathrm{d}t = 2\pi i.$$

Substitute this result into the equation (12.7), we see that

$$\lim_{\substack{R\to\infty \\ \epsilon\to 0}} \int_{\gamma_1} e^z \log z \, \mathrm{d}z = 2\pi i - \lim_{\substack{R\to\infty \\ \epsilon\to 0}} \int_{\gamma_3} e^z \log z \, \mathrm{d}z. \tag{12.10}$$

Finally, if $z = \gamma_3(t) = \epsilon e^{it}$, where $t \in [\theta_R, \theta_{-R}] \subseteq [-\pi, \pi]$, then we know that $|\gamma_3(t)| \le 2\pi\epsilon$ and

$$|f(z)| = |\exp(\epsilon e^{it}) \log(\epsilon e^{it})| = e^{\epsilon \cos t} |\log \epsilon + it| \le e^1 (|\log \epsilon| + 2\pi).$$

Since f is obviously continuous on γ_3, Theorem 4.10 (The M-L Formula) implies that

$$\left| \int_{\gamma_3} f(z) \, \mathrm{d}z \right| \le 2\pi e^1 (|\log \epsilon| + 2\pi)\epsilon \to 0$$

as $\epsilon \to 0$. Hence we conclude from the expression (12.10) that

$$\int_{\gamma} e^z \log z \, \mathrm{d}z = 2\pi i$$

which ends the proof of the problem.

Problem 12.4

Bak and Newman Chapter 12 Exercise 4.

Proof. Using exactly the same argument as the proof of [4, Example 3, pp. 163, 164], we have

$$\sum_{k=0}^{n} (-1)^n (C_k^n)^{\frac{1}{3}} = \frac{1}{2\pi i} \int_C [f(z)]^{\frac{1}{3}} \cdot \frac{\pi}{\sin \pi z} \, \mathrm{d}z, \tag{12.11}$$

where C is any contour in $-1 < \operatorname{Re} z < n+1$ and

$$f(z) = \frac{\sin \pi z}{\pi z (1 - z)(1 - \frac{z}{2}) \cdots (1 - \frac{z}{n})}.$$

Furthermore, we can split the integral (12.11) as

$$\sum_{k=0}^{n} (-1)^n (C_k^n)^{\frac{1}{3}} = \frac{1}{2\pi i} \left[\int_{-\frac{1}{2}+i\infty}^{-\frac{1}{2}-i\infty} + \int_{n+\frac{1}{2}-i\infty}^{n+\frac{1}{2}+i\infty} [f(z)]^{\frac{1}{3}} \cdot \frac{\pi}{\sin \pi z} \, \mathrm{d}z \right], \tag{12.12}$$

If $z = n - \omega$, then we have

$$\begin{aligned}
f(n-\omega) &= \frac{n! \sin \pi(n-\omega)}{\pi(n-\omega)[(\omega+1-n)(\omega+2-n) \cdots (\omega-1)(\omega-0)]} \\
&= \frac{(-1)^{n+1} n! \sin \pi\omega}{\pi(n-\omega)(-1)^n [-\omega(1-\omega) \cdots (n-2-\omega)(n-1-\omega)]} \\
&= \frac{n! \sin \pi\omega}{\pi\omega[(1-\omega)(2-\omega) \cdots (n-2-\omega)(n-1-\omega)(n-\omega)]} \\
&= \frac{\sin \pi\omega}{\pi\omega\left[\frac{(1-\omega)}{1} \cdot \frac{2-\omega}{2} \cdots \frac{(n-2-\omega)}{n-2} \cdot \frac{(n-1-\omega)}{n-1} \cdot \frac{(n-\omega)}{n}\right]} \\
&= \frac{\sin \pi\omega}{\pi\omega(1-\omega)(1-\frac{\omega}{2}) \cdots (1-\frac{\omega}{n})}
\end{aligned}$$

$$= f(\omega).$$

Therefore, we see that

$$\left| \int_{n+\frac{1}{2}-i\infty}^{n+\frac{1}{2}+i\infty} [f(z)]^{\frac{1}{3}} \cdot \frac{\pi}{\sin \pi z}\, \mathrm{d}z \right| = \left| \int_{-\frac{1}{2}+i\infty}^{-\frac{1}{2}-i\infty} [f(\omega)]^{\frac{1}{3}} \cdot \frac{\pi}{\sin \pi \omega}\, \mathrm{d}\omega \right|$$

and when $\operatorname{Re} z = -\frac{1}{2}$, we deduce from the expression (12.12) that

$$\left| \sum_{k=0}^{\infty} (-1)^n (C_k^n)^{\frac{1}{3}} \right| \le \frac{1}{\pi} \left| \int_{-\frac{1}{2}+i\infty}^{-\frac{1}{2}-i\infty} [f(z)]^{\frac{1}{3}} \cdot \frac{\pi}{\sin \pi z}\, \mathrm{d}z \right|$$

$$\le \frac{1}{\pi} \times \pi^{\frac{2}{3}} \times \left(\frac{2}{\sqrt{n+1}} \right)^{\frac{1}{3}} \int_{\operatorname{Re} z=-\frac{1}{2}} \left| \frac{\mathrm{d}z}{(\sin \pi z)^{\frac{2}{3}}} \right|$$

$$= \frac{\sqrt[3]{2}}{\sqrt[3]{\pi} \cdot \sqrt[6]{n+1}} \int_{\operatorname{Re} z=-\frac{1}{2}} \left| \frac{\mathrm{d}z}{(\sin \pi z)^{\frac{2}{3}}} \right|. \qquad (12.13)$$

Note that if $z = -\frac{1}{2} + iy$, then the definition of $\sin z$ gives

$$\left| \sin \pi z \right| = \left| \frac{1}{2i} (\mathrm{e}^{-\frac{\pi i}{2} - \pi y} - \mathrm{e}^{\frac{\pi i}{2} + \pi y}) \right| = \frac{1}{2} (\mathrm{e}^{-\pi y} + \mathrm{e}^{\pi y}).$$

Consequently, if $y \in [0, \infty)$, then $\left| \sin \pi z \right| \ge \frac{1}{2}\mathrm{e}^{\pi y}$. If $y \in (-\infty, 0)$, then we have $\left| \sin \pi z \right| \ge \frac{1}{2}\mathrm{e}^{-\pi y}$. Now these imply that

$$\int_{\operatorname{Re} z=-\frac{1}{2}} \left| \frac{\mathrm{d}z}{(\sin \pi z)^{\frac{2}{3}}} \right| = \sqrt[3]{4} \cdot \left[\int_{-\infty}^{0} \mathrm{e}^{\frac{2\pi}{3} y}\, \mathrm{d}y + \int_{0}^{\infty} \mathrm{e}^{-\frac{2\pi}{3} y}\, \mathrm{d}y \right]$$

$$= \sqrt[3]{4} \cdot \left[\frac{3}{2\pi} \mathrm{e}^{\frac{2\pi}{3} y} \Big|_{-\infty}^{0} - \frac{3}{2\pi} \mathrm{e}^{-\frac{2\pi}{3} y} \Big|_{0}^{\infty} \right]$$

$$= \sqrt[3]{4} \cdot \frac{3}{\pi}.$$

Hence it follows from the inequality (12.13) that

$$\left| \sum_{k=0}^{\infty} (-1)^n (C_k^n)^{\frac{1}{3}} \right| \le \frac{\sqrt[3]{2}}{\sqrt[3]{\pi} \cdot \sqrt[6]{n+1}} \cdot \sqrt[3]{4} \cdot \frac{3}{\pi} = \frac{6}{\pi \sqrt[3]{\pi} \cdot \sqrt[6]{n+1}}$$

which guarantees

$$\sum_{k=0}^{\infty} (-1)^n (C_k^n)^{\frac{1}{3}} \to 0$$

as $n \to \infty$, completing the proof of the problem. $\qquad \blacksquare$

Problem 12.5

Bak and Newman Chapter 12 Exercise 5.

Proof.

(a) Using the same argument as in the proof of [4, Example 3, pp. 163, 164], we know that

$$\sum_{k=0}^{\infty} (-1)^k \sqrt{C_k^n} = \frac{1}{2\pi i} \left[\int_{-\frac{3}{4}+i\infty}^{-\frac{3}{4}-i\infty} + \int_{n+\frac{3}{4}-i\infty}^{n+\frac{3}{4}+i\infty} \sqrt{f(z)} \cdot \frac{\pi}{\sin \pi z}\, \mathrm{d}z \right]. \qquad (12.14)$$

As we have shown in Problem 12.4, the integrand $\sqrt{f(z)} \cdot \frac{\pi}{\sin \pi z}$ is invariant under the substitution $z \mapsto n - z$, it suffices to consider the first integral in the expression (12.14). On the line $\operatorname{Re} z = -\frac{3}{4}$, we have $z = -\frac{3}{4} + iy$ and

$$
\begin{aligned}
\left| z(1-z)\left(1 - \frac{z}{2}\right) \cdots \left(1 - \frac{z}{n}\right) \right| &= |z| \cdot |1 - z| \cdot \left| 1 - \frac{z}{2} \right| \cdots \left| 1 - \frac{z}{n} \right| \\
&\geq \frac{3}{4} \cdot \left(1 + \frac{3}{4}\right) \cdot \left(1 + \frac{3}{4 \cdot 2}\right) \cdots \left(1 + \frac{3}{4n}\right) \\
&= \frac{3}{4} \cdot \prod_{k=1}^{n} \left(1 + \frac{3}{4k}\right).
\end{aligned}
\tag{12.15}
$$

Recall the binomial theorem ([22, Exercise 22, p. 201]) that

$$
(1 + x)^\alpha = \sum_{n=0}^{\infty} C_n^\alpha x^n = \sum_{n=0}^{\infty} C_n^\alpha = \frac{\alpha(\alpha - 1) \cdots (\alpha - n + 1)}{n!} x^n.
$$

If we put $x = \frac{1}{k}$ and $\alpha = \frac{3}{4}$, then we obtain

$$
\left(1 + \frac{1}{k}\right)^{\frac{3}{4}} = 1 + \frac{3}{4k} - \frac{3}{32k^2} + \frac{5}{128k^3} - \cdots \leq 1 + \frac{3}{4k}
\tag{12.16}
$$

because $|C_n^{\frac{3}{4}}|$ is a decreasing function of n. Combining the inequalities (12.15) and (12.16), we see immediately that

$$
\begin{aligned}
\left| z(1-z)\left(1 - \frac{z}{2}\right) \cdots \left(1 - \frac{z}{n}\right) \right| &\geq \frac{3}{4} \cdot \prod_{k=1}^{n} \left(1 + \frac{1}{k}\right)^{\frac{3}{4}} \\
&= \frac{3}{4} \cdot \left[\prod_{k=1}^{n} \left(\frac{k+1}{k}\right) \right]^{\frac{3}{4}} \\
&= \frac{3}{4} \cdot (n + 1)^{\frac{3}{4}}.
\end{aligned}
\tag{12.17}
$$

Hence it concludes that

$$
\begin{aligned}
\left| \int_{-\frac{3}{4}+i\infty}^{-\frac{3}{4}-i\infty} \sqrt{f(z)} \cdot \frac{\pi}{\sin \pi z}\, dz \right| &\leq \frac{1}{\pi} \cdot \int_{-\frac{3}{4}+i\infty}^{-\frac{3}{4}-i\infty} \frac{\sqrt{|\sin \pi z|}}{\sqrt{|\pi z(1-z)(1 - \frac{z}{2}) \cdots (1 - \frac{z}{n})|}} \cdot \left| \frac{\pi\, dz}{\sin \pi z} \right| \\
&\leq \frac{1}{\pi} \cdot \int_{-\frac{3}{4}+i\infty}^{-\frac{3}{4}-i\infty} \frac{2\sqrt{\pi}}{\sqrt{3}(n+1)^{\frac{3}{8}}} \cdot \left| \frac{dz}{\sqrt{\sin \pi z}} \right| \\
&= \frac{2}{\sqrt{3\pi}(n+1)^{\frac{3}{8}}} \int_{-\frac{3}{4}+i\infty}^{-\frac{3}{4}-i\infty} \left| \frac{dz}{\sqrt{\sin \pi z}} \right| \\
&\leq An^{-\frac{3}{8}}
\end{aligned}
$$

for some constant $A > 0$.

(b) Let $\delta > 0$ be small. Suppose that $-t = -1 + \delta$ so that $t \in (0, 1)$. Now our integration becomes along $\operatorname{Re} z = -t$ and $\operatorname{Re} z = n + t$. On $\operatorname{Re} z = -t$, since $t \in (0, 1)$, instead of the inequality (12.17), we establish

$$
\left| z(1-z)\left(1 - \frac{z}{2}\right) \cdots \left(1 - \frac{z}{n}\right) \right| \geq t \prod_{k=1}^{n} \left(1 + \frac{t}{k}\right)
$$

$$\geq t\left[\prod_{k=1}^{n}\left(1+\frac{1}{k}\right)\right]^{t}$$

$$= (1-\delta)(n+1)^{1-\delta}$$

$$\geq (1-\delta)n^{1-\delta}. \tag{12.18}$$

Recall the facts from the proof of Problem 12.4 that $|\sin \pi z| \geq \frac{1}{2}e^{\pi y}$ if $y \in [0,\infty)$ and $|\sin \pi z| \geq \frac{1}{2}e^{-\pi y}$ if $y \in (-\infty, 0)$, where $y = \operatorname{Im} z$.

Now instead of (12.14), we have

$$\sum_{k=0}^{\infty}(-1)^{k}\sqrt{C_{k}^{n}} = \frac{1}{2\pi i}\left[\int_{-1+\delta+i\infty}^{-1+\delta-i\infty} + \int_{n+1-\delta-i\infty}^{n+1-\delta+i\infty} \sqrt{f(z)} \cdot \frac{\pi}{\sin \pi z}\,dz\right].$$

Applying similar argument we have employed in part (a), we follow from the inequality (12.18) that

$$\left|\sum_{k=0}^{\infty}(-1)^{k}\sqrt{C_{k}^{n}}\right| \leq \frac{1}{\pi}\int_{-1+\delta+i\infty}^{-1+\delta-i\infty}\left|\sqrt{f(z)} \cdot \frac{\pi\,dz}{\sin \pi z}\right|$$

$$= \frac{1}{\sqrt{\pi}}\int_{-1+\delta+i\infty}^{-1+\delta-i\infty}\left|\frac{dz}{\sqrt{\sin \pi z} \times \sqrt{z(1-z)(1-\frac{z}{2})\cdots(1-\frac{z}{n})}}\right|$$

$$\leq \frac{1}{\sqrt{\pi(1-\delta)n^{1-\delta}}}\int_{\operatorname{Re}z=-1+\delta}\left|\frac{dz}{\sqrt{\sin \pi z}}\right|$$

$$\leq \frac{A}{\sqrt{(1-\delta)n^{1-\delta}}}$$

for some positive constant A.

We have completed the proof of the problem.   ▧

Problem 12.6

Bak and Newman Chapter 12 Exercise 6.

Proof. Suppose that $x > 0$ and $h(t) = \frac{1}{t^{t-1}} > 0$ on $[0,\infty)$. Since $e^{xt} \geq 1 + xt$, we observe from the definition that

$$g(x + 2\pi i) = f(x) = \int_{0}^{\infty}\frac{e^{xt}}{t^{t}}\,dt \geq x\int_{0}^{\infty}\frac{dt}{t^{t-1}} = x\left[\int_{0}^{1}h(t)\,dt + \int_{1}^{\infty}h(t)\,dt\right]. \tag{12.19}$$

Since t^{1-t} is bounded on $[0,1]$, the integral

$$\int_{0}^{1}h(t)\,dt = \int_{0}^{1}t^{1-t}\,dt$$

is convergent and then the convergence of the integral

$$\int_{0}^{\infty}\frac{dt}{t^{t-1}}$$

depends on the convergence of the second integral inside the brackets (12.19). To this end, we study the convergence of the series

$$\sum_{n=1}^{\infty}h(n) = \sum_{n=1}^{\infty}\frac{1}{n^{n-1}}. \tag{12.20}$$

Let $a_n = \frac{1}{n^{n-1}}$. Then it is easy to see that

$$\alpha = \limsup_{n\to\infty} \sqrt[n]{a_n} = \lim_{n\to\infty} \frac{1}{n^{1-\frac{1}{n}}} = 0,$$

so [27, Theorem 6.7, p. 76] ensures that the series (12.20) converges. Since $h(t) \geq 0$ and h decreases monotonically on $[1,\infty)$, it follows from [28, Theorem 1.10, p. 37] that the second integral in the brackets (12.19) is convergent. Therefore, there is a positive constant M such that

$$g(x + 2\pi i) \geq Mx$$

which shows certainly that $g(x + 2\pi i) \to \infty$ as $x \to \infty$. This completes the analysis of the problem.

CHAPTER **13**

Introduction to Conformal Mapping

Problem 13.1

Bak and Newman Chapter 13 Exercise 1.

Proof. Suppose that $z_0 \neq 0$ and $z_1, z_2 \in \mathbb{C}$ are distinct numbers such that $f(z_1) = f(z_2)$. If $z_2 = 0$, then it forces that $z_1 = 0$. Therefore, without loss of generality, we may assume that $z_2 \neq 0$. Then we can write

$$f(z_1) - f(z_2) = z_2^k \left[\left(\frac{z_1}{z_2} \right)^k - 1 \right]$$

which means that $f(z_1) = f(z_2)$ if and only if $z_1 = z_1(n) = z_2 \exp(\frac{2n\pi i}{k})$ if and only if

$$\operatorname{Arg} z_1 - \operatorname{Arg} z_2 = \frac{2n\pi}{k} \tag{13.1}$$

for some $n \in \{1, 2, \ldots, k-1\}$. Hence the equation (13.1) insures that if $\delta > 0$ is chosen such that $D(z_0; \delta)$ lies in the sector $S = \{z \in \mathbb{C} \,|\, \alpha < \operatorname{Arg} z < \beta\}$, where $0 < \beta - \alpha < \frac{2\pi}{k}$ (see Figure 13.1 below), then f is locally 1-1 at z_0 by Definition 13.3.

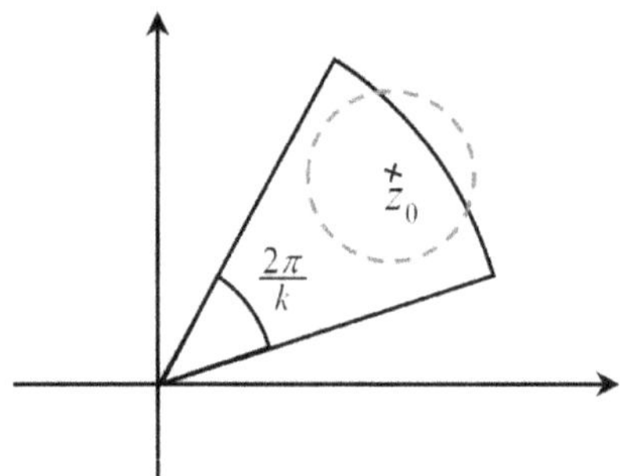

Figure 13.1: The sector S and the disc $D(z_0; \delta)$.

More precisely, if $\theta = \operatorname{Arg} z_0$, then we may take $\alpha = \theta - \frac{\pi}{k}$ and $\beta = \theta + \frac{\pi}{k}$ so that the ray connecting the origin and the point z_0 bisects the sector S. This completes the proof of the problem. ∎

Problem 13.2

Bak and Newman Chapter 13 Exercise 2.

Proof. If $x = c$, then we have

$$|\omega| = |\exp(c + iy)| = e^c$$

which is a circle centred at 0 with radius e^c. Next, if $y = c$, then $\omega = e^x \cdot e^{ic}$ which is a ray with angle c and $e^x > 0$. (Note that when x runs through $\mathbb{R}$, e^x runs through $(0, \infty)$.) We have completed the proof of the problem. $\blacksquare$

Problem 13.3

Bak and Newman Chapter 13 Exercise 3.

Proof. We remark that throughout this chapter, we denote $\mathbb{H}$ to be the upper half-plane.

(i) Note that $f_1(z) = z + 2$ maps the strip S onto the strip $S_1 = \{z \mid 0 < \operatorname{Re} z < 3\}$ conformally. Next, the mapping $f_2(z) = \frac{\pi}{3} f_1(z)$ maps S_1 onto the strip $S_2 = \{z \mid 0 < \operatorname{Re} z < \pi\}$ conformally. Furthermore, the mapping $f_3(z) = i f_2(z)$ sends S_2 onto the vertical strip $S_4 = \{z \mid 0 < \operatorname{Im} z < \pi\}$ and the mapping $f_4(z) = e^{f_3(z)}$ maps S_4 onto $\mathbb{H}$ conformally. By Theorem 13.16, the mapping $f_5(z) = \frac{f_4(z) - i}{f_4(z) + i}$ conformally maps $\mathbb{H}$ onto T. Hence a required conformal mapping f is given by

$$f(z) = \frac{\exp(\frac{\pi i}{3}(z + 2)) - i}{\exp(\frac{\pi i}{3}(z + 2)) + i}.$$

(ii) Theorem 13.23 implies that the (unique) bilinear transformation f mapping $-2, 0, 2$ into $-1, 0, 2$ is given by

$$\frac{(f(z) - 0)(2 + 1)}{(f(z) + 1)(2 - 0)} = \frac{(z - 0)(2 + 2)}{(z + 2)(2 - 0)}$$

which reduces to

$$f(z) = \frac{4z}{6 - z}.$$

(iii) The mapping $f_1(z) = z^4$ maps S onto $\mathbb{H}$ conformally. Next, the mapping $f_2(z) = \log f_1(z)$ maps $\mathbb{H}$ onto the strip $\{z \mid 0 < \operatorname{Im} z < \pi\}$ conformally. Finally, $f_3(z) = \frac{1}{\pi} f_2(z)$ conformally maps the strip $\{z \mid 0 < \operatorname{Im} z < \pi\}$ onto the strip $\{z \mid 0 < \operatorname{Im} z < 1\}$. Hence a required conformal mapping f is given by

$$f(z) = \frac{4}{\pi} \log z.$$

(iv) The mapping $f_1(z) = iz^{\frac{1}{2}}$ sends S onto the upper semi-disc $S_1 = \{z \mid |z| = 1 \text{ and } \operatorname{Im} z > 0\}$ conformally. By [4, Example 1, p. 180], the mapping $f_2(z) = -\frac{(z-1)^2}{4(z+1)^2}$ sends S_1 conformally onto $\mathbb{H}$. According to Theorem 13.6, $f_3(z) = \frac{z-i}{z+i}$ is a conformal mapping of $\mathbb{H}$ onto the T. Therefore, a required conformal mapping f is given by

$$f(z) = \left[-\frac{(i\sqrt{z} - 1)^2}{4(i\sqrt{z} + 1)^2} - i \right] \times \left[-\frac{(i\sqrt{z} - 1)^2}{4(i\sqrt{z} + 1)^2} + i \right]^{-1}.$$

We complete the analysis of the problem. $\blacksquare$

Problem 13.4

*Bak and Newman Chapter 13 Exercise 4.**

Proof. Let the region between the two circles be G. The two circles touch internally at 2, so [4, Example 1, p. 181] indicates that the mapping $f_1(z) = \frac{1}{z-2}$ sends G conformally onto two parallel lines. Since $f_1(0) = -\frac{1}{2}$ and $f_1(-2) = -\frac{1}{4}$, the two parallel lines are $\operatorname{Re} z = -\frac{1}{4}$ and $\operatorname{Re} z = -\frac{1}{2}$. Then the mapping $f_2(z) = \frac{1}{z-2} + \frac{1}{2} = \frac{z}{2(z-2)}$ maps G onto the infinite strip $S = \{z \in \mathbb{C} \mid 0 < \operatorname{Re} z < \frac{1}{4}\}$ conformally. Next, the mapping $f_3(z) = 4\pi i f_2(z) = \frac{2\pi i z}{z-2}$ sends G conformally onto the strip $T = \{z \in \mathbb{C} \mid 0 < \operatorname{Im} z < \pi\}$ and so the mapping

$$f_4(z) = \exp f_3(z) = \exp\left(\frac{2\pi i z}{z-2}\right)$$

sends G onto $\mathbb{H}$ conformally. Finally, Theorem 13.16 shows that the mapping f given by

$$f(z) = \frac{\exp(\frac{2\pi i}{z-2}) - i}{\exp(\frac{2\pi i}{z-2}) + i}$$

maps G onto the unit disc $D(0;1)$ conformally, completing the proof of the problem. $\blacksquare$

Problem 13.5

*Bak and Newman Chapter 13 Exercise 5.**

Proof. Let $S = \{z = x + iy \in \mathbb{C} \mid x > 0 \text{ and } 0 < y < 1\}$. Then $f_1(z) = \pi i z$ sends S conformally onto $S_1 = \{z = x + iy, \mid -\pi < x < 0 \text{ and } y > 0\}$. Next, $f_2(z) = \pi i z + \frac{\pi}{2}$ will map S onto the semi-infinite strip

$$S_2 = \left\{z \in \mathbb{C} \,\middle|\, -\frac{\pi}{2} < \operatorname{Re} z < \frac{\pi}{2} \text{ and } \operatorname{Im} z > 0\right\}$$

conformally. We know that the Schwarz-Christoffel transformation $f(z) = \sin z$ maps S_2 onto $\mathbb{H}$, so the composition $f_3 = f \circ f_2$ maps S onto $\mathbb{H}$ conformally. Since

$$f_3(z) = f(f_2(z)) = \sin\left(\frac{\pi}{2} + \pi i z\right) = \cos(\pi i z),$$

we observe from Theorem 13.16 that the mapping

$$g(z) = \frac{\cos(\pi i z) - i}{\cos(\pi i z) + i}$$

sends S conformally onto the unit disc $D(0;1)$. This ends the proof of the problem. $\blacksquare$

Problem 13.6

*Bak and Newman Chapter 13 Exercise 6.**

Proof. By [4, Example 1, p. 180], the mapping f defined by

$$f(z) = -\frac{(z-1)^2}{4(z+1)^2}$$

sends S onto $\mathbb{H}$ conformally. Now Theorem 13.16 ensures that the mapping g given by

$$g(z) = \frac{f(z) - i}{f(z) + i} = \left[-\frac{(z-1)^2}{4(z+1)^2} - i \right] \times \left[-\frac{(z-1)^2}{4(z+1)^2} + i \right]^{-1}$$

maps S conformally onto the unit disc $D(0; 1)$ and we complete the proof of the problem.   ▨

Problem 13.7

*Bak and Newman Chapter 13 Exercise 7.**

Proof. Suppose that U, V, W are regions and there exist conformal mappings $f : U \to V$, $g : V \to W$ such that $f(U) = V$ and $g(V) = W$. In other words, we have $U \sim V$ and $V \sim W$.

- **Reflexive Property:** The identity map id $: U \to U$ is certainly bijective. Furthermore, it satisfies $\text{id}'(z) = 1$ for all $z \in U$, so Theorem 13.4 implies that id is conformal, i.e., $U \sim U$.

- **Symmetric Property:** Since $f : U \to V$ is a conformal mapping, Theorem 13.8 shows that $F = f^{-1} : V \to U$ is also a conformal mapping. Since f is bijective, F is also bijective. Thus we conclude that $V \sim U$.

- **Transitive Property:** We know that $h = g \circ f : U \to W$ is bijective because f and g are bijective. Since f and g are analytic, Problem 3.3 ensures that h is also analytic. By Definition 13.9, h is a conformal mapping which means that $U \sim W$.

According to the definition, "conformal equivalence" is an equivalence relation which completes the proof of the problem.   ▨

Problem 13.8

Bak and Newman Chapter 13 Exercise 8.

Proof.

(a) Suppose that $R = \{z = x + iy \,|\, a \le x \le b \text{ and } c \le y \le d\}$ and $\partial R = \gamma_1 + \gamma_2 + \gamma_3 + \gamma_4$, where γ_1 and γ_3 are the horizontal lines and γ_2 and γ_4 are the vertical sides.[a] Let $f(z) = az + b$, where $a \neq 0$. It is clear that $f : \partial R \to f(\partial R)$ is bijective. By Theorem 13.11, $f(\gamma_1)$, $f(\gamma_2)$, $f(\gamma_3)$ and $f(\gamma_4)$ are lines. Since f is conformal at the four vertices, $f(\gamma_1)$ and $f(\gamma_3)$ are parallel lines such that they are perpendicular to both $f(\gamma_2)$ and $f(\gamma_4)$. Consequently, $f(\partial R)$ is also a polygon.

(b) Let R and $R' = f(R)$ be the said rectangles. The hypotheses imply that f is nonconstant. By a rotation and a translation if necessary, we may assume that points z of R have the form $0 \le \operatorname{Re} z \le a$ and $0 \le \operatorname{Im} z \le b$ for some positive constants a and b. Denote $0, v_1, v_2$ and v_3 to be its vertices. Similarly, we may also assume that if $z \in R'$, then $0 \le \operatorname{Re} z \le c$ and $0 \le \operatorname{Im} z \le d$ for some positive constants c and d with vertices $0, v_1', v_2'$ and v_3'. Figure 13.2 shows the setting for this problem.

---

[a] If not, then we can consider the rectangle $R' = e^{i\theta} R$ for some θ whose sides are either horizontal or vertical.

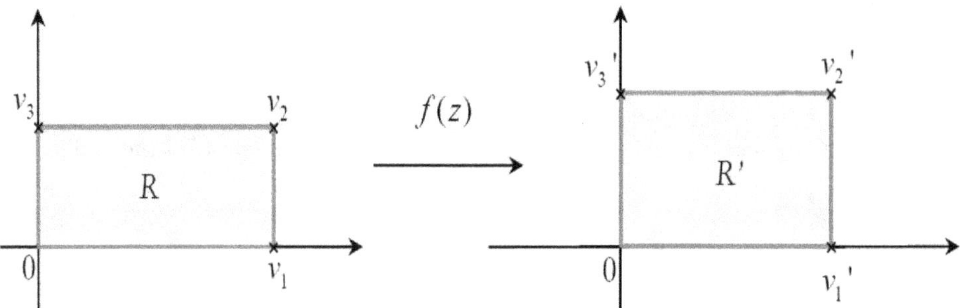

Figure 13.2: The rectangles R and R'.

Suppose that $f(z_0) \in \{0, v_1', v_2', v_3'\}$. By Problem 7.3, we have $z_0 \in \partial R$. We claim that if $z_0 \notin \{0, v_1, v_2, v_3\}$, then $f(z_0) \notin \{0, v_1', v_2', v_3'\}$. Otherwise, we assume, without loss of generality, that $f(z_0) = v_2'$. Now we want to compare $\angle \gamma_1, \gamma_2$ and $\angle f(\gamma_1), f(\gamma_2)$, see Figure 13.3 below for an illustration:

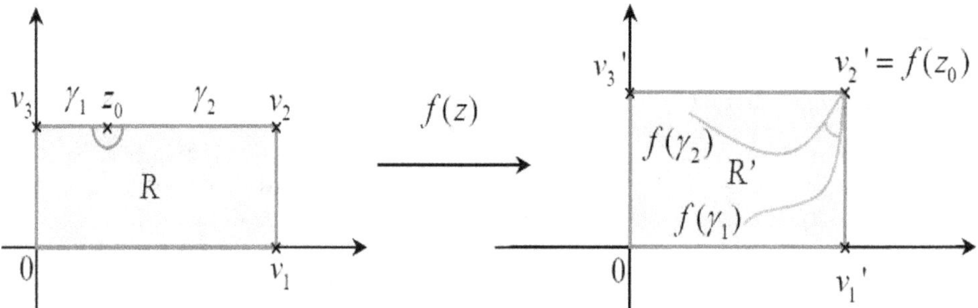

Figure 13.3: The angles of $\angle \gamma_1, \gamma_2$ and $\angle f(\gamma_1), f(\gamma_2)$.

By Theorems 13.4 and 13.7, we have $\angle \gamma_1, \gamma_2 = k\pi$, where k is the least positive integer such that $f^{(k)}(z_0) \neq 0$. However, as the interior angle at v_2' is $\frac{\pi}{2}$, so we will have

$$\angle \gamma_1, \gamma_2 \neq \angle f(\gamma_1), f(\gamma_2),$$

a contradiction. Thus this proves our claim.

By Theorem 7.1 (The Open Mapping Theorem), we have the observation that interior points of R are mapped by f to interior points of R'. Assume that there was a $\omega \in R^\circ$ such that $f(\omega) \in \partial R$. For each $n \in \mathbb{N}$, the continuity of f ensures that there exists a $\delta_n > 0$ such that $z_n \in D(\omega; \delta_n) \subseteq R^\circ$ implies

$$|\omega - f(z_n)| < \frac{1}{n}.$$

Thus we can find a bounded sequence $\{z_n\}$ in R° such that $f(z_n) \to \omega$ as $n \to \infty$. By the Bolzano-Weierstrass Theorem, $\{z_n\}$ has a convergent subsequence with limit z. Now $z \in R^\circ$ is impossible because it contradicts to our mentioned observation. Hence f maps ∂R to $\partial R'$.

Finally, since a rectangle is also a parallelogram, we conclude from Problem 7.23 that f is a linear polynomial.

We have completed the proof of the problem.

Problem 13.9

Bak and Newman Chapter 13 Exercise 9.

Proof. We check the definition [8, pp. 16, 17]. Suppose that $f_1(z) = \frac{a_1 z + b_1}{c_1 z + d_1}$ and $f_2(z) = \frac{a_2 z + b_2}{c_2 z + d_2}$, where $a_1 d_1 - b_1 c_1 \neq 0$ and $a_2 d_2 - b_2 c_2 \neq 0$. Then it is easy to check that

$$
\begin{aligned}
f_1(f_2(z)) &= \frac{a_1 f_2(z) + b_1}{c_1 f_2(z) + d_1} \\
&= \frac{a_1 \cdot \frac{a_2 z + b_2}{c_2 z + d_2} + b_1}{c_1 \cdot \frac{a_2 z + b_2}{c_2 z + d_2} + d_1} \\
&= \frac{a_1 (a_2 z + b_2) + b_1 (c_2 z + d_2)}{c_1 (a_2 z + b_2) + d_1 (c_2 z + d_2)} \\
&= \frac{(a_1 a_2 + b_1 c_2) z + (a_1 b_2 + b_1 d_2)}{(c_1 a_2 + d_1 c_2) z + (c_1 b_2 + d_1 d_2)}
\end{aligned}
$$

and

$$
\begin{aligned}
&(a_1 a_2 + b_1 c_2)(c_1 b_2 + d_1 d_2) - (a_1 b_2 + b_1 d_2)(c_1 a_2 + d_1 c_2) \\
&= a_1 d_1 (a_2 d_2 - b_2 c_2) + b_1 c_1 (c_2 b_2 - d_2 a_2) \\
&= (a_1 d_1 - b_1 c_1)(a_2 d_2 - b_2 c_2) > 0.
\end{aligned}
$$

Thus $f_1 \circ f_2$ is also a bilinear transformation.

If f_1, f_2 and f_3 are bilinear transformations, then it is true that

$$(f_1 \circ f_2) \circ f_3 = f_1 \circ (f_2 \circ f_3)$$

as compositions of functions. Obviously, the identity $\operatorname{id}(z) = z$ is a bilinear transformation satisfying

$$f \circ \operatorname{id} = \operatorname{id} \circ f = f.$$

Finally, we notice from [4, p. 177] that if $f(z) = \frac{az+b}{cz+d}$, then $f^{-1}(z) = \frac{dz-b}{-cz+a}$ and therefore,

$$f \circ f^{-1} = f^{-1} \circ f = \operatorname{id}.$$

Hence bilinear mappings form a group under composition which completes the analysis0 of the problem. $\blacksquare$

Problem 13.10

Bak and Newman Chapter 13 Exercise 10.

Proof. The mappings are bilinear, so Theorem 13.11 ensures that they map the unit circle onto a circle or a line.

(a) Suppose that $z = e^{i\theta}$. Then we have $\omega = \operatorname{cis}(-\theta)$ which means that $|\omega| = 1$.

(b) Since $\omega(-1) = -\frac{1}{2}$ and ω has a pole at 1, its image is the vertical line $\operatorname{Im} z = -\frac{1}{2}$.

(c) Notice that $\omega = \omega_2 \circ \omega_1$, where $\omega_1(z) = z - 2$ and $\omega_2 = \frac{1}{z}$. It is easily seen that

$$\omega_1(C(0;1)) = C(-2;1).$$

Since $1 \neq |-2|$, it follows from the proof of Lemma 13.10 that $\omega_2(C(-2;1)) = C(\beta;R)$, where

$$\beta = \frac{-2}{|-2|^2 - 1^2} = -\frac{2}{3} \quad \text{and} \quad R = \frac{1}{|-2|^2 - 1} = \frac{1}{3}.$$

Hence we have completed the proof of the problem.

> ### Problem 13.11
>
> Bak and Newman *Chapter 13 Exercise 11.*

Proof. We deduce from Theorem 13.15 that f has the form

$$f(z) = e^{i\theta}\left(\frac{z - \alpha}{1 - \overline{\alpha}z}\right),$$

where $|\alpha| < 1$. Since $0 = f(0) = e^{i\theta}\alpha$, we have $\alpha = 0$ and thus $f(z) = e^{i\theta}z$. Since $f'(0) = e^{i\theta} > 0$, $\theta = 0$ which implies that $f(z) = z$ as desired. This completes the proof of the problem.

> ### Problem 13.12
>
> Bak and Newman *Chapter 13 Exercise 12.*

Proof. Note that $f_1, f_2 : D \to D(0;1)$ are bijective conformal mappings. Then the mapping $f_2^{-1} : D(0;1) \to D$ is also bijective and conformal. By Problem 13.7, we see that the mapping $f = f_1 \circ f_2^{-1} : D(0;1) \to D(0;1)$ is an automorphism. Clearly, $f(0) = f_1(f_2^{-1}(0)) = f_1(z_0) = 0$. By Problem 3.3 and Proposition 3.5, we obtain

$$f'(0) = f_1'(f_2^{-1}(0)) \cdot (f_2^{-1})'(0) = f_1'(f_2^{-1}(0)) \cdot \frac{1}{f_2'(f_2^{-1}(0))} = \frac{f_1'(z_0)}{f_2'(z_0)} > 0.$$

Hence f satisfies the hypotheses of Problem 13.11 so that $f(z) \equiv z$, i.e., $f_1 \equiv f_2$ which ends the proof of the problem.

> ### Problem 13.13
>
> Bak and Newman *Chapter 13 Exercise 13.*

Proof. Suppose that $D(z_1;r_1), D(z_2;r_2)$ are discs and U_1, U_2 are half-planes, where $r_1, r_2 > 0$. Let U be $\mathbb{H}$. As what we have used in Problem 13.7, $D_1 \sim D_2$ means that D_1 and D_2 are conformally equivalent.

- **Case (i):** $D(z_1;r_1) \sim D(z_2;r_2)$. Let f be this conformal mapping. Now the mapping $g_k(z) = \frac{1}{r_k}(z - z_k)$ maps $D(z_k;r_k)$ conformally onto the unit disc $D(0;1)$, where $k = 1, 2$. It is obvious that each g_k is a bilinear transformation. Since $f = g_2^{-1} \circ g_1$, it follows from Problem 13.9 that f is also in the form of a bilinear transformation.

- **Case (ii):** $D(z_1; r_1) \sim U_1$. Let f be this conformal mapping. It is easily seen that there exist constants θ and b such that the mapping $h(z) = e^{i\theta}(z + b)$ sends U_1 conformally onto U. It deduces from Theorem 13.16 that the mapping $\phi(z) = \frac{z-i}{z+i}$ is a conformal mapping of U onto $D(0; 1)$. We know that $f = h^{-1} \circ \phi^{-1} \circ g$, where $g(z) = \frac{1}{r_1}(z - z_1)$. Since the inverses h^{-1} and ϕ^{-1} are bilinear transformations, Problem 13.9 guarantees that our f is in the form of a bilinear transformation.

- **Case (iii):** $U_1 \sim U_2$. Let f be this conformal mapping. Suppose that $h_1(z) = e^{i\theta_1}(z + b_1)$ and $h_2(z) = e^{i\theta_2}(z + b_2)$ map U_1 and U_2 conformally onto U respectively, where $\theta_1, \theta_2, b_1, b_2$ are some constants. Then we have $f = h_2^{-1} \circ h_1$ which is a bilinear transformation by Problem 13.9.

Consequently, any conformal mapping of a half-plane or disc onto another half-plane or disc must be in the form of a bilinear transformation. This completes the analysis of the problem.

Problem 13.14

Bak and Newman Chapter 13 Exercise 14.

Proof. Let $g = -f$. Then we have

$$g(z) = \frac{(-a)z + (-b)}{cz + d}$$

and $(-a)d - (-b)c = -(ad - bc) > 0$. It follows from Theorem 13.17 that g is an automorphism of $\mathbb{H}$. Therefore, f maps $\mathbb{H}$ conformally onto the lower half-plane, completing the proof of the problem.

Problem 13.15

Bak and Newman Chapter 13 Exercise 15.

Proof. Let Π_1 be the first quadrant. Now Theorem 13.17 shows that an automorphism of $\mathbb{H}$ is of the form

$$h(z) = \frac{az + b}{cz + d},$$

where $ad - bc > 0$. We note that $f(z) = \sqrt{z}$ is a conformal mapping of $\mathbb{H}$ onto Π_1. Similarly, $g(z) = z^2$ is a conformal mapping of Π_1 onto $\mathbb{H}$. Since the composition of two conformal mappings is also a conformal mapping, the mapping $F = f \circ h \circ g : \Pi_1 \to \Pi_1$ is an automorphism. Hence we conclude that

$$F(z) = f(h(g(z))) = \sqrt{\frac{az^2 + b}{cz^2 + d}}.$$

We complete the proof of the problem.

Problem 13.16

Bak and Newman Chapter 13 Exercise 16.

Proof. We follow the given hint. Let $h_1 = f^{-1} \circ e^{i\theta} \circ f$ and $h_2 = f^{-1} \circ e^{-i\theta} g \circ f$, where $f(z) = \frac{z-i}{z+i}$ and $g(z) = e^{i\theta}(\frac{z-\alpha}{1-\overline{\alpha}z})$ with $|\alpha| < 1$. By the properties of composition of functions, we obtain

$$h = f^{-1} \circ g \circ f = f^{-1} \circ e^{i\theta} \circ e^{-i\theta} g \circ f = (f^{-1} \circ e^{i\theta} \circ f) \circ (f^{-1} \circ e^{-i\theta} g \circ f) = h_1 \circ h_2. \quad (13.2)$$

Since $f(z) = \frac{z-i}{z+i}$, we have $f^{-1}(z) = \frac{i(z+1)}{-z+1}$. Therefore, we see that

$$e^{-i\theta} g(f(z)) = \frac{\frac{z-i}{z+i} - \alpha}{1 - \overline{\alpha}(\frac{z-i}{z+i})} = \frac{(1-\alpha)z - (1+\alpha)i}{(1-\overline{\alpha})z + (1+\overline{\alpha})i}$$

and direct computation shows

$$\begin{aligned}
h_2(z) &= i \cdot \frac{\frac{(1-\alpha)z-(1+\alpha)i}{(1-\overline{\alpha})z+(1+\overline{\alpha})i} + 1}{-\frac{(1-\alpha)z-(1+\alpha)i}{(1-\overline{\alpha})z+(1+\overline{\alpha})i} + 1} \\
&= i \cdot \frac{(1-\alpha)z - (1+\alpha)i + (1-\overline{\alpha})z + (1+\overline{\alpha})i}{-(1-\alpha)z + (1+\alpha)i + (1-\overline{\alpha})z + (1+\overline{\alpha})i} \\
&= i \cdot \frac{[2 - (\alpha + \overline{\alpha})]z + (\overline{\alpha} - \alpha)i}{(\alpha - \overline{\alpha})z + [2 + (\alpha + \overline{\alpha})]i} \\
&= \frac{(1 - \operatorname{Re}\alpha)z + \operatorname{Im}\alpha}{(\operatorname{Im}\alpha)z + (1 + \operatorname{Re}\alpha)}.
\end{aligned}$$

Recall that $|\alpha| < 1$, so $(1 - \operatorname{Re}\alpha)(1 + \operatorname{Re}\alpha) - (\operatorname{Im}\alpha)^2 = 1 - [(\operatorname{Re}\alpha)^2 + (\operatorname{Im}\alpha)^2] = 1 - |\alpha|^2 > 0$. In other words, h_2 is a bilinear transformation.

If $\theta = \pi$, then $h_1(z) = -z$ and we obtain from the formula (13.2) that $h = -h_2$ and thus this insures that h is in the form of a bilinear transformation. Suppose that $\theta \neq \pi$. Then we have $\cos\theta \neq -1$ and $\cos\frac{\theta}{2} \neq 0$. Now simple algebra shows that

$$\begin{aligned}
h_1(z) &= \frac{e^{i\theta}(\frac{z-i}{z+i}) + 1}{-e^{i\theta}(\frac{z-i}{z+i}) + 1} \\
&= i \cdot \frac{e^{i\theta}(z - i) + z + i}{-e^{i\theta}(z - i) + z + i} \\
&= i \cdot \frac{(1 + e^{i\theta})z + (1 - e^{i\theta})i}{(1 - e^{i\theta})z + (1 + e^{i\theta})i} \\
&= i \cdot \frac{(2\cos\frac{\theta}{2})z - (2i\sin\frac{\theta}{2})i}{(-2i\sin\frac{\theta}{2})z + (2\cos\frac{\theta}{2})i} \\
&= \frac{(1 + \cos\theta)z + \sin\theta}{(-\sin\theta)z + (1 + \cos\theta)}.
\end{aligned}$$

As $(2 + \cos\theta)^2 + \sin^2\theta = 5 + 4\cos\theta > 0$, h_1 is in the form of a bilinear transformation. By Problem 13.9, h is also a bilinear transformation in this case and we have completed the proof of the problem. $\blacksquare$

Problem 13.17

Bak and Newman Chapter 13 Exercise 17.

Proof.

(a) Now $\frac{z-1}{z+1} = z$ if and only if $z^2 = -1$ if and only if $z = \pm i$.

(b) Similarly, $\frac{z}{z+1} = z$ if and only if $z^2 = 0$ if and only if $z = 0$.

This completes the proof of the problem.

Proof. By Lemma 13.20, we know that

$$T(z) = \frac{(z - z_2)(z_3 - z_1)}{(z - z_1)(z_3 - z_2)}$$

is the (unique) bilinear mapping sending z_1, z_2 and z_3 to $\infty, 0$ and 1 respectively. Next, Theorem 13.11 implies that the image of the circle or line containing z_1, z_2 and z_3 under T is either a circle or a line. Since $\infty, 0$ and 1 lie on the real axis, the image must be the real axis. Therefore, $T(z_4)$ is real if and only if z_4 lies on the circle or the line containing z_1, z_2 and z_3, completing the proof of the problem.

Proof.

(a) By Theorem 13.23, we conclude that

$$\frac{2(\omega - i)}{(\omega + 1)(1 - i)} = \frac{-2(z - i)}{(z - 1)(-1 - i)}.$$

After simplification, we conclude that $\omega = -\frac{1}{z}$.

(b) By Theorem 13.23 again, we get

$$\frac{2i(\omega - i)}{i\omega} = \frac{2iz}{i(z + i)}$$

which gives $\omega = z + i$.

(c) We know that

$$\lim_{\omega_1 \to \infty} \frac{(\omega - 0)(\frac{1}{3} - \omega_1)}{(\omega - \omega_1)(\frac{1}{3} - 0)} = \lim_{\omega_1 \to \infty} \frac{\omega(\frac{1}{3\omega_1} - 1)}{\frac{1}{3}(\frac{\omega}{\omega_1} - 1)} = 3\omega,$$

so Theorem 13.23 implies that

$$3\omega = \frac{(z - i)(2i + i)}{(z + i)(2i - i)} = 3 \cdot \frac{z - i}{z + i}$$

which reduces to $\omega = \frac{z-i}{z+i}$.

We have ended the proof of the problem.

Proof. Figure 13.4 shows the conformal mapping f of the region between the two circles (the light blue part) and the annulus (the light orange part).

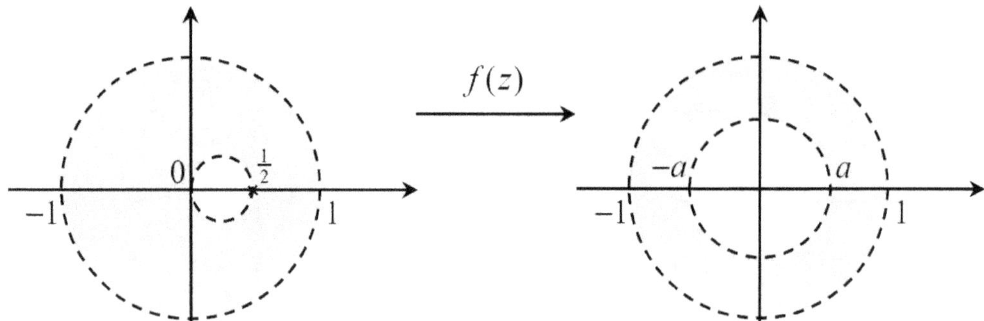

Figure 13.4: The conformal mapping of the region between the two circles and the annulus.

We know from Theorem 13.15 that

$$f(z) = e^{i\theta} \cdot \frac{z - \alpha}{1 - \overline{\alpha}z} \tag{13.3}$$

is an automorphism of $D(0;1)$, where $|\alpha| < 1$ and $\theta \in \mathbb{R}$. Now it remains to determine α so that f maps $D(\frac{1}{4};\frac{1}{4})$ conformally onto $D(0;a)$. Obviously, $\alpha \neq 0$; otherwise, $f = e^{i\theta} \cdot \mathrm{id}$ which is impossible. Now the hypothesis and the fact that f maps $C(0;1)$ onto itself certainly force that our f must map $C(\frac{1}{4};\frac{1}{4})$ onto the circle $C(0;a)$. By choosing θ appropriately in the formula (13.3), we may assume that $f(\frac{1}{2}) = a$. Since f is conformal, we have $f(0) = -a$ so that

$$\frac{\frac{1}{2} - \alpha}{1 - \frac{\alpha}{2}} = e^{-i\theta}a = \alpha$$

$$4\alpha = 1 + |\alpha|^2$$

and thus α must be real. Consequently, we have $\alpha^2 - 4\alpha + 1 = 0$ which implies that $\alpha = 2 \pm \sqrt{3}$. Since $|\alpha| < 1$, we obtain $\alpha = 2 - \sqrt{3}$ which gives the desired conformal mapping. This completes the proof of the problem. ∎

Problem 13.21

*Bak and Newman Chapter 13 Exercise 21.**

Proof. If $\zeta = \cosh z$, then we can show easily from Proposition 3.5 or from [16, Eqn. (8), p. 201] directly that

$$\frac{\mathrm{d}}{\mathrm{d}z}(\cosh^{-1} z) = \frac{1}{\frac{\mathrm{d}}{\mathrm{d}\zeta}(\cosh \zeta)} = \frac{1}{\sinh \zeta} = \frac{1}{\sqrt{\cosh^2 \zeta - 1}} = \frac{1}{\sqrt{z^2 - 1}}.$$

Therefore, we conclude that

$$f(z) = \cosh^{-1} z = \int_0^z \frac{\mathrm{d}\zeta}{\sqrt{\zeta^2 - 1}}.$$

By [16, Eqn. (8), p. 201] again, we know that $\cosh^{-1} z = i \cos^{-1} z$. Using [3, Example 1.8.8, p. 91; Exercise 48, p. 93] we can derive the identity $\sin^{-1} z + \cos^{-1} z = \frac{\pi}{2}$, so we establish

$$f(z) = \cosh^{-1} z = \frac{\pi i}{2} - i \sin^{-1} z. \tag{13.4}$$

We note from [4, pp. 187, 188] that $\sin^{-1} z$ maps $\mathbb{H}$ conformally onto the semi-infinite strip $\{z \in \mathbb{C} \mid -\frac{\pi}{2} < \operatorname{Re} z < \frac{\pi}{2} \text{ and } \operatorname{Im} z > 0\}$, thus the equation (13.4) ensures that our mapping f sends $\mathbb{H}$ conformally onto the semi-infinite strip $\{z \in \mathbb{C} \mid \operatorname{Re} z > 0 \text{ and } 0 < \operatorname{Im} z < \pi\}$.[b] We have completed the proof of the problem.

> **Problem 13.22**
>
> *Bak and Newman Chapter 13 Exercise 22.**

Proof. Applying the Schwarz-Christoffel transformation and idea as well as the notations of [4, Sect. III, pp. 191, 192], we choose $a_1 = -1$, $a_2 = 1$ and $a_3 > 1$. Since the exterior angles of an isosceles right triangle are $\frac{3\pi}{4}, \frac{3\pi}{4}$ and $\frac{\pi}{2}$, we obtain

$$\alpha_1 \pi = \frac{3\pi}{4}, \quad \alpha_2 \pi = \frac{3\pi}{4} \quad \text{and} \quad \alpha_3 \pi = \frac{\pi}{2}$$

which give $\alpha_1 = \alpha_2 = \frac{3}{4}$ and $\alpha_3 = \frac{1}{2}$. Thus the mapping

$$f(z) = \int_0^z \frac{d\zeta}{(\zeta+1)^{\frac{3}{4}} (\zeta-1)^{\frac{3}{4}} (\zeta-a_3)^{\frac{1}{2}}} \tag{13.5}$$

is a conformal mapping sends $\mathbb{H}$ onto an isosceles right triangle. In fact, we can omit the factor $(\zeta - a_3)^{\frac{1}{2}}$ in the formula (13.5), so we can reduce the formula (13.5) to

$$f(z) = \int_0^z \frac{d\zeta}{(\zeta^2 - 1)^{\frac{3}{4}}}$$

which is our required mapping. This ends the proof of the problem.

> **Problem 13.23**
>
> *Bak and Newman Chapter 13 Exercise 23.**

Proof. Take $a = -1$, $b = 0$, $c = 1$ and replace d by ∞ in [4, Eqn. (3), p. 189] to get

$$f(z) = \int_0^z \frac{d\zeta}{\sqrt{\zeta(1-\zeta^2)}},$$

which sends $\mathbb{H}$ conformally onto a rectangle, where the branch of $\sqrt{\zeta(1-\zeta^2)}$ is taken to be the one that is positive when $0 < \zeta < 1$, see Figure 13.5 below:

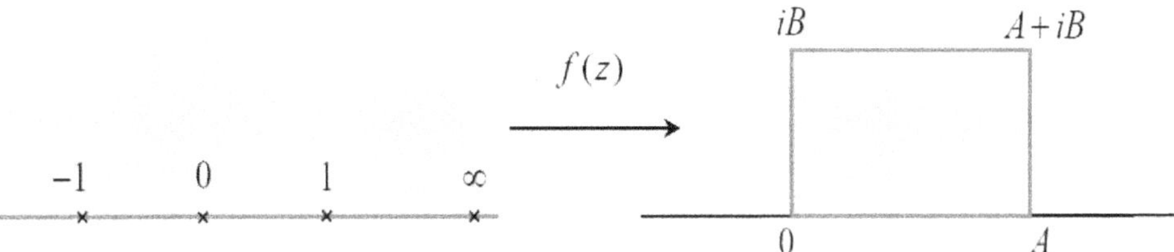

Figure 13.5: The conformal mapping of $\mathbb{H}$ onto a rectangle.

[b]The reader is suggested to refer to [13, Maps 5.53, 5.54, pp. 173, 174].

Recall from [4, p. 191] that the point at infinity is mapped by f onto one of the vertices of the rectangle. We claim that this rectangle is actually a square. To this end, we know that $f(0) = 0$ and

$$f'(z) = \frac{1}{\sqrt{z(1 - z^2)}} > 0$$

on the interval $0 < z < 1$, so f maps the interval $[0, 1]$ onto the interval $[0, A]$, i.e., $f(1) = A$ or

$$A = \int_0^1 \frac{\mathrm{d}x}{\sqrt{x(1 - x^2)}}.$$

As z crosses over the point 1, the argument of $f'(z)$ will increase by $\frac{\pi}{2}$. Next, we consider the path of integration $[0, \infty]$. As z crosses "over" the point ∞ (from ∞ to $-\infty$), the argument of $f'(z)$ will also increase by $\frac{\pi}{2}$ so that $f(\infty) = A + iB$. Since

$$A + iB = f(\infty) = f(1) + \int_1^\infty \frac{\mathrm{d}x}{\sqrt{x(1 - x^2)}},$$

we get

$$iB = \int_1^\infty \frac{\mathrm{d}x}{\sqrt{x(x^2 - 1)}}. \tag{13.6}$$

Applying the substitution $x = \frac{1}{y}$ to the integral (13.6), it can be shown easily that $B = A$ which means that the rectangle is in fact a square. This completes the proof of the problem. ∎

CHAPTER 14

The Riemann Mapping Theorem

Problem 14.1

Bak and Newman Chapter 14 Exercise 1.

Proof. By the discussion on [4, p. 195], we know that

$$\int_C \overline{g(\zeta)}\, \mathrm{d}\zeta = 0 \tag{14.1}$$

for any closed curve C in D, so by using similar argument as in the proof of Theorem 8.5, we conclude that our F is well-defined and analytic in D.

Let $z_1, z_2 \in \mathbb{C}$ and γ be a (smooth) curve from z_1 to z_2. Furthermore, suppose that γ_1 and γ_2 are curves from z_0 to z_1 and z_2 respectively. Then $\gamma_1 + \gamma - \gamma_2$ forms a closed curve and so we observe from the equation (14.1) that

$$\int_{\gamma_1} \overline{g(\zeta)}\, \mathrm{d}\zeta - \int_{\gamma_2} \overline{g(\zeta)}\, \mathrm{d}\zeta = -\int_{\gamma} \overline{g(\zeta)}\, \mathrm{d}\zeta$$

or equivalently,

$$F(z_2) - F(z_1) = \int_{\gamma} \overline{g(\zeta)}\, \mathrm{d}\zeta. \tag{14.2}$$

If we write $g = u + iv$ and $z(t) = x(t) + iy(t)$, then we apply Definition 4.3 to the expression (14.2) to get

$$
\begin{aligned}
F(z_2) - F(z_1) &= \int_{t_1}^{t_2} (u - iv)\dot{z}(t)\, \mathrm{d}t \\
&= \int_{t_1}^{t_2} (u - iv)(\,\mathrm{d}x + i\,\mathrm{d}y) \\
&= \int_{t_1}^{t_2} (u\,\mathrm{d}x + v\,\mathrm{d}y) + i \int_{t_1}^{t_2} (u\,\mathrm{d}y - v\,\mathrm{d}x),
\end{aligned}
\tag{14.3}
$$

where t_1 and t_2 are the values such that $z_1 = z(t_1)$ and $z_2 = z(t_2)$.

If $\operatorname{Re} F(z) = C_1$ for some constant C_1 on γ, then the representation (14.3) implies that

$$\int_{t_1}^{t_2} (u\,\mathrm{d}x + v\,\mathrm{d}y) = 0 \tag{14.4}$$

187

and if we represent $g = u + iv$ in the vector form (u, v), then the result (14.4) means that the velocity vector (u, v) is orthogonal to the tangle vector $(x'(t), y'(t))$, completing the proof of the problem.

Problem 14.2

Bak and Newman Chapter 14 Exercise 2.

Proof. It follows from the representation (14.3) that if $\operatorname{Im} F(z) = C_2$ for a constant C_2, then we have

$$\int_{t_1}^{t_2} (u\,dy - v\,dx) = 0$$

which means that the velocity vector (u, v) is orthogonal to the vector $(-y'(t), x'(t))$. Since $(x'(t), y'(t)) \cdot (-y'(t), x'(t)) = 0$, the flow g is parallel to the tangent vector $(x'(t), y'(t))$, i.e., g is tangent to $\operatorname{Im} F(z) = C_2$. This completes the proof of the problem,

---

Problem 14.3

Bak and Newman Chapter 14 Exercise 3.

Proof.

(a) By Problem 14.1, we take $z_0 = 0$ to get

$$F(z) = \int_0^z \overline{g}(\zeta)\,d\zeta = \int_0^z \zeta\,d\zeta = \frac{\zeta^2}{2}\Big|_0^z = \frac{z^2}{2}. \tag{14.5}$$

Let $c \in \mathbb{R}$. If $z = x + iy$, then $z^2 = x^2 - y^2 + 2ixy$. By Problem 14.2, the streamlines of g is $\operatorname{Im} F(z) = c$. Therefore, we follow from the expression (14.5) that the streamlines of g are

$$xy = c.$$

(b) For simplicity, we take $z_0 = 1$ so that

$$F(z) = \int_1^z \overline{g}(\zeta)\,d\zeta = \int_1^z \frac{d\zeta}{\zeta} = \log z = \log|z| + i\operatorname{Arg} z.$$

In this case, $\operatorname{Im} F(z) = \operatorname{Arg} z$. Hence the streamlines of g are rays from the origin. We end the proof of the problem.

Problem 14.4

Bak and Newman Chapter 14 Exercise 4.

Proof. We follow the given hint. Let C be a real constant. Let D be the exterior of $D(0; 1)$ and for simplicity, we pick $I = [0, 1]$ to be the interval for consideration. Suppose that $F : D \to \mathbb{C} \setminus I$ is a conformal mapping with $F(z) \sim z$ as $z \to \infty$. Thus we may suppose that

$$F(z) = z + A_0 + \frac{A_1}{z} + \frac{A_2}{z^2} + \cdots,$$

where $A_k \in \mathbb{C}$ for all $k = 0, 1, 2, \ldots$. If $z = e^{i\theta}$ with $\theta \in [-\pi, \pi]$, then we notice that

$$\begin{aligned}
A_k z^{-k} &= (\operatorname{Re} A_k + i\operatorname{Im} A_k)(\cos k\theta - i\sin k\theta) \\
&= (\operatorname{Re} A_k \cos k\theta + \operatorname{Im} A_k \sin k\theta) + i(-\operatorname{Re} A_k \sin k\theta + \operatorname{Im} A_k \cos k\theta).
\end{aligned} \tag{14.6}$$

Now $\operatorname{Im} F(e^{i\theta}) = C$, so the expression (14.6) implies that

$$\sin\theta + \operatorname{Im} A_0 + \sum_{k=1}^{\infty}(-\operatorname{Re} A_k \sin k\theta + \operatorname{Im} A_k \cos k\theta) = C. \tag{14.7}$$

By the identities $\sin z = \frac{1}{2i}(e^{iz} - e^{-iz})$ and $\cos z = \frac{1}{2}(e^{iz} + e^{-iz})$, we can rewrite (14.7) as

$$\sum_{-\infty}^{\infty} c_k e^{ik\theta} = C, \tag{14.8}$$

where

$$c_k = \begin{cases}
\dfrac{\operatorname{Im} A_k + i\operatorname{Re} A_k}{2}, & \text{if } k \geq 2; \\[2mm]
\dfrac{\operatorname{Im} A_1 - (1 - \operatorname{Re} A_1)i}{2} & \text{if } k = 1; \\[2mm]
\operatorname{Im} A_0 & \text{if } k = 0; \\[2mm]
\dfrac{\operatorname{Im} A_1 + (1 - \operatorname{Re} A_1)i}{2} & \text{if } k = -1; \\[2mm]
\dfrac{\operatorname{Im} A_{|k|} - i\operatorname{Re} A_{|k|}}{2}, & \text{if } k \leq -2.
\end{cases} \tag{14.9}$$

We apply the facts[a]

$$\frac{1}{2\pi}\int_{-\pi}^{\pi} e^{inx}\, dx = \begin{cases} 1, & \text{if } n = 0; \\ 0, & \text{if } n = \pm 1, \pm 2, \ldots \end{cases}$$

to the representation (14.8) to get

$$c_k = \begin{cases} C, & \text{if } k = 0; \\ 0, & \text{if } k \in \mathbb{Z} \setminus \{0\}. \end{cases}$$

Therefore, the definition (14.9) asserts that $\operatorname{Re} A_1 = 1$, $\operatorname{Re} A_k = 0$ for all $k \geq 2$ and $\operatorname{Im} A_k = 0$ for all $k \geq 1$. In other words, we have established that

$$F(z) = z + A_0 + \frac{1}{z}.$$

This completes the proof of the problem. ∎

> **Remark 14.1**
>
> In fact, the mapping in Problem 14.5 is classically called the **Joukowsky transform**, see [3, Example 7.1.8, pp. 408-410] or [9, Exercise 15, pp. 66, 67].

---

[a] See [22, Eqn. (61) & (62), pp. 185, 186]

Problem 14.5

Bak and Newman Chapter 14 Exercise 5.

Proof.

(a) Let $f(z) = 2z + \frac{1}{z}$ and let D and D' be the exteriors of $D(0;1)$ and the said ellipse respectively. Since $f'(z) = 2 - \frac{1}{z^2}$, $f'(z) = 0$ if and only if $z = \pm\frac{1}{\sqrt{2}} \notin D$. Thus f is conformal throughout D by Theorem 13.4. Take $z = e^{i\theta}$, so

$$f(e^{i\theta}) = 2(\cos\theta + i\sin\theta) + (\cos\theta - i\sin\theta) = 3\cos\theta + i\sin\theta$$

which means that

$$f(C(0;1)) = \left\{(x,y)\,\middle|\,\frac{x^2}{9} + y^2 = 1\right\}.$$

As a conformal map, f is continuous on the connected set D so that $f(D)$ is connected. In addition, since $f(z) \to \infty$ as $z \to \infty$, f must map D into D'. It remains to show that the mapping $f : D \to D'$ is surjective, but it is easy to see because if $(x,y) \in D'$ satisfies $\frac{x^2}{9} + y^2 > 1$, then the point $z = \frac{x}{3} + iy$ lies in D. Hence the f is actually a surjective conformal mapping.

(b) Without loss of generality, we let the real line segment be $[-2,2]$. By [4, Example 2, p. 197], $g(z) = z + \frac{1}{z}$ maps D conformally onto $\mathbb{C}\setminus[-2,2]$. Thus the composition

$$g \circ f^{-1} : D' \to \mathbb{C}\setminus[-2,2]$$

is a desired conformal mapping.

Consequently, we have completed the proof of the problem. ■

Problem 14.6

Bak and Newman Chapter 14 Exercise 6.

Proof. Since $f(z_0) \in D(0;1)$, Theorem 13.15 implies that the mapping of the form

$$h(z) = e^{i\theta}\left(\frac{z - f(z_0)}{1 - \overline{f(z_0)}z}\right) \tag{14.10}$$

is an automorphism of $D(0;1)$, where $\theta \in \mathbb{R}$. By Problem 13.7, the composition $g = h \circ f$ is a conformal mapping of R onto $D(0;1)$. Obviously, the expression (14.10) gives

$$g(z_0) = h(f(z_0)) = 0.$$

Furthermore, direct computation gives

$$g'(z) = e^{i\theta} \cdot \frac{[1 - \overline{f(z_0)} \cdot f(z)]f'(z) - [f(z) - f(z_0)] \cdot [-\overline{f(z_0)} \cdot f'(z)]}{[1 - \overline{f(z_0)} \cdot f(z)]^2}$$

so that

$$g'(z_0) = e^{i\theta} \cdot \frac{f'(z_0)}{1 - |f(z_0)|^2}.$$

Now if we take $\theta = -\operatorname{Arg} f'(z_0)$, then we obtain the desired result that $g'(z_0) > 0$. We end the proof of the problem. ■

Proof. We define $g : R \to U$ by $g(z) = \overline{f(\overline{z})}$. We claim that g is analytic in R. To this end, let $z = x + iy \in R$ and $f(z) = u(x,y) + iv(x,y)$, where u and v are real-valued functions. Then we have

$$g(z) = \overline{f(\overline{z})} = \overline{u(x,-y) + iv(x,-y)} = u(x,-y) - iv(x,-y) = U(x,y) + iV(x,y),$$

where $U(x,y) = u(x,-y)$ and $V(x,y) = -v(x,-y)$. Since f is analytic in R, Theorem 6.6 and Corollary 2.10 together imply that u and v have C^1 partial derivatives. In particular, $g_x = U_x + iV_x$ and $g_y = U_y + iV_y$ are continuous. By Proposition 3.1, we have

$$\frac{\partial u(x,y)}{\partial x} = \frac{\partial v(x,y)}{\partial y} \quad \text{and} \quad \frac{\partial u(x,y)}{\partial y} = -\frac{\partial v(x,y)}{\partial x}$$

which give

$$\frac{\partial U(x,y)}{\partial x} = \frac{\partial u(x,-y)}{\partial x} = \frac{\partial v(x,-y)}{\partial(-y)} = \frac{\partial V(x,y)}{\partial y}$$

and

$$\frac{\partial V(x,y)}{\partial x} = \frac{\partial(-v(x,-y))}{\partial x} = \frac{\partial u(x,-y)}{\partial(-y)} = -\frac{\partial U(x,y)}{\partial y}$$

Hence it follows from Proposition 3.2 that g is analytic at z, so the claim is true.

Since R is symmetric respect to $\mathbb{R}$, it is easy to see that the map $z \mapsto \overline{z}$ is bijective in R. Since $f : R \to U$ is the Riemann mapping, it is bijective. Finally, since g is the composition of bijective maps, it is also bijective. By Definition 13.9, g is a conformal mapping. Recall that z_0 is real, thus we obtain

$$g(z_0) = \overline{f(\overline{z_0})} = \overline{f(z_0)} = 0.$$

Furthermore, for every $\omega \in R$, the definition of analyticity of g guarantees that

$$\begin{aligned}
g'(\omega) &= \lim_{z \to \omega} \frac{g(z) - g(\omega)}{z - \omega} \\
&= \lim_{z \to \omega} \frac{\overline{f(\overline{z})} - \overline{f(\overline{\omega})}}{z - \omega} \\
&= \lim_{z \to \omega} \frac{\overline{f(\overline{z})} - \overline{f(\overline{\omega})}}{\overline{\overline{z} - \overline{\omega}}} \\
&= \overline{\left[\lim_{z \to \omega} \frac{f(\overline{z}) - f(\overline{\omega})}{\overline{z} - \overline{\omega}} \right]} \\
&= \overline{\left[\lim_{\zeta \to \overline{\omega}} \frac{f(\zeta) - f(\overline{\omega})}{\zeta - \overline{\omega}} \right]} \\
&= \overline{f'(\overline{\omega})}.
\end{aligned} \tag{14.11}$$

Particularly, if we take $\omega = z_0$, then we deduce from the expression (14.11) and the hypotheses that $g'(z_0) = \overline{f'(\overline{z_0})} = f'(z_0) > 0$. By the Riemann Mapping Theorem, the mapping f is unique, so we conclude that $f(z) = g(z)$ which means $f(\overline{z}) = \overline{f(z)}$ for all $z \in R$, as desired. This completes the proof of the problem. ∎

Proof. By Theorem 13.16, every conformal mapping f of $\mathbb{H}$ onto $D(0;1)$ is of the form

$$f(z) = e^{i\theta}\left(\frac{z - \alpha}{z - \overline{\alpha}}\right), \tag{14.12}$$

where $\operatorname{Im}\alpha > 0$.

(a) Since $f(-1) = 1$, $f(0) = i$ and $f(1) = -1$, we deduce from the formula (14.12) that

$$e^{i\theta}\left(\frac{1 + \alpha}{1 + \overline{\alpha}}\right) = 1, \quad e^{i\theta}\cdot\frac{\alpha}{\overline{\alpha}} = i \quad \text{and} \quad e^{i\theta}\left(\frac{1 - \alpha}{1 - \overline{\alpha}}\right) = -1. \tag{14.13}$$

Suppose that $\alpha = a + ib$. Then the first and the third equations (14.13) yield that $a^2 + b^2 = 1$. Besides, the first two equations (14.13) imply that

$$a - b + 1 = 0. \tag{14.14}$$

Finally, the last two equations (14.13) establish that

$$a + b - 1 = 0. \tag{14.15}$$

Solving the equations (14.14) and (14.15), we get $a = 0$ and $b = 1$, i.e., $\alpha = i$. Using the second expression (14.13), we find that $e^{i\theta} = -i$. Hence the desired conformal mapping is given by

$$f(z) = -i\cdot\frac{z - i}{z + i} = -\frac{iz + 1}{z + i}.$$

(b) The condition $f(i) = 0$ implies that $\alpha = i$. Next, the condition $f(1) = 1$ implies that $e^{i\theta} = i$. Hence the required conformal mapping is

$$f(z) = i\cdot\frac{z - i}{z + i} = \frac{iz + 1}{z + i}.$$

We have completed the proof of the problem. $\blacksquare$

> **Problem 14.9**
>
> *Bak and Newman Chapter 14 Exercise 9.*

Proof. If $R = \mathbb{C}$, then the mapping $f : \mathbb{C} \to \mathbb{C}$ defined by $f(z) = z - z_1 + z_2$ is conformal at every $z \in \mathbb{C}$ because $f'(z) = 1$ by Theorem 13.4. Since $f(z_1) = z_2$, it satisfies our requirements.

Next, we suppose that $R \neq \mathbb{C}$. Since R is simply connected, the Riemann Mapping Theorem ensures that there exist conformal mapping $\varphi_1 : R \to D(0;1)$ and $\varphi_2 : R \to D(0;1)$ such that $\varphi_1(z_1) = \varphi_2(z_2) = 0$. By Problem 13.7, the mapping $\varphi = \varphi_2^{-1} \circ \varphi_1 : R \to R$ is a conformal mapping and we have

$$\varphi = \varphi_2^{-1}(\varphi_1(z_1)) = \varphi_2^{-1}(0) = z_2,$$

completing the analysis of the problem. $\blacksquare$

> **Problem 14.10**
>
> *Bak and Newman Chapter 14 Exercise 10.*

Proof. Assume that $f : \mathbb{C} \to R$ was a conformal mapping. By the Riemann Mapping Theorem, there exists a conformal mapping $g : R \to D(0;1)$. By Problem 13.7, $h = g \circ f : \mathbb{C} \to D(0;1)$ is conformal, i.e., h is a bounded entire function which certainly contradicts Theorem 5.10 (Liouville's Theorem). This completes the proof of the problem. $\blacksquare$

Problem 14.11

Bak and Newman Chapter 14 Exercise 11.

Proof.

(a) Note that $G = \{g : R \to U \mid g$ is analytic and $g'(z_0) > 0\}$. By the proof of Part (B) of the Riemann Mapping Theorem [4, p. 202], for every $g \in G$, we see easily that

$$g'(z_0) = |g'(z_0)| < \frac{1}{\delta},$$

where δ is a positive number independent of g such that $D(z_0;\delta) \subset R$. In conclusion, we find that

$$\sup_{g \in G} g'(z_0) = M^* \le \frac{1}{\delta}.$$

(b) Following the same argument as the proof of Part (B), we can verify that there exists a function $\Phi \in G$ such that $\Phi'(z_0) = M^*$. Next, if $\Phi(z_0) = \alpha$ with $0 < |\alpha| < 1$, then the map $\widehat{g} : R \to U$ defined by

$$\widehat{g}(z) = \frac{\Phi(z) - \alpha}{1 - \overline{\alpha}\Phi(z)}$$

is clearly analytic, i.e., $\widehat{g} \in G$. Besides, we have

$$\widehat{g}'(z_0) = \frac{\Phi'(z_0)}{1 - |\alpha|^2} > \Phi'(z_0),$$

a contradiction. Therefore, we conclude that $\Phi(z_0) = 0$. Let $\varphi : R \to U$ be the Riemann mapping function with $\varphi(z_0) = 0$ and $\varphi'(z_0) = M > 0$. Notice that $\mathcal{F} \subseteq G$, so it happens that $\Phi'(z_0) = M^* \ge M > 0$. Now we consider the map

$$f = \Phi \circ \varphi^{-1} : U \to U.$$

As a composition of two analytic functions, f is also analytic in U and $|f(z)| < 1$ on U. Simple algebra shows that

$$f(0) = \Phi(\varphi^{-1}(0)) = \Phi(z_0) = 0.$$

Consequently, our f satisfies the hypotheses of Theorem 7.2 (Schwarz's Lemma), so we see that $|f(z)| \le |z|$ and $|f'(0)| \le 1$.

On the other hand, by combining Proposition 3.5 and Problem 3.3, we know that

$$f'(0) = \frac{\Phi'(z_0)}{\varphi'(z_0)} = \frac{M^*}{M} \ge 1.$$

Hence according to Theorem 7.2 (Schwarz's Lemma), it is true that $f(z) = z$, i.e., $f = \mathrm{id}$. By the definition, we finally get

$$\Phi = \varphi$$

so that Φ is injective.

We complete the analysis of the problem. $\blacksquare$

Problem 14.12

*Bak and Newman Chapter 14 Exercise 12.**

Proof.

(a) By similar idea as the proof of Problem 13.6, the mapping $f : S \to U$ defined by

$$
\begin{aligned}
f(z) &= \frac{-(1+i)z^2 + 2(1-i)z - (1+i)}{-(1-i)z^2 + 2(1+i)z - (1-i)} \\[2mm]
&= \frac{z^2 - 2(\frac{1-i}{1+i})z + 1}{\frac{1-i}{1+i}z^2 - 2z + \frac{1-i}{1+i}} \\[2mm]
&= \frac{z^2 + 2iz + 1}{-iz^2 - 2z - i} \\[2mm]
&= i \cdot \frac{z^2 + 2iz + 1}{z^2 - 2iz + 1}
\end{aligned}
\tag{14.16}
$$

is conformal. Since S and U are Jordan regions by Definition 14.2, we follow from Theorem 14.3 (The Carathéodory-Osgood Theorem) that f can be extended to a homeomorphism between $\overline{S}$ and $\overline{U}$.

(b) Since the roots of $z^2 - 2i + 1 = 0$ are $z = (1 \pm \sqrt{2})i \notin \overline{S}$, the representation (14.16) ensures that f is analytic on $\overline{S}$. However, it follows from the representation (14.16) that the inverse $f^{-1} : U \to S$ has the form

$$
f^{-1}(z) = \frac{(-1 + iz) + \sqrt{2 - 2z^2}}{i - z}
$$

which has a simple pole at $-i \in \overline{U}$, so f^{-1} is not analytic on $\overline{U}$.

This completes the proof of the problem. $\blacksquare$

Problem 14.13

*Bak and Newman Chapter 14 Exercise 13.**

Proof. Figure 14.1 shows the shape of a Norman window N. Without loss of generality, we may suppose that 0 lies on ∂N.

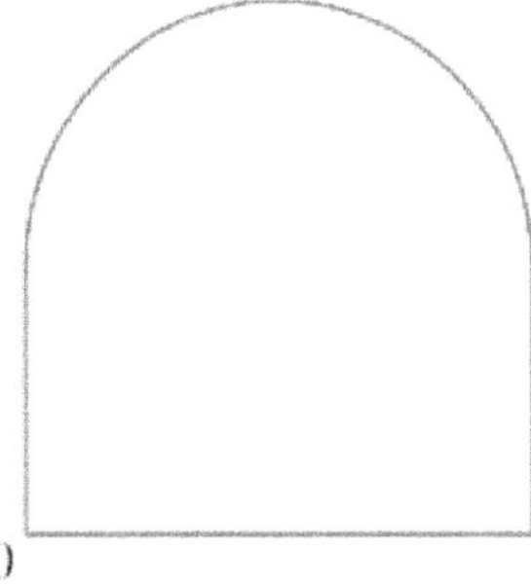

Figure 14.1: The shape of a Norman window N.

Assume that $f : \overline{U} \to N$ was analytic and surjective. By Theorem 7.1 (The Open Mapping Theorem)[b], f will map a $z_0 \in C(0;1)$ onto the origin. Now we may further suppose that $z_0 = 1$. If $f'(1) \neq 0$, then Theorem 13.4 implies that f maps the rays $I_1 = \{1 + it \mid 0 \leq t \leq \infty\}$ and $I_2 = \{1 - it \mid 0 \leq t \leq \infty\}$ onto two curves whose tangent lines form a straight angle,

$$\angle f(I_1), f(I_2) = \pi, \tag{14.17}$$

but it is impossible because the angle at 0 is just $\frac{\pi}{2}$. Next, if $f'(1) = 0$, then Theorem 13.7 will ensure that, instead of the value (14.17), we have

$$\angle f(I_1), f(I_2) = k\pi,$$

where k is the least positive integer such that $f^{(k)}(1) \neq 0$. However, this is also impossible. Hence no such analytic and surjective map f exists. This completes the proof of the problem.

---

[b]See also the proof of Problem 13.8.

Maximum-Modulus Theorems for Unbounded Domains

Problem 15.1

Bak and Newman Chapter 15 Exercise 1.[*]

Proof. Without loss of generality, we may assume that f is non-constant. We basically follow the proof of Theorem 15.1. Suppose first that $D = \{z \in \mathbb{C} \mid \operatorname{Re} z > 0\}$, $f(z) \ll 1$ on ∂D, $f(z) \ll \log z$ for all $z \in D$ and $z_0 \in D$. Consider the auxiliary function

$$h(z) = \frac{f^N(z)}{z+1}, \tag{15.1}$$

where $N \in \mathbb{N}$. By the hypothesis, we have $|h(z)| \leq 1$ along the imaginary axis. In addition, if $z = x + iy \in D$, then $x > 0$ and so

$$|z+1| = \sqrt{x^2 + y^2 + 2x + 1} \geq \sqrt{x^2 + y^2} = |z|.$$

Therefore, for all $z \in D$ with $|z| = R > 0$, we have

$$|h(z)| \leq \frac{|\log z|^N}{R} \leq \frac{[(\log R)^2 + (\operatorname{Arg} z)^2]^{\frac{N}{2}}}{R}. \tag{15.2}$$

Since $-\pi \leq \operatorname{Arg} z \leq \pi$, we can take R large enough so that the bound (15.2) reduces to

$$|h(z)| \leq \frac{\sqrt{2^N}(\log R)^N}{R} \leq 1 \tag{15.3}$$

for all z in the right semi-disc $D_R = \{z \in D \mid |z| \leq R\}$. Since h is certainly analytic in D_R and D_R is compact, Theorem 6.13 (The Maximum Modulus Theorem) implies that $|h(z_0)| \leq 1$. By the definition (15.1), we achieve

$$\left| \frac{f^N(z_0)}{z_0 + 1} \right| \leq 1$$

or equivalently,

$$|f(z_0)| \leq |1 + z_0|^{\frac{1}{N}}.$$

Since N is arbitrary, we take $N \to \infty$ to conclude that $|f(z_0)| \leq 1$ as desired.

Next, we suppose that D is an arbitrary region. Then we consider

$$g(z) = \frac{f(z) - f(a)}{z - a},$$

where a is a fixed point in D. By Proposition 6.7, g is C-analytic in D. Furthermore, the hypothesis implies that, for large $|z|$,

$$|g(z)| = \frac{|f(z) - f(a)|}{|z - a|} \le \frac{|\log z| + |\log a|}{|z| - |a|}$$

so that $g(z) \to 0$ as $z \to \infty$. Consequently, there exists a constant $M > 0$ such that $|g(z)| \le M$ on $\overline{D}$. Similar to the previous paragraph, we set $D_R = \{z \in D \mid |z| \le R\}$ and $h(z) = f^N(z)g(z)$, where $R > 0$ and $N \in \mathbb{N}$. Thus it must be true that $|h(z)| \le M$ on ∂D_R for sufficiently large R. Hence it follows from Theorem 6.13 (The Maximum Modulus Theorem) that $|h(z_0)| \le M$ for every $z_0 \in D$. If $g(z_0) \ne 0$, then we get

$$|f(z_0)| \le \left| \frac{h(z)}{g(z)} \right|^{\frac{1}{N}} \le \frac{M^{\frac{1}{N}}}{|g(z_0)|^{\frac{1}{N}}}.$$

By taking $N \to \infty$, we get $|f(z_0)| \le 1$. According to Theorem 6.9 (The Uniqueness Theorem), since f is nonconstant, the set $S = \{z \in D \mid g(z) = 0\}$ is discrete, so we must have

$$|f(z)| \le 1 \tag{15.4}$$

on $D \setminus S$. Finally, the analyticity of f ensures that the bound (15.4) holds in D. Hence our conclusion of Theorem 15.1 is still valid in this case.

We see that our argument works because of the inequality (15.3). Thus if $f(z) \ll P(\log z)$ in D, where $P(\omega)$ is a polynomial in ω of degree $n \ge 1$, then there exists a positive constant M such that

$$|h(z)| \le \frac{M^N (\log R)^{nN}}{R} \le 1$$

holds in $D_R = \{z \in D \mid |z| \le R\}$ for sufficiently large enough R. This completes the proof of the problem. ◼

> **Problem 15.2**
>
> *Bak and Newman Chapter 15 Exercise 2.*

Proof. We consider

$$D_1 = \{z \in \mathbb{C} \mid -\frac{\pi}{4} < \operatorname{Arg} z < \frac{\pi}{4}\} \quad \text{and} \quad F(z) = \exp(z^2).$$

On $z = r\exp(\pm\frac{\pi}{4}i)$, we see that $z^2 = \pm ir^2$ so that $|F(z)| = |\exp(\pm ir^2)| = 1$, i.e., F is bounded on ∂D_1. However, it is easy to check that

$$F(x) = e^{x^2} \to \infty$$

as $x \to \infty$. Consequently, F is unbounded in D_1.

Suppose that $\mathscr{F}$ is the class of all functions f analytic in $\overline{D_1}$ such that f is bounded on ∂D_1 but $f(z)$ is unbounded in D_1. Let $f(z)$ be an analytic function "smaller" than $F(z)$ in D_1 in the sense that for every $\epsilon > 0$, there exists a constant A_ϵ such that

$$|f(z)| \le A_\epsilon \exp(\epsilon|z|^2)$$

for all $z \in D_1$. Now Corollary 15.5 with $\alpha = \frac{\pi}{2}$ implies immediately that this function f must be bounded in the region D_1. Combining this fact and the observation in the previous paragraph, we conclude that $F \in \mathscr{F}$ but $f \notin \mathscr{F}$. Hence $F(z) = e^{z^2}$ is the "smallest" analytic function in this sense.

By replacing z by $e^{-i\frac{3\pi}{4}} z$, we see at once that the function

$$G(z) = F(e^{-i\frac{3\pi}{4}} z) = \exp(e^{-i\frac{3\pi}{2}} z^2) = e^{iz^2}$$

is the desired "smallest" non-constant analytic function in D, completing the proof of the problem.

> **Problem 15.3**
>
> *Bak and Newman Chapter 15 Exercise 3.*

Proof. Let $F(z) = \exp(e^z)$ and $D = \{x + iy \mid x \in \mathbb{R}$ and $-\frac{\pi}{2} < y < \frac{\pi}{2}\}$. Clearly, F is C-analytic in D. If $z = x \pm i\frac{\pi}{2}$, then we have

$$e^z = \pm ie^x.$$

In other words, e^z maps ∂D onto the imaginary axis which implies that

$$|F(z)| = |\exp(e^z)| = |\exp(\pm ie^x)| = 1$$

on ∂D, proving the first assertion. However, we note that

$$F(\log x) = \exp(e^{\log x}) = e^x \to \infty$$

as $x \to \infty$. In other words, F is unbounded in the region D.

Suppose that $f(z)$ is an analytic function "smaller" than $F(z)$ in D in the sense that for every $\epsilon > 0$, there exists a constant A_ϵ such that

$$|f(z)| \le A_\epsilon \exp(\epsilon |e^z|)$$

for all $z \in D$. In this case, the analytic function $g(z) = f(\log z)$ satisfies

$$|g(z)| \le A_\epsilon e^{\epsilon |z|}$$

in the region D. By Theorem 15.4 (The Phragmén-Lindelöf Theorem), $g(z)$ and then $f(z)$ are also bounded throughout $\overline{D}$.

Suppose that $\mathscr{F}$ is the class of all functions f analytic in $\overline{D}$ such that f is bounded on ∂D but $f(\log z)$ is unbounded in D. Then the first assertion says immediately that $F \in \mathscr{F}$ but the previous paragraph means that any analytic function f "smaller" than F implies that $f \notin \mathscr{F}$. This completes the proof of the problem.

> **Problem 15.4**
>
> *Bak and Newman Chapter 15 Exercise 4.*

Proof. Assume that there were constants A and B such that

$$|g(z)| \le A \exp(|z|^B) \tag{15.5}$$

in $\mathbb{C}$.[a] As given by the hint, we choose a positive integer N such that $N > 2B$ and then we divide the plane into N wedges of equal angles $\alpha = \frac{\pi}{N}$. Let $W_1, W_2, \dots, W_N$ be these wedges. Since g is entire, it is C-analytic in each W_k. Next, the hypothesis ensures that

$$|g(z)| \le M_k \tag{15.6}$$

on ∂W_k for some positive constant M_k. Furthermore, it is clear from the assumption (15.5) that

$$|g(z)| \le A\exp(|z|^B) \le A\exp(|z|^{\frac{N}{2}}) = A\exp(|z|^{\frac{\pi}{2\alpha}}). \tag{15.7}$$

Given $\epsilon > 0$. Then the inequality $|z|^{\frac{\pi}{2\alpha}} \le \epsilon |z|^{\frac{\pi}{\alpha}}$ holds if and only if $|z| \ge \epsilon^{-\frac{2\alpha}{\pi}}$. Thus the inequality (15.7) gives

$$|g(z)| \le A\exp(\epsilon |z|^{\frac{\pi}{\alpha}}) \tag{15.8}$$

if $z \in W_k$ and $|z| \ge \epsilon^{-\frac{2\alpha}{\pi}}$. If $z \in W_k$ and $|z| < \epsilon^{-\frac{2\alpha}{\pi}}$, then since e^x is strictly increasing, it is easy to see from the inequality (15.7) that

$$|g(z)| \le A\exp(\epsilon^{-1}). \tag{15.9}$$

Let $A_\epsilon = \max(A, A\exp(\epsilon^{-1}))$. We conclude from the inequalities (15.8) and (15.9) that

$$|g(z)| \le A_\epsilon \exp(\epsilon |z|^{\frac{\pi}{\alpha}})$$

for all $z \in W_k$. In other words, our function g satisfies the hypotheses of Corollary 15.5. Hence the inequality (15.6) holds throughout W_k and then g is bounded in $\mathbb{C}$. Therefore, Theorem 5.10 (Liouville's Theorem) implies that g is constant which contradicts our hypothesis. This completes the analysis of the problem.

---

[a] By Definition 16.12, g is of finite order.

Harmonic Functions

Problem 16.1

Bak and Newman Chapter 16 Exercise 1.

Proof. Since f is analytic, Theorem 16.2 says that u and v are harmonic, i.e, $u_{xx} + u_{yy} = 0$ and $v_{xx} + v_{yy} = 0$. Since

$$(u + v)_{xx} + (u + v)_{yy} = (u_{xx} + u_{yy}) + (v_{xx} + v_{yy}) = 0,$$

$u+v$ is harmonic by Definition 16.1. By direct computation, we have $(uv)_{xx} = u_{xx}v + 2u_x v_x + uv_{xx}$ and $(uv)_{yy} = u_{yy}v + 2u_y v_y + uv_{yy}$ whose sum is

$$(uv)_{xx} + (uv)_{yy} = 2(u_x v_x + u_y v_y). \tag{16.1}$$

By Proposition 3.1, the equation (16.1) reduces to 0, so uv is also harmonic by Definition 16.1, completing the proof of the problem. ∎

Problem 16.2

Bak and Newman Chapter 16 Exercise 2.

Proof. Suppose that u is a harmonic function. By Theorem 16.3, u_x is the real part of an analytic function f. As an analytic function, f is infinitely differentiable. Hence u is also infinitely differentiable. Let $g = u_x$. Then we deduce from [28, Theorem 15.12, p. 146] that

$$g_{xx} + g_{yy} = \frac{\partial}{\partial x}u_{xx} + u_{yyx} = \frac{\partial}{\partial x}u_{xx} + u_{xyy} = \frac{\partial}{\partial x}(u_{xx} + u_{yy}) = 0.$$

Consequently, u_x is harmonic. Similarly, u_y is also harmonic and we have completed the proof of the problem. ∎

Problem 16.3

Bak and Newman Chapter 16 Exercise 3.

Proof. Let $f = u^2$. Obviously, we have $f_x = 2uu_x$ and $f_{xx} = 2(uu_{xx} + u_x^2)$. Similarly, we also have $f_{yy} = 2(uu_{yy} + u_y^2)$. Thus we obtain

$$f_{xx} + f_{yy} = 2u(u_{xx} + u_{yy}) + 2(u_x^2 + u_y^2) = 2(u_x^2 + u_y^2) \geq 0$$

so that $f_{xx} + f_{yy} = 0$ if and only if $u_x = u_y = 0$ if and only if u is a constant by Theorem 1.10, a contradiction. This ends the proof of the problem. $\blacksquare$

> ### Problem 16.4
>
> *Bak and Newman Chapter 16 Exercise 4.*

Proof. Let $u = \log(x^2 + y^2)$. Then we get $u_x = \frac{2x}{x^2+y^2}$ and

$$u_{xx} = \frac{2(y^2 - x^2)}{(x^2 + y^2)^2}. \tag{16.2}$$

Similarly, we get

$$u_{yy} = \frac{2(x^2 - y^2)}{(x^2 + y^2)^2}. \tag{16.3}$$

Thus the sum of the formulas (16.2) and (16.3) certainly imply that $u_{xx} + u_{yy} = 0$, i.e., u is harmonic in $D = \mathbb{C} \setminus \{0\}$.

Assume that u was the real part of an analytic function $f = u + iv$ in D. We note that $u = \log(x^2 + y^2) = 2\log|z|$, so we write $f = 2g$, where $g = \log|z| + i\frac{v}{2}$ is analytic in D. Therefore, g is an analytic branch of $\log z$ up to an imaginary constant in D. Next, recall from Problem 8.8 that an analytic branch of $\log z$ can be defined with $0 < \operatorname{Arg} z < 2\pi$ in the plane $\mathbb{C} \setminus [0, \infty)$. This fact forces that g is not continuous on $[0, \infty)$ which contradicts the assumption that g is analytic in D. Hence no such g and then f exists and we complete the analysis of the problem. $\blacksquare$

> ### Problem 16.5
>
> *Bak and Newman Chapter 16 Exercise 5.*

Proof.

(a) Note that $x = r\cos\theta$ and $y = r\sin\theta$, so we obtain

$$\frac{\partial x}{\partial r} = \cos\theta, \quad \frac{\partial x}{\partial \theta} = -r\sin\theta, \quad \frac{\partial y}{\partial r} = \sin\theta \quad \text{and} \quad \frac{\partial y}{\partial \theta} = r\cos\theta.$$

These imply that

$$\frac{\partial u}{\partial r} = \frac{\partial u}{\partial x} \cdot \frac{\partial x}{\partial r} + \frac{\partial u}{\partial y} \cdot \frac{\partial y}{\partial r} = \cos\theta \frac{\partial u}{\partial x} + \sin\theta \frac{\partial u}{\partial y} \tag{16.4}$$

and thus

$$\frac{\partial^2 u}{\partial r^2} = \cos^2\theta \frac{\partial^2 u}{\partial x^2} + 2\cos\theta\sin\theta \frac{\partial^2 u}{\partial x \partial y} + \sin^2\theta \frac{\partial^2 u}{\partial y^2}. \tag{16.5}$$

Similarly, we have

$$\frac{\partial u}{\partial \theta} = -r\sin\theta \frac{\partial u}{\partial x} + r\cos\theta \frac{\partial u}{\partial y}$$

so that

$$\frac{\partial^2 u}{\partial \theta^2} = -r\left(\cos\theta\frac{\partial u}{\partial x} + \sin\theta\frac{\partial u}{\partial y}\right) + r^2\left(\sin^2\theta\frac{\partial^2 u}{\partial x^2} - 2\cos\theta\sin\theta\frac{\partial^2 u}{\partial x\partial y} + \cos^2\theta\frac{\partial^2 u}{\partial y^2}\right). \tag{16.6}$$

Combining the expressions (16.4), (16.5) and (16.6), we conclude that

$$u_{rr} + \frac{1}{r}u_r + \frac{1}{r^2}u_{\theta\theta} = u_{xx} + u_{yy}. \tag{16.7}$$

Now if $u(r,\theta)$ is a harmonic function depending on r alone, then $u_{\theta\theta} = 0$ so that Laplace's equation (16.7) becomes

$$u_{rr} + \frac{1}{r}u_r = 0 \tag{16.8}$$

as desired.

(b) The differential equation (16.8) can be written in the form $(ru_r)_r = 0$ which means that $ru_r = a$ for some constant a, or equivalently $u_r = \frac{a}{r}$. By integration, we establish

$$u(r,\theta) = a\log r + b$$

for some constant b.

We complete the proof of the problem. ◾

Problem 16.6

Bak and Newman Chapter 16 Exercise 6.

Proof. We start with the following form of Poisson Formula [1, Theorem 22, p. 168]:

$$U(a) = \frac{1}{2\pi}\int_{|\zeta|=1}\frac{1-|a|^2}{|\zeta-a|^2}U(\zeta)\frac{\mathrm{d}\zeta}{i\zeta} \tag{16.9}$$

where U is C-harmonic in $D(0;1)$ and $a \in D(0;1)$. By Theorem 13.16, the map $\varphi : \mathbb{H} \to D(0;1)$ given by

$$\varphi(z) = \frac{z-i}{z+i} \tag{16.10}$$

is conformal and surjective. Besides, it is easy to check that φ maps $\partial\mathbb{H} = \mathbb{R}$ onto $C(0;1) \setminus \{1\}$. Now we consider the map $U = u \circ \varphi^{-1} : D(0;1) \to \mathbb{R}$. Since $\mathbb{H}$ is simply connected, Theorem 16.3 ensures that $u = \operatorname{Re} f$ for some analytic function f on $\mathbb{H}$. Write $f = u + iv$ so that

$$f \circ \varphi^{-1} = u \circ \varphi^{-1} + iv \circ \varphi^{-1}.$$

Since both f and φ^{-1} are analytic, its composition $f \circ \varphi^{-1} : D(0;1) \to \mathbb{C}$ is also analytic in $D(0;1)$ and we follow from Theorem 16.2 that $u \circ \varphi^{-1}$ is harmonic in $D(0;1)$. Since u is continuous on $\mathbb{R}$ and bounded, $u \circ \varphi^{-1}$ can be made to be continuous at 1 so that it is actually continuous on $C(0;1)$. In other words, $u \circ \varphi^{-1}$ is C-harmonic in $D(0;1)$. Thus we put $U = u \circ \varphi^{-1}$ into the formula (16.9) to get

$$u(\varphi^{-1}(a)) = \frac{1}{2\pi}\int_{|\zeta|=1}\frac{1-|a|^2}{|\zeta-a|^2}u(\varphi^{-1}(\zeta))\frac{\mathrm{d}\zeta}{i\zeta}. \tag{16.11}$$

Put $\zeta = \varphi(t)$ and $a = \varphi(z)$ into the formula (16.11) and then using the expression (16.10) and the fact $\varphi'(t) = \frac{2i}{(t+i)^2}$ to establish

$$u(x+iy) = u(z) = u(\varphi^{-1}(a)) = \frac{1}{2\pi}\int_{\mathbb{R}}\frac{1-|\varphi(z)|^2}{|\varphi(t)-\varphi(z)|^2}\cdot u(t)\cdot\frac{2}{t^2+1}\,\mathrm{d}t. \tag{16.12}$$

Since $\left|\frac{z-i}{z+i}\right|^2 = \frac{x^2+(y-1)^2}{x^2+(y+1)^2}$ and

$$|\varphi(t) - \varphi(z)|^2 = \left|\frac{t-i}{t+i} - \frac{z-i}{z+i}\right|^2 = \frac{|2y + 2(t-x)i|^2}{(1+t^2)[x^2+(y+1)^2]} = \frac{4[(t-x)^2 + y^2]}{(1+t^2)[x^2+(y+1)^2]},$$

the expression (16.12) becomes

$$
\begin{aligned}
u(x+iy) &= \frac{1}{2\pi} \int_{-\infty}^{\infty} \left[1 - \frac{x^2+(y-1)^2}{x^2+(y+1)^2}\right] \cdot \frac{(1+t^2)[x^2+(y+1)^2]}{4[(t-x)^2+y^2]} \cdot u(t) \cdot \frac{2}{t^2+1}\, dt \\
&= \frac{1}{4\pi} \int_{-\infty}^{\infty} \frac{(y+1)^2 - (y-1)^2}{(t-x)^2+y^2} u(t)\, dt \\
&= \frac{1}{\pi} \int_{-\infty}^{\infty} \frac{y \cdot u(t)}{(t-x)^2+y^2}\, dt
\end{aligned}
$$

which is our desired result. We have completed the proof of the problem. ▨

Problem 16.7

Bak and Newman Chapter 16 Exercise 7.

Proof. Following the idea of [4, Example i, p. 232], since z^3 is analytic in $D(0;1)$, $\mathrm{Re}\,(z^3)$ is harmonic in $D(0;1)$ by Theorem 16.2. Note that

$$\mathrm{Re}\,(z^3) = x^3 - 3xy^2. \tag{16.13}$$

Since $x^2 + y^2 = 1$, the equation (16.13) reduces to $\mathrm{Re}\,(z^3) = 4x^3 - 3x$ on the boundary $C(0;1)$. Hence if we choose

$$u(x,y) = \frac{1}{4}[\mathrm{Re}\,(z^3) + 3x] = \frac{1}{4}(x^3 - 3xy^2 + 3x),$$

then it is harmonic in $D(0;1)$ by Definition 16.1 and satisfies $u(x,y) = x^3$ on $C(0;1)$. This ends the proof of the problem. ▨

Problem 16.8

Bak and Newman Chapter 16 Exercise 8.

Proof. By [4, Example ii, pp. 232, 233], we know that

$$u(z) = \frac{3}{2} - \frac{1}{\pi}\mathrm{Arg}\left(\frac{z-1}{z+1}\right),$$

where $|z| \le 1$. If $u(x,y) = k$, where $k \in [0,1]$, then we get

$$\theta_k = \mathrm{Arg}\left(\frac{z-1}{z+1}\right) = \pi\left(\frac{3}{2} - k\right). \tag{16.14}$$

It is clear that as k runs through $[0,1]$, the formula (16.14) indicates that θ_k runs from $\frac{3\pi}{2}$ to $\frac{\pi}{2}$. Suppose that $z = x + iy$. Simple algebra gives

$$\frac{z-1}{z+1} = \frac{x^2+y^2-1}{(x+1)^2+y^2} + i\frac{2y}{(x+1)^2+y^2}.$$

By the definition, we know that

$$\tan \theta_k = \frac{2y}{x^2 + y^2 - 1}$$

and then the value (16.14) implies that

$$\frac{2y}{x^2 + y^2 - 1} = \frac{1}{\tan k\pi}$$

or equivalently

$$x^2 + y^2 - (2 \tan k\pi)y - 1 = 0. \tag{16.15}$$

It is trivial that $(-1, 0)$ and $(1, 0)$ always lie on the locus represented by the equation (16.15). Furthermore, if $k = \frac{1}{2}$, then we conclude from the equation (16.15) that $y = 0$ and $x = \pm 1$. If $k = 0$ or 1, then the equation (16.15) reduces to the unit circle $C(0; 1)$.

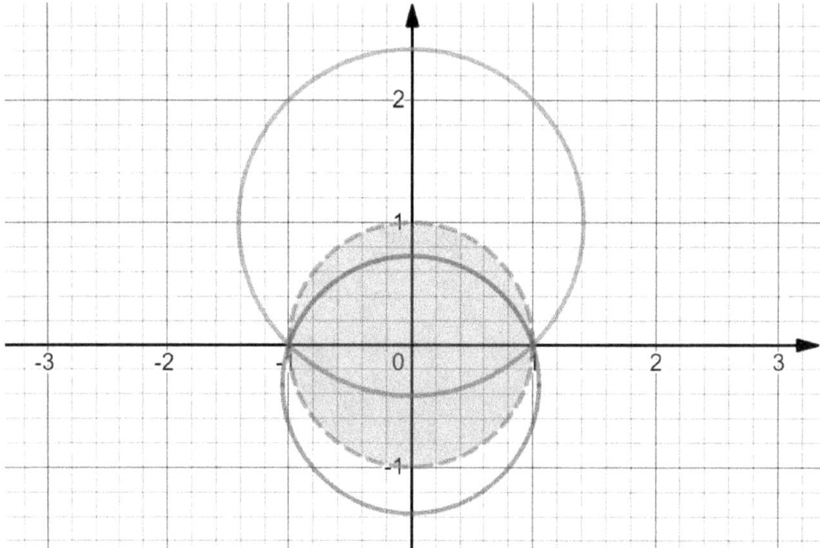

Figure 16.1: The level curves of $u(x, y) = k$ for $k \in (0, \frac{1}{2}) \cup (\frac{1}{2}, 1)$.

Next, suppose that $k \in (0, \frac{1}{2}) \cup (\frac{1}{2}, 1)$. Then the centre and the radius of the circle (16.15) are given by

$$p = (0, \tan k\pi) \quad \text{and} \quad |\sec k\pi|$$

respectively. In fact, for $0 < k < \frac{1}{2}$, we know that $\tan k\pi > 0$ and thus the centre lies on the positive imaginary axis which implies that the level curve $u(x, y) = k$ is the lower circular segment of the circle $C(p, \sec k\pi)$. This is exactly the dark red arc in Figure 16.1 lying inside the unit disc $D(0; 1)$. Similarly, for $\frac{1}{2} < k < \pi$, then $\tan k\pi < 0$ so that its centre lies on the negative imaginary axis. In this case, the level curve $u(x, y) = k$ is the upper circular segment of the circle $C(p, |\sec k\pi|)$, see the purple arc lying inside the unit disc $D(0; 1)$ in Figure 16.1. We have completed the analysis of the problem. ∎

Problem 16.9

Bak and Newman Chapter 16 Exercise 9.

Proof. By Problem 16.6 with the function

$$u(t) = \begin{cases} 1, & \text{if } t > 0; \\ \\ 0, & \text{if } t < 0, \end{cases}$$

we get

$$u(x + iy) = \frac{1}{\pi} \int_0^\infty \frac{y}{(t - x)^2 + y^2}\, dt. \tag{16.16}$$

Using the substitution $y \tan u = t - x$ and the identity $\tan^{-1}(-x) = -\tan^{-1} x$ for all $x \in \mathbb{R}$, we can reduce the expression (16.16) to the following form

$$u(x + iy) = \frac{1}{2} + \frac{1}{\pi} \tan^{-1} \frac{x}{y}$$

which is our desired harmonic function. This ends the proof of the problem.

Problem 16.10

Bak and Newman Chapter 16 Exercise 10.

Proof. Let S be the semi-infinite strip $\{z = x + iy \mid -\frac{\pi}{2} < x < \frac{\pi}{2} \text{ and } y > 0\}$. The graph of the the temperature problem with the prescribed boundary values is shown in Figure 16.2. We note that this is a Dirichlet problem for the region S with its boundary in the z plane. Our method of solution is to obtain a new Dirichlet problem for the upper half-plane $\mathbb{H}$ with its boundary in the ω plane.

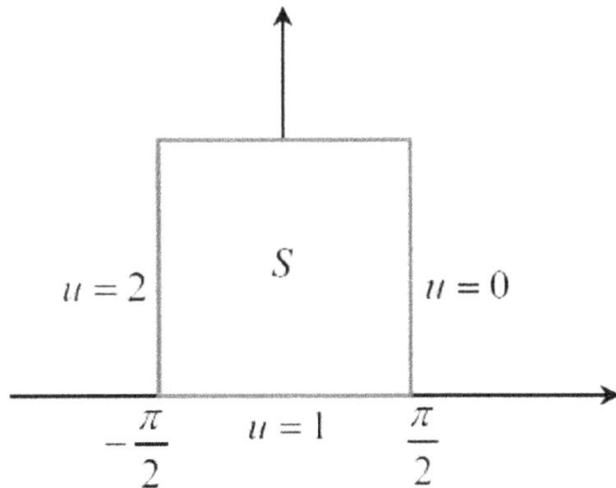

Figure 16.2: The temperature problem with the prescribed boundary values.

Recall from §13.3 that the function $\omega = f(z) = \sin z$ maps S conformally onto $\mathbb{H}$ and the interval $[-\frac{\pi}{2}, \frac{\pi}{2}]$ is mapped onto $[-1, 1]$. On $z = -\frac{\pi}{2} + iy$ with $y > 0$, we have

$$\sin z = -\cos(iy) = -\frac{1}{2}(e^y + e^{-y}).$$

Therefore, f maps the vertical line $z = -\frac{\pi}{2} + iy$ onto $(-\infty, -1)$. Similarly, it can be shown easily that $\sin z$ maps the vertical line $z = \frac{\pi}{2} + iy$ with $y > 0$ onto $(1, \infty)$. Consequently, the mapping $f(z) = \sin z$ transforms this boundary value problem into another boundary value problem in $\mathbb{H}$.

Stimulating by [4, Example ii, pp. 232, 233], we consider the mapping

$$\zeta = \log(\omega^2 - 1) = \log|\omega^2 - 1| + i\operatorname{Arg}(\omega^2 - 1)$$

which is analytic in $\mathbb{H}$. By the definition, we know that

$$\operatorname{Arg}(\omega^2 - 1) = \operatorname{Arg}(\omega + 1) + \operatorname{Arg}(\omega - 1) = \theta_2 + \theta_1,$$

where θ_1 and θ_2 are shown in Figure 16.3 below.

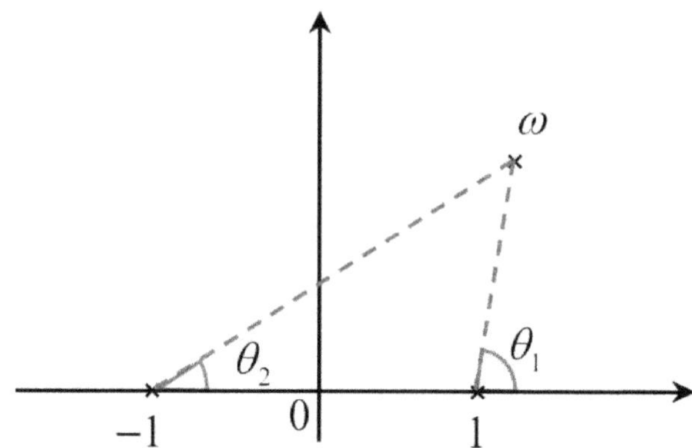

Figure 16.3: The angle $\operatorname{Arg}(\omega^2 - 1)$.

Clearly, the function $\operatorname{Im}\zeta = \operatorname{Arg}(\omega^2 - 1)$ is harmonic in $\mathbb{H}$ by Theorem 16.2. In addition, since $\operatorname{Arg}(\omega^2 - 1) = \operatorname{Arg}(\omega + 1) + \operatorname{Arg}(\omega - 1)$, we actually obtain

$$\operatorname{Arg}(\omega^2 - 1) = \begin{cases} 2\pi, & \text{if } \omega \in (-\infty, -1); \\ \pi, & \text{if } \omega \in (-1, 1); \\ 0, & \text{if } \omega \in (1, \infty). \end{cases} \tag{16.17}$$

These facts show that, in $\mathbb{H}$, $\operatorname{Im}\zeta = \operatorname{Arg}(\omega^2 - 1)$ is harmonic with the prescribed boundary values (16.17). Hence the composition function

$$u(x, y) = \frac{1}{\pi}\operatorname{Arg}(\sin^2 z - 1)$$

is harmonic in S with the prescribed boundary values

$$u(x, y) = \begin{cases} 2, & \text{if } z = -\frac{\pi}{2} + iy \text{ with } y > 0; \\ 1, & \text{if } z \in (-1, 1); \\ 0, & \text{if } z = \frac{\pi}{2} + iy \text{ with } y > 0. \end{cases}$$

This completes the proof of the problem. $\blacksquare$

Problem 16.11

Bak and Newman Chapter 16 Exercise 11.

Proof. If the number of zeros of $f(z) = e^z - P(z)$ was finite, then we deduce from Theorem 16.13 that f has the form

$$e^z - P(z) = P_1(z)e^{P_2(z)} \tag{16.18}$$

for some polynomials P_1 and P_2. If $\deg P_2 \geq 2$, then the equation (16.18) implies that

$$\lim_{z \to \infty} P_1(z) = \lim_{z \to \infty} \frac{e^z - P(z)}{e^{P_2(z)}} = 0.$$

By Problem 1.26, $P_1(z) \equiv 0$. In this case, the expression (16.18) becomes $e^z = P(z)$ which is impossible. If $\deg P_2 = 0$, then P_2 is a constant and so the expression (16.18) yields that $e^z = Q(z)$ for some polynomial Q, a contradiction. Hence we must have $\deg P_2 = 1$ and furthermore, $P_2(z) = z$. Now we rewrite the expression (16.18) as

$$[1 - P_1(z)]e^z = P(z). \tag{16.19}$$

If $P_1(z) \neq 1$, then we follow from the expression (16.19) that e^z is a rational function, a contradiction. Thus $P_1(z) \equiv 1$ and so $P(z) = 0$ by the expression (16.19), but this contradicts to our hypothesis.

For the entire function $g(z) = \sin z - P(z)$, if $\sin z - P(z) = P_1(z)e^{P_2(z)}$ for some polynomials P_1 and P_2, then we have

$$\lim_{\substack{z \to \infty \\ z \in \mathbb{R}}} P_1(z) = \lim_{\substack{z \to \infty \\ z \in \mathbb{R}}} \frac{\sin z - P(z)}{e^{P_2(z)}} = 0$$

which contradicts Problem 1.26. Thus we must have $P_1(z) \equiv 0$ and then $\sin z - P(z) = 0$ which is another contradiction. Now we complete the analysis of the problem. $\blacksquare$

Problem 16.12

Bak and Newman Chapter 16 Exercise 12.

Proof. Suppose that f does not have infinitely many zeros. By Theorem 16.13, f is in the form

$$f(z) = Q(z)e^{P(z)}, \tag{16.20}$$

where P and Q are polynomials. Since f is non-vanishing, the representation (16.20) and Theorem 5.12 (The Fundamental Theorem of Algebra) force that Q is a non-zero constant and thus we may assume further that $Q \equiv 1$, i.e., $f(z) = e^{P(z)}$.

Let $S = \big\{k \in \mathbb{R} \,\big|\, \lim_{z \to \infty} |f(z)| \exp(-|z|^k) = 0\big\}$ and $P(z) = a_n z^n + a_{n-1} z^{n-1} + \cdots + a_0$, where $a_n \neq 0$. Then it is easy to see that

$$\lim_{z \to \infty} \frac{|P(z)|}{|a_n| \cdot |z|^n} = 1$$

so that

$$\lim_{z \to \infty} \frac{|f(z)|}{\exp(|a_n| \cdot |z|^n)} = 1. \tag{16.21}$$

Assume that $n > j$. Since $j = \inf S$, the definition of infimum means that n is *not* a lower bound of S, i.e., there exists a $k \in S$ such that $n > k$. Therefore, we have $\exp(|z|^{-n}) < \exp(|z|^{-k})$ for large z which gives

$$\lim_{z \to \infty} \frac{|f(z)|}{\exp(|z|^n)} = \lim_{z \to \infty} \frac{|f(z)|}{\exp(|z|^k)} = 0$$

and this means that $n \in S$, but it implies the contrary result $n > j > n$. In other words, we must have $n \leq j$. If $n < j$, then we can find a $k_0 \in (n, j)$. Combining this fact and the limit (16.21), it is trivial to see that

$$\lim_{z \to \infty} \frac{|f(z)|}{\exp(|z|^{k_0})} = \lim_{z \to \infty} \frac{|f(z)|}{\exp(|a_n| \cdot |z|^n)} \times \frac{\exp(|a_n| \cdot |z|^n)}{\exp(|z|^{k_0})} = 0.$$

Consequently, it yields that $k_0 \in S$ which implies that $k_0 \geq j$, a contradiction. Hence it must been that $n = j$ and this completes the analysis of the problem. ∎

Problem 16.13

*Bak and Newman Chapter 16 Exercise 13.**

Proof. Assume that $\sin z - z \neq c$ for some $c \in \mathbb{C}$ and for all $z \in \mathbb{C}$. On the one hand, we have

$$\sin z - z \neq c + 2\pi \tag{16.22}$$

on $\mathbb{C}$. Otherwise, if $\sin z_0 - z_0 - 2\pi = c$ for some $z_0 \in \mathbb{C}$, then we have

$$\sin(z_0 + 2\pi) - (z_0 + 2\pi) = \sin z_0 - z_0 - 2\pi = c,$$

a contradiction. On the other hand, since the function $f(z) = \sin z - z$ is clearly an entire function of finite order, the Little Picard Theorem ensures that f assumes other values infinitely many times, but this contradicts the observation (16.22). Hence we obtain the desired result that $\sin z - z = c$ has a solution. This completes the proof of the problem. ∎

Problem 16.14

*Bak and Newman Chapter 16 Exercise 14.**

Proof. Since $g_0(0) = 0$, we only consider the function

$$F(z) = \sum_{k=1}^{\infty} \frac{f_k(z)}{g_k(k)} \quad \text{and} \quad F_n(z) = \sum_{k=1}^{n} \frac{f_k(z)}{g_k(k)},$$

where $n \in \mathbb{N}$. Since each f_k is entire, every F_n is also entire. Let K be a compact set of $\mathbb{C}$ and $z \in K$. Then it can be shown by induction easily that

$$|f_k(z)| \leq \underbrace{\exp(\exp(\cdots \exp(|\operatorname{Re} z|)))}_{k \text{ exponential factors}} \quad \text{and} \quad g_k(k) = \underbrace{k^{k^{\cdot^{\cdot^{k}}}}}_{k+1 \text{ terms}}. \tag{16.23}$$

Since K is compact, the set $\{|\operatorname{Re} z| \,|\, z \in K\}$ must be bounded by a positive constant M. Suppose that N is a positive integer such that $N \geq 2\max(\mathrm{e}, M)$. In this case, N must satisfy

$$\frac{M}{N} \leq \frac{1}{2}$$

so that

$$\mathrm{e}^M \leq \mathrm{e}^{\frac{N}{2}} \leq N^{\frac{N}{2}}.$$

Consequently, we achieve

$$\left| \frac{f_1(z)}{g_1(z)} \right| \leq \frac{\mathrm{e}^M}{N^N} \leq \frac{1}{N^{\frac{N}{2}}} \leq \frac{1}{2}$$

and because $N \geq 2$, so

$$\left| \frac{f_2(z)}{g_2(z)} \right| \leq \frac{e^{e^M}}{N^{N^N}} \leq \frac{\exp(\frac{1}{2}N^N)}{N^{N^N}} \leq \frac{N^{\frac{1}{2}N^N}}{N^{N^N}} = \frac{1}{\sqrt{N^{N^N}}} \leq \frac{1}{2^2}.$$

By induction, it can be shown that $k \geq N$ implies

$$\left| \frac{f_k(z)}{g_k(k)} \right| \leq \frac{1}{2^k}.$$

Now Theorem 1.9 (The Weierstrass M-Test) can be applied to conclude that $F_n \to F$ uniformly in K. Finally, since K is taken arbitrary, it follows from Theorem 7.6 that F is entire. This proves the first assertion.

For the second assertion, we note that if $z = x > 0$, then $F(x)$ is a series of positive terms which gives

$$|F(z)| = F(x) \geq \frac{f_n(x)}{g_n(n)}$$

for every $n \in \mathbb{N}$. Given that $k \in \mathbb{N}$, since $\{g_n(n)\}$ is a strictly increasing sequence of positive integers, the binomial theorem implies that for all large positive integers N, we have

$$
\begin{aligned}
e^{g_{N+k}(N+k)} &= (1+\alpha)^{g_{N+k}(N+k)} \\
&\geq \frac{[g_{N+k}(N+k)-1]\alpha^2}{2} \cdot g_{N+k}(N+k) \\
&> \frac{3}{2}k g_{N+k}(N+k),
\end{aligned}
\tag{16.24}
$$

where $\alpha = e - 1 > 1$. Now, for each fixed $j \in \mathbb{N}$, we know from the representation (16.23) that $g_j(j) \geq 1$, so we may take $x_j = e^{g_j(j)} > 1$. Combining the inequality (16.24) and the fact

$$f_{j+2}(x_{N+k}) = \underbrace{\exp(\exp(\exp\cdots\exp(x_{N+k})))}_{(j+2) \text{ terms}} = \underbrace{\exp(\exp(\exp\cdots\exp[g_{N+k}(N+k)]))}_{(j+3) \text{ terms}},$$

we obtain

$$F(x_{N+k}) = \frac{f_{j+2}(x_{N+k})}{g_{j+2}(j+2)} \geq \underbrace{\exp(\exp(\exp\cdots\exp[g_{N+k}(N+k)]))}_{(j+2) \text{ terms}}$$

for all large positive integers N. Therefore, the result

$$\underbrace{\log(\log(\log\cdots\log[F(x_{N+k})]))}_{j \text{ terms}} \geq \exp(\exp[g_{N+k}(N+k)])$$

$$
\begin{aligned}
&> \exp\left[\frac{3}{2}k g_{N+k}(N+k)\right] \\
&= \exp[k g_{N+k}(N+k)] \cdot \exp\left[\frac{k}{2}g_{N+k}(N+k)\right] \\
&> \exp[k g_{N+k}(N+k)] \\
&= x_{N+k}^k
\end{aligned}
$$

holds for all large positive integers N. By the definition, F is *not* of j-fold exponential order for any positive integer j. This proves the second assertion and then we complete the proof of the problem.

Different Forms of Analytic Functions

Problem 17.1

Bak and Newman Chapter 17 Exercise 1.

Proof. By Definition 17.1, we get

$$
\begin{aligned}
P_N &= \prod_{k=2}^{N} \left(1 - \frac{1}{k^2}\right) \\
&= \prod_{k=2}^{N} \left[\frac{(k-1)(k+1)}{k^2}\right] \\
&= \frac{1 \cdot 3}{2 \cdot 2} \times \frac{2 \cdot 4}{3 \cdot 3} \times \frac{3 \cdot 5}{4 \cdot 4} \times \cdots \times \frac{(N-3)(N-1)}{(N-2)(N-2)} \times \frac{(N-2)N}{(N-1)(N-1)} \times \frac{(N-1)(N+1)}{N \cdot N} \\
&= \frac{1}{2} \times \frac{N+1}{N}
\end{aligned}
$$

so that $P_N \to \frac{1}{2}$ as $N \to \infty$, completing the proof of the problem.

Problem 17.2

Bak and Newman Chapter 17 Exercise 2.

Proof. By Definition 17.1, we see that

$$
\begin{aligned}
P_N &= \prod_{k=2}^{N} \left[1 + \frac{(-1)^k}{k}\right] \\
&= \prod_{k=2}^{N} \left[\frac{k + (-1)^k}{k}\right] \\
&= \underbrace{\frac{3}{2} \times \frac{2}{3} \times \frac{5}{4} \times \frac{4}{5} \times \cdots \times \frac{N-1+(-1)^{N-1}}{N-1} \times \frac{N+(-1)^N}{N}}_{(N-1) \text{ terms}}.
\end{aligned}
\tag{17.1}
$$

Obviously, we know that

$$\frac{N + (-1)^N}{N} = \begin{cases} \dfrac{N-1}{N}, & \text{if } N \text{ is odd;} \\[2ex] \dfrac{N+1}{N}, & \text{otherwise.} \end{cases}$$

Hence the expression (17.1) reduces to

$$P_N = \begin{cases} 1, & \text{if } N \text{ is odd;} \\[2ex] \dfrac{N+1}{N}, & \text{otherwise.} \end{cases}$$

Now we conclude that $P_N \to 1$ as $N \to \infty$. This completes the proof of the problem.

> **Problem 17.3**
>
> *Bak and Newman Chapter 17 Exercise 3.*[*]

Proof. For every $N \in \mathbb{N}$, we consider

$$\sum_{n=1}^{N} \log\left(1 + \frac{i}{n}\right) = \sum_{n=1}^{N}\left[\log\left|1 + \frac{i}{n}\right| + i\operatorname{Arg}\left(1 + \frac{i}{n}\right)\right]$$
$$= \sum_{n=1}^{N} \log\left|1 + \frac{i}{n}\right| + i\sum_{n=1}^{N} \arctan\frac{1}{n}. \tag{17.2}$$

Since $e^x \geq 1 + x$ for $x \geq 0$, we obtain $\log(1 + x) \leq x$ and then

$$\log\left|1 + \frac{i}{n}\right| = \log\sqrt{1 + \frac{1}{n^2}} = \frac{1}{2}\log\left(1 + \frac{1}{n^2}\right) \leq \frac{1}{2n^2}. \tag{17.3}$$

Applying the Comparison Test [27, Theorem 6.6, p. 76] to the inequality (17.3), we conclude that

$$\sum_{n=1}^{\infty} \log\left|1 + \frac{i}{n}\right| \tag{17.4}$$

converges. By Proposition 17.2, the product $\displaystyle\prod_{n=1}^{\infty}\left|1 + \frac{i}{n}\right|$ converges.

Assume that the product $\displaystyle\prod_{n=1}^{\infty}\left(1 + \frac{i}{n}\right)$ was convergent. Then Proposition 17.2 tells us that the series $\displaystyle\sum_{n=1}^{\infty} \log\left(1 + \frac{i}{n}\right)$ converges. Applying this fact with the convergence of the series (17.4) to the sum (17.2), we see that the series

$$\sum_{n=1}^{\infty} \arctan\frac{1}{n} \tag{17.5}$$

is also convergent. However, since $\arctan\frac{1}{n} \geq \frac{\pi}{4n}$ for all $n \geq 1$, the Comparison Test implies that the series (17.5) is divergent, a contradiction. Hence the product $\displaystyle\prod_{n=1}^{\infty}\left(1 + \frac{i}{n}\right)$ is divergent. This ends the proof of the problem.

> **Problem 17.4**
>
> *Bak and Newman Chapter 17 Exercise 4.*

Proof. Using similar argument as in the proof of Proposition 17.3, we take $N \in \mathbb{N}$ such that $k > N$ implies $|z_k| < \frac{1}{2}$. Then, for every $k > N$, we have

$$\log(1 + z_k) - z_k = -\frac{z_k^2}{2} + \frac{z_k^3}{3} - \cdots = z_k^2\left(-\frac{1}{2} + \frac{z_k}{3} - \cdots\right) \tag{17.6}$$

so that

$$
\begin{aligned}
|\log(1 + z_k) - z_k| &= |z_k|^2 \cdot \left| -\frac{1}{2} + \frac{z_k}{3} - \cdots \right| \\
&\leq |z_k|^2 \cdot \left(\frac{1}{2} + \frac{|z_k|}{3} + \frac{|z_k|^2}{4} + \cdots\right) \\
&\leq |z_k|^2 \cdot \left(\frac{1}{2} + \frac{1}{4} + \frac{1}{8} + \cdots\right) \\
&= |z_k|^2.
\end{aligned}
$$

Hence the series $\displaystyle\sum_{k=N+1}^{\infty} [\log(1 + z_k) - z_k]$ and then the series $\displaystyle\sum_{k=1}^{\infty}[\log(1 + z_k) - z_k]$ are convergent. Since $\displaystyle\sum_{k=1}^{\infty} z_k$ converges, we conclude that the series

$$\sum_{k=1}^{\infty} \log(1 + z_k)$$

is convergent. By Proposition 17.2, the product $\displaystyle\prod_{k=1}^{\infty}(1 + z_k)$ converges which completes the proof of the problem. $\blacksquare$

> **Problem 17.5**
>
> *Bak and Newman Chapter 17 Exercise 5.*

Proof. Let $z_k = \frac{(-1)^k}{\sqrt{k}}$, where $k = 2, 3, \ldots$. Assume that the product $\displaystyle\prod_{k=2}^{\infty}(1 + z_k)$ was convergent. By Proposition 17.2, the series

$$\sum_{k=2}^{\infty} \log(1 + z_k) \tag{17.7}$$

is convergent. Using the formula (17.6), if $k \geq 4$, then we have[a]

$$
\begin{aligned}
\log(1 + z_k) - z_k &= \frac{1}{k} \cdot \left[-\frac{1}{2} + \frac{(-1)^k}{3\sqrt{k}} - \frac{1}{4k} + \cdots \right] \\
&\leq \frac{1}{k} \cdot \left(-\frac{1}{2} + \frac{1}{3\sqrt{k}} \right)
\end{aligned}
$$

---

[a] Notice that all z_k and $1 + z_k$ are real numbers.

$$\leq \frac{1}{k} \cdot \left(-\frac{1}{2} + \frac{1}{6} \right)$$
$$\leq -\frac{1}{3k}$$

or equivalently,

$$z_k - \log(1 + z_k) \geq \frac{1}{3k}. \tag{17.8}$$

Thus the convergence of $\sum_{k=2}^{\infty} z_k$ and the series (17.7) imply that the series $\sum_{k=4}^{\infty} [z_k - \log(1 + z_k)]$ is convergent, so it follows from the inequality (17.8) and the Comparison Test that the series

$$\sum_{k=4}^{\infty} \frac{1}{k}$$

is convergent, a contradiction. Consequently, the product $\prod_{k=2}^{\infty}(1 + z_k)$ must be divergent, completing the proof of the problem. ◼

Problem 17.6

Bak and Newman Chapter 17 Exercise 6.

Proof. Now we claim that

$$P_N = (1 + z)(1 + z^2) \cdots (1 + z^{2^{N-1}}) = 1 + z + z^2 + \cdots + z^{2^N - 1}$$

for every $N \in \mathbb{N}$. The case for $N = 1$ is trivial. Suppose that

$$P_k = (1 + z)(1 + z^2) \cdots (1 + z^{2^{k-1}}) = 1 + z + z^2 + \cdots + z^{2^k - 1} \tag{17.9}$$

for some $k \in \mathbb{N}$. If $N = k + 1$, then the assumption step (17.9) implies that

$$P_{k+1} = [(1 + z)(1 + z^2) \cdots (1 + z^{2^{k-1}})](1 + z^{2^k})$$
$$= (1 + z + z^2 + \cdots + z^{2^k - 1})(1 + z^{2^k})$$
$$= 1 + z + z^2 + \cdots + z^{2^k - 1} + z^{2^k} + z^{2^k + 1} + \cdots + z^{2^{k+1} - 1}.$$

Hence our claim follows from induction.

If K is a compact subset of $D(0; 1)$, then there exists a $0 < \delta < 0$ such that $K \subseteq D(0; \delta)$. In $D(0; \delta)$, we see that the series $\sum_{k=0}^{\infty} z^k$ converges uniformly to $\frac{1}{1-z}$,[b] so it yields that

$$P_N \rightarrow \sum_{k=0}^{\infty} z^k = \frac{1}{1 - z}$$

uniformly on $D(0; \delta)$ and in particularly, on K. Since K is arbitrary, we know from Definition 7.5 that the product $\prod_{k=0}^{\infty}(1 + z^{2^k})$ converges uniformly on compacta to $\frac{1}{1-z}$ in $|z| < 1$. This completes the proof of the problem. ◼

---

[b]See also Problem 2.19 and [4, p. 27].

Problem 17.7

Bak and Newman Chapter 17 Exercise 7.

Proof. Since $\lambda_k \to \infty$, Theorem 17.7 (The Weierstrass Product Theorem) ensures the existence of such an entire function g. Since $\sum_{k=1}^{\infty} \frac{1}{k^2}$ converges, we deduce from the discussion on [4, p. 245] that the function

$$g(z) = \prod_{k=1}^{\infty} \left(1 - \frac{z}{k^2}\right) \tag{17.10}$$

satisfies our requirements, completing the analysis of the problem. ∎

Problem 17.8

Bak and Newman Chapter 17 Exercise 8.

Proof. Recall from Problem 3.22 that

$$\sin z = z - \frac{z^3}{3!} + \frac{z^5}{5!} - \cdots = \sum_{k=0}^{\infty} \frac{(-1)^k}{(2k+1)!} z^{2k+1},$$

so we obtain

$$\frac{\sin(\pi\sqrt{z})}{\pi\sqrt{z}} = \sum_{k=0}^{\infty} \frac{(-1)^k \pi^{2k}}{(2k+1)!} z^k. \tag{17.11}$$

If $a_k = \frac{(-1)^k \pi^{2k}}{(2k+1)!}$, then

$$\lim_{k\to\infty} \frac{a_{k+1}}{a_k} = \lim_{k\to\infty} \frac{-\pi^2}{(2k+3)(2k+2)} = 0.$$

By Problem 2.13 and Theorem 2.8, the power series (17.11) converges everywhere, i.e., the function $\frac{\sin(\pi\sqrt{z})}{\pi\sqrt{z}}$ is in fact entire. Since the zeros of $\sin z$ are $k\pi$ for all $k \in \mathbb{Z}$, the zeros of $\sin(\pi\sqrt{z})$ are k^2 for all $k \in \mathbb{N}$. In other words, one solution to Problem 17.7 is given by the entire function (17.11). This ends the proof of the problem. ∎

Problem 17.9

Bak and Newman Chapter 17 Exercise 9.

Proof. By the inspiration of [4, Example 3, p. 247], an entire function with a single zero at every $k + \frac{1}{2}$ with $k \in \mathbb{Z}$ is given by

$$f(z) = \left\{ \prod_{k=0}^{\infty} \left[\left(1 - \frac{z}{k+\frac{1}{2}}\right) \exp\left(\frac{z}{k+\frac{1}{2}}\right)\right] \right\} \times \left\{ \prod_{k=0}^{\infty} \left[\left(1 - \frac{z}{-k-\frac{1}{2}}\right) \exp\left(\frac{z}{-k-\frac{1}{2}}\right)\right] \right\}$$

$$= \prod_{k=0}^{\infty} \left[\left(1 - \frac{z}{k+\frac{1}{2}}\right) \exp\left(\frac{z}{k+\frac{1}{2}}\right)\left(1 + \frac{z}{k+\frac{1}{2}}\right) \exp\left(-\frac{z}{k+\frac{1}{2}}\right)\right]$$

$$= \prod_{k=0}^{\infty} \left[1 - \frac{4z^2}{(2k+1)^2}\right].$$

Note that $\cos \pi z$ is an entire function having simple zeros at every $k+\frac{1}{2}$, where $k \in \mathbb{Z}$. Therefore, we have

$$\cos \pi z = C \prod_{k=0}^{\infty} \left[1 - \frac{4z^2}{(2k+1)^2}\right] \tag{17.12}$$

for some nonzero constant C. Put $z = 0$ into the expression (17.12), we get immediately that $C = 1$. Hence we have completed the proof of the problem.

> ### Problem 17.10
>
> *Bak and Newman Chapter 17 Exercise 10.*

Proof.

(a) Applying similar idea of [4, Example 1, p. 246], an entire function g with zeros at every positive integer is given by

$$g(z) = \prod_{k=1}^{\infty} \left(1 - \frac{z}{k}\right) e^{\frac{z}{k}}.$$

Next, we define

$$f(z) = g\left(\frac{1}{1-z}\right) = \prod_{k=1}^{\infty} \left(1 - \frac{1}{k(1-z)}\right) e^{\frac{1}{k(1-z)}}$$

which is clearly analytic in $|z| < 1$ and $f(z) = 0$ if and only if $z = 1 - \frac{1}{k}$, where $k \in \mathbb{N}$.

(b) Suppose that $\{z_k\}$ is a sequence of distinct numbers such that $z_k \to z_0$ as $k \to \infty$. Then the sequence $\{\lambda_k = \frac{1}{z_k - z_0}\}$ satisfies $\lambda_k \to \infty$ as $k \to \infty$. By Theorem 17.7 (The Weierstrass Product Theorem), one can find an entire function g such that $g(z) = 0$ if and only if $z = \lambda_k$ for every $k \in \mathbb{N}$. Now if we define

$$f(z) = g\left(\frac{1}{z - z_0}\right),$$

then f is analytic in $|z| < |z_0|$ and $f(z) = 0$ if and only if $z = z_1, z_2, \ldots$.

We complete the proof of the problem.

> ### Problem 17.11
>
> *Bak and Newman Chapter 17 Exercise 11.*

Proof. Since $f(z) = \varphi(z, t)$ satisties the hypotheses of Theorem 17.9, the function $F(z)$ is analytic in a region D. Thus we get

$$F'(z) = \frac{1}{2\pi i} \int_C \frac{F(\zeta)}{(\zeta - z)^2} \, d\zeta = \frac{1}{2\pi i} \int_C \left(\int_a^b \frac{f(\zeta)}{(\zeta - z)^2} \, dt\right) d\zeta, \tag{17.13}$$

where C is a disc contained in D with centre z. As suggested by the question, we switch the order of integration in the formula (17.13) to obtain

$$F'(z) = \int_a^b \left(\frac{1}{2\pi i} \int_C \frac{f(\zeta)}{(\zeta - z)^2} \, d\zeta\right) dt. \tag{17.14}$$

Using the formula at the bottom of [4, p. 80], we see that

$$\varphi_z(z,t) = f'(z) = \frac{1}{2\pi i} \int_C \frac{f(\zeta)}{(\zeta - z)^2}\, d\zeta.$$

After the substitution of this formula into the expression (17.14), we reduce it to

$$F'(z) = \int_a^b \varphi_z(z,t)\, dt$$

which is our desired formula and we end the proof of the problem.

Problem 17.12

Bak and Newman Chapter 17 Exercise 12.

Proof. Given that $\epsilon > 0$ which will be determined later. We write

$$\int_\alpha^\beta \frac{h(u)y}{(u-x)^2 + y^2}\, du = \int_\alpha^{x-\epsilon} \frac{h(u)y}{(u-x)^2 + y^2}\, du + \int_{x-\epsilon}^{x+\epsilon} \frac{h(u)y}{(u-x)^2 + y^2}\, du$$
$$+ \int_{x+\epsilon}^\beta \frac{h(u)y}{(u-x)^2 + y^2}\, du, \tag{17.15}$$

where h is a continuous function on $[\alpha, \beta]$ and $x \in (\alpha, \beta)$. Since $[\alpha, \beta]$ is compact, h is bounded by a positive constant M. Furthermore, on $[\alpha, x-\epsilon]$, it is easy to see that $(u-x)^2 + y^2 \geq \epsilon^2 + y^2 \geq \epsilon^2$. Therefore, we get

$$\left| \int_\alpha^{x-\epsilon} \frac{h(u)y}{(u-x)^2 + y^2}\, du \right| \leq \int_\alpha^{x-\epsilon} \left| \frac{h(u)y}{(u-x)^2 + y^2} \right|\, du \leq \int_\alpha^{x-\epsilon} \frac{M|y|}{\epsilon^2}\, du \leq \frac{M|y|(\beta - \alpha)}{\epsilon^2}. \tag{17.16}$$

Similarly, it is true that

$$\left| \int_{x+\epsilon}^\beta \frac{h(u)y}{(u-x)^2 + y^2}\, du \right| \leq \frac{M|y|(\beta - \alpha)}{\epsilon^2}. \tag{17.17}$$

Finally, since the function

$$g(u) = \frac{y}{(u-x)^2 + y^2}$$

is integrable and does not change sign on $[x - \epsilon, x + \epsilon]$, we observe from the Weighted Mean Value Theorem for Integrals [5, Exercise 17, p. 215] that

$$\int_{x-\epsilon}^{x+\epsilon} \frac{h(u)y}{(u-x)^2 + y^2}\, du = h(\xi) \int_{x-\epsilon}^{x+\epsilon} \frac{y\, du}{(u-x)^2 + y^2} \tag{17.18}$$

for some $\xi \in [x - \epsilon, x + \epsilon]$. We apply the substitution $u = x + y\tan\theta$ to the integral on the right-hand side of the expression (17.18), we have

$$\int_{x-\epsilon}^{x+\epsilon} \frac{h(u)y}{(u-x)^2 + y^2}\, du = h(\xi) \int_{\tan^{-1}(-\epsilon/y)}^{\tan^{-1}(\epsilon/y)} d\theta = 2h(\xi) \tan^{-1} \frac{\epsilon}{y}. \tag{17.19}$$

Finally, we take $\epsilon = \sqrt[4]{y}$ and we put the results (17.16), (17.17) and (17.19) into the expression (17.15) to get

$$\lim_{y\to 0} \int_\alpha^\beta \frac{h(u)y}{(u-x)^2 + y^2}\, du = \lim_{y\to 0} 2h(\xi) \tan^{-1} \frac{1}{y^{\frac{3}{4}}}.$$

Since $\tan^{-1}(y^{-\frac{3}{4}}) \to \frac{\pi}{2}$ and $\xi \to x$ as $y \to 0$, we conclude that

$$\lim_{y \to 0} \int_\alpha^\beta \frac{h(u)y}{(u-x)^2 + y^2}\, du = \pi h(x),$$

completing the proof of the problem.

Problem 17.13

Bak and Newman Chapter 17 Exercise 13.

Proof. According to direct integration, it is clear that

$$f(z) = \int_0^1 \frac{dt}{1 - zt} = -\frac{\log(1 - z)}{z}. \tag{17.20}$$

Recall from [4, p. 115] that $\log z$ is analytic in $\mathbb{C} \setminus (-\infty, 0]$, so $\log(1 - z)$ and then f is analytic in $D = \mathbb{C} \setminus [1, \infty)$. Suppose that γ is a simple closed curve encircling the point $z = 1$. Since the function $F(z) = 1 - z$ is analytic inside and on γ and $F(z) \neq 0$ on γ, Corollary 10.9 (The Argument Principle) implies that

$$\frac{1}{2\pi} \Delta \mathrm{Arg}\,(1 - z) = \frac{1}{2\pi i} \int_\gamma \frac{F'(z)}{F(z)}\, dz = 1$$

which means that $\Delta \mathrm{Arg}\,(1 - z) = 2\pi i$ as z traverses along γ. Hence it follows from the representation (17.20) that f has a "jump" of $\frac{2\pi i}{x}$ as z crosses from the upper half-plane to the lower half-plane through any point $x > 1$. This completes the analysis of the problem.

Problem 17.14

*Bak and Newman Chapter 17 Exercise 14.**

Proof.

(a) By the definition, $|\phi(n)| = \phi(n) \leq n$. Since $|n^z| = n^{\mathrm{Re}\,z}$, we obtain

$$\sum_{n=1}^\infty \left|\frac{\phi(n)}{n^z}\right| = \sum_{n=1}^\infty \frac{\phi(n)}{|n^z|} \leq \sum_{n=1}^\infty \frac{n}{n^{\mathrm{Re}\,z}} = \sum_{n=1}^\infty \frac{1}{n^{\mathrm{Re}\,z - 1}}.$$

By [27, Theorems 6.6, 6.10, pp. 76, 77], the series

$$\sum_{n=1}^\infty \frac{\phi(n)}{n^z}$$

converges absolutely for $\mathrm{Re}\,z - 1 > 1$ or equivalently, $\mathrm{Re}\,z > 2$.

(b) By [4, Example, pp. 254, 255], we have

$$\zeta(z) \sum_{n=1}^\infty \frac{\phi(n)}{n^z} = \sum_{n=1}^\infty \frac{1}{n^z} \cdot \sum_{n=1}^\infty \frac{\phi(n)}{n^z} = \sum_{n=1}^\infty \frac{c_n}{n^z},$$

where

$$c_n = \sum_{d|n} \phi\left(\frac{n}{d}\right) = \sum_{d|n} \phi(n) = n.$$

Consequently, we note that

$$\zeta(z) \sum_{n=1}^{\infty} \frac{\phi(n)}{n^z} = \sum_{n=1}^{\infty} \frac{1}{n^{z-1}} = \zeta(z-1)$$

for $\operatorname{Re} z > 2$.

Hence we end the proof of the problem.

Analytic Continuation; The Gamma and Zeta Functions

Problem 18.1

Bak and Newman Chapter 18 Exercise 1.

Proof. Let $D_1 = \mathbb{C} \setminus \{(x,0) \mid x \leq 0\}$ and $D_2 = \mathbb{C} \setminus \{(0,y) \mid y \leq 0\}$. Define $g_1 : D_1 \to \mathbb{C}$ by

$$g_1(z) = \log |z| + i\operatorname{Arg} z, \tag{18.1}$$

where $\operatorname{Arg} z \in (-\pi, \pi)$. Similarly, we define $g_2 : D_2 \to \mathbb{C}$ by

$$g_2(z) = \log |z| + i\operatorname{Arg} z, \tag{18.2}$$

where $\operatorname{Arg} z \in (-\frac{\pi}{2}, \frac{3\pi}{2})$. Obviously, we have

$$e^{g_1(z)} = e^{g_2(z)} = z,$$

so g_1 and g_2 are analytic branches of $\log z$ by Definition 8.7. If we let D be the first quadrant and define $f : D \to \mathbb{C}$ by

$$f(z) = \log |z| + i\operatorname{Arg} z,$$

where $\operatorname{Arg} z \in (0, \frac{\pi}{2})$, then each of g_1 and g_2 is an analytic continuation of f. Suppose that ζ is a point lying in the third quadrant. On the one hand, we know from the definition (18.1) that

$$\operatorname{Im} g_1(\zeta) = \operatorname{Arg} \zeta \in \left(-\pi, -\frac{\pi}{2}\right).$$

On the other hand, the definition (18.2) gives

$$\operatorname{Im} g_2(\zeta) = \operatorname{Arg} \zeta \in \left(\pi, \frac{3\pi}{2}\right).$$

Therefore, we establish that $g_1(\zeta) \neq g_2(\zeta)$, completing the proof of the problem. $\blacksquare$

Problem 18.2

*Bak and Newman Chapter 18 Exercise 2.**

Proof.

(a) Assume that there was a point $|z_0| = 1$ such that the series $\sum_{n=0}^{\infty} a_n z_0^n$ converges. Without loss of generality, we may assume that $z_0 = 1$. Otherwise, if $z_0 = e^{i\theta_0}$ for some $\theta_0 \in [0, 2\pi]$, then we have

$$\sum_{n=0}^{\infty} a_n z_0^n = \sum_{n=0}^{\infty} (a_n e^{in\theta_0}) \cdot 1^n.$$

Thus it must be true that $a_n \to 0$ as $n \to \infty$. Given $\epsilon > 0$. Then there exists an $N \in \mathbb{N}$ such that $|a_n| < \epsilon$ for all $n \geq N$. Therefore, for every $z \in D(0; 1)$, we see that

$$|f(z)| \leq \sum_{n=0}^{\infty} |a_n| \cdot |z|^n = \sum_{n=0}^{N-1} |a_n| \cdot |z|^n + \sum_{n=N}^{\infty} |a_n| \cdot |z|^n \leq \sum_{n=0}^{N-1} |a_n| + \frac{\epsilon}{1 - |z|}$$

which gives, for $0 < x < 1$,

$$\lim_{\substack{x \to 1^- \\ x \in \mathbb{R}}} (1 - x) f(x) \leq \epsilon.$$

Since ϵ is arbitrary, we actually have $(1 - x) f(x) \to 0$ as $x \to 1^-$ and $x \in \mathbb{R}$. However, this contradicts the hypothesis that f has a pole at $z = 1$. Hence the power series diverges at every point on $C(0; 1)$.

(b) Suppose that the radius of convergence of

$$f(z) = \sum_{n=0}^{\infty} a_n z^n \tag{18.3}$$

is $R > 0$ and f has a pole at $z_0 = Re^{i\theta_0}$. Then the function

$$g(z) = f(Rz) = \sum_{n=0}^{\infty} (a_n R^n) z^n \tag{18.4}$$

has radius of convergence 1 and g has a pole at $z = e^{i\theta_0}$. By the argument in part (a), we may assume that the pole of g is at $z = 1$. Hence part (a) implies that the power series (18.4) diverges at every point on $C(0; 1)$ which is equivalent to saying that the power series (18.3) diverges at all points on $C(0; R)$.

This completes the proof of the problem. ∎

Problem 18.3

Bak and Newman Chapter 18 Exercise 3.

Proof. Define $f(z) = \sum_{n=0}^{\infty} (-1)^n a_n z^n$ and its radius of convergence to be $R < \infty$. Then we have

$$f(-z) = \sum_{n=0}^{\infty} a_n z^n$$

with R as its radius of convergence. Now the function $f(-z)$ satisfies the hypotheses of Theorem 18.3, so it has a singularity at $z = R$. Hence we conclude that $f(z)$ has a singularity at $z = -R$ which ends the proof of the problem. ∎

> **Problem 18.4**
>
> *Bak and Newman Chapter 18 Exercise 4.*

Proof.

(a) By the application of Problem 18.5 in advance, we have

$$\frac{1}{\sqrt[3]{n}} = \frac{1}{\Gamma(\frac{1}{3})} \int_0^\infty e^{-nt} t^{-\frac{2}{3}} \, dt.$$

If $|z| < 1$, then we derive

$$\sum_{n=1}^\infty \frac{z^n}{\sqrt[3]{n}} = \sum_{n=1}^\infty \left[\frac{1}{\Gamma(\frac{1}{3})} \int_0^\infty z^n e^{-nt} t^{-\frac{2}{3}} \, dt \right]$$

$$= \frac{1}{\Gamma(\frac{1}{3})} \int_0^\infty \left[\sum_{n=1}^\infty (ze^{-t})^n t^{-\frac{2}{3}} \right] dt$$

$$= \frac{1}{\Gamma(\frac{1}{3})} \int_0^\infty \frac{t^{-\frac{2}{3}} \cdot ze^{-t}}{1 - ze^{-t}} \, dt$$

$$= \frac{z}{\Gamma(\frac{1}{3})} \int_0^\infty \frac{t^{-\frac{2}{3}}}{e^t - z} \, dt \tag{18.5}$$

$$= \frac{z}{\Gamma(\frac{1}{3})} \lim_{\substack{\epsilon \to 0 \\ N \to \infty}} \int_\epsilon^N \frac{t^{-\frac{2}{3}}}{e^t - z} \, dt. \tag{18.6}$$

Let $f(t) = e^t$ and $g(t) = t^{-\frac{2}{3}}$. Clearly, f and g are continuous real-valued functions on $[\epsilon, N]$. Furthermore, $f'(t) = e^t > 0$ is also continuous on $[\epsilon, N]$. By Proposition 17.10, the integral (18.6) is analytic in $\mathbb{C} \setminus [e^\epsilon, e^N]$. Since $\epsilon \to 0$ and $N \to \infty$, the integral (18.5) is actually analytic in $\mathbb{C} \setminus [1, \infty)$ as required.

(b) Now we apply the second well-known formula on [4, p. 262], we have

$$\frac{1}{n^2 + 1} = \int_0^\infty e^{-nt} \sin t \, dt$$

which show that if $|z| < 1$, then

$$\sum_{n=0}^\infty \frac{z^n}{n^2 + 1} = \sum_{n=0}^\infty \left[\int_0^\infty (ze^{-t})^n \sin t \, dt \right]$$

$$= \int_0^\infty \left[\sum_{n=0}^\infty (ze^{-t})^n \sin t \right] dt$$

$$= \int_0^\infty \frac{e^t \sin t}{e^t - z} \, dt \tag{18.7}$$

$$= \lim_{\substack{\epsilon \to 0 \\ N \to \infty}} \int_\epsilon^N \frac{e^t \sin t}{e^t - z} \, dt. \tag{18.8}$$

Again, if we let $f(t) = e^t$ and $g(t) = e^t \sin t$, then both are continuous real-valued functions on $[\epsilon, N]$ and $f' > 0$ is also continuous on $[\epsilon, N]$. Hence we derive from Proposition 17.10 that the integral (18.8) is analytic outside $[e^\epsilon, e^N]$. Finally, by letting $\epsilon \to 0$ and $N \to \infty$, the integral (18.7) is analytic outside $[1, \infty)$.

We have ended the proof of the problem. ∎

Bak and Newman Chapter 18 Exercise 5.

Proof. Suppose that $x = nt$. Thus it follows from the definition [4, Eqn. (1), p. 265] that

$$\int_0^\infty e^{-nt} t^{p-1}\, \mathrm{d}t = \int_0^\infty e^{-x} \cdot \frac{x^{p-1}}{n^{p-1}} \cdot \frac{\mathrm{d}x}{n} = \frac{1}{n^p} \int_0^\infty e^{-x} x^{p-1}\, \mathrm{d}x = \frac{\Gamma(p)}{n^p}$$

for $p > 0$. This completes the proof of the problem.

Bak and Newman Chapter 18 Exercise 6.

Proof. By the substitution $x = t^2$, we deduce from the definition [4, Eqn. (1), p. 265] and then the fact $\Gamma(\frac{1}{2}) = \sqrt{\pi}$ (see [4, p. 268]) that

$$\int_0^\infty e^{-t^2}\, \mathrm{d}t = \frac{1}{2}\int_0^\infty e^{-x} x^{-\frac{1}{2}}\, \mathrm{d}x = \frac{1}{2}\Gamma\left(\frac{1}{2}\right) = \frac{\sqrt{\pi}}{2}$$

as required, completing the proof of the problem.

Bak and Newman Chapter 18 Exercise 7.

Proof. Define

$$\Gamma_n(z) = \int_0^n t^{z-1}\left(1 - \frac{t}{n}\right)^n \mathrm{d}t,$$

where $\operatorname{Re} z > 0$ and $n \in \mathbb{N}$. Then we have

$$\Gamma(z) - \Gamma_n(z) = \int_0^n \left[e^{-t} - \left(1 - \frac{t}{n}\right)^n\right] t^{z-1}\, \mathrm{d}t + \int_n^\infty e^{-t} t^{z-1}\, \mathrm{d}t. \tag{18.9}$$

It is easy to see that the second integral in the expression (18.9) tends to 0 as $n \to \infty$.

Let $t \le n$, i.e., $\frac{t}{n} \le 1$. By the power series expansion of e^{-x}, we see that

$$e^{-\frac{t}{n}} = 1 - \frac{t}{n} + \frac{t^2}{2!n^2} - \frac{t^3}{3!n^3} + \cdots$$

so that on the one hand,

$$e^{-\frac{t}{n}} - \left(1 - \frac{t}{n}\right) = \frac{t^2}{2!n^2} - \frac{t^3}{3!n^3} + \cdots = \frac{t^2}{n^2}\left(\frac{1}{2!} - \frac{t}{3!n} + \cdots\right) \le \frac{t^2}{2n^2}.$$

We derive from the given identity with $a = \exp(-\frac{t}{n})$ and $b = 1 - \frac{t}{n}$ that

$$\left|e^{-t} - \left(1 - \frac{t}{n}\right)^n\right| \le n \exp\left[-\frac{(n-1)t}{n}\right] \cdot \left|e^{-\frac{t}{n}} - \left(1 - \frac{t}{n}\right)\right| \le ne^{-t} \cdot e^{\frac{t}{n}} \cdot \frac{t^2}{2n^2} \le e^{-t} \cdot \frac{et^2}{2n}.$$

Consequently, it implies that

$$\left|\int_0^n \left[e^{-t} - \left(1 - \frac{t}{n}\right)^n\right] t^{z-1}\, \mathrm{d}t\right| = \int_0^n |t^{z-1}| \cdot \left|e^{-t} - \left(1 - \frac{t}{n}\right)^n\right| \mathrm{d}t \le \frac{e}{2n} \int_0^n e^{-t} t^{\operatorname{Re} z+1}\, \mathrm{d}t.$$

By the definition, we know that

$$\lim_{n\to\infty} \int_0^n e^{-t} t^{\operatorname{Re} z+1}\,\mathrm{d}t = \int_0^\infty e^{-t} t^{\operatorname{Re} z+1}\,\mathrm{d}t = \Gamma(\operatorname{Re} z + 2) < \infty$$

so that

$$\lim_{n\to\infty} \left| \int_0^n \left[e^{-t} - \left(1 - \frac{t}{n}\right)^n \right] t^{z-1}\,\mathrm{d}t \right| = 0$$

and then it yields from the expression (18.9) immediately that

$$\Gamma(z) = \lim_{n\to\infty} \Gamma_n(z),$$

ending the proof of the problem.

> **Problem 18.8**
>
> *Bak and Newman Chapter 18 Exercise 8.**

Proof. Suppose that $z = x > 0$. If we take the logarithm on both sides of the product formula

$$\Gamma^{-1}(x) = xe^{\gamma x} \prod_{k=1}^{\infty} \left(1 + \frac{x}{k}\right) e^{-\frac{x}{k}},$$

then we arrive at

$$-\log\Gamma(x) = \log x + \gamma x + \sum_{k=1}^{\infty}\left[\log\left(1 + \frac{x}{k}\right) - \frac{x}{k}\right]$$

and thus[a]

$$-\frac{\Gamma'(x)}{\Gamma(x)} = \frac{1}{x} + \gamma + \frac{\mathrm{d}}{\mathrm{d}x}\sum_{k=1}^{\infty}\left[\log\left(1 + \frac{x}{k}\right) - \frac{x}{k}\right]. \tag{18.10}$$

Now termwise differentiation is permitted for the series (18.10) provided that the differentiated series is uniformly convergent. To see this, we consider the series

$$\sum_{k=1}^{\infty} \frac{\mathrm{d}}{\mathrm{d}x}\left[\log\left(1 + \frac{x}{k}\right) - \frac{x}{k}\right] = \sum_{k=1}^{\infty}\left(\frac{1}{x+k} - \frac{1}{k}\right) = -\sum_{k=1}^{\infty} \frac{x}{k(x+k)}.$$

Let $x \in [\frac{1}{2}, 1]$. Then we have $k(x+k) \geq k^2$ which gives

$$\sum_{k=1}^{\infty} \frac{x}{k(x+k)} \leq \sum_{k=1}^{\infty} \frac{1}{k^2}.$$

By the Comparison Test, the differentiated series is uniformly convergent on $[\frac{1}{2}, 1]$. Hence it follows from [28, Theorem 10.8, p. 4] that we can take term by term differentiation to get

$$-\frac{\Gamma'(x)}{\Gamma(x)} = \frac{1}{x} + \gamma - \sum_{k=1}^{\infty} \frac{x}{k(x+k)}. \tag{18.11}$$

Put $x = 1$ into the expression (18.11) and use the fact $\Gamma(1) = 1$, we conclude that

$$-\Gamma'(1) = 1 + \gamma - 1 = \gamma$$

which means that $\Gamma'(1) = -\gamma$, completing the proof of the problem.

---

[a]The logarithm derivative $\psi(z) = (\log\Gamma(z))'$ is called the **digamma function**.

Problem 18.9

Bak and Newman Chapter 18 Exercise 9.

Proof. Recall from the fact[b]

$$\left(1 - \frac{1}{2^z}\right)\zeta(z) = 1 + \frac{1}{3^z} + \frac{1}{5^z} + \cdots$$

that

$$\begin{aligned}
f(z) &= \left(1 - \frac{2}{2^z}\right)\zeta(z) \\
&= \left(1 - \frac{1}{2^z}\right)\zeta(z) - \frac{1}{2^z}\zeta(z) \\
&= \left(1 + \frac{1}{3^z} + \frac{1}{5^z} + \cdots\right) - \frac{1}{2^z} - \frac{1}{4^z} - \frac{1}{6^z} - \cdots \\
&= 1 - \frac{1}{2^z} + \frac{1}{3^z} - \frac{1}{4^z} + \cdots.
\end{aligned}$$

By Theorem 18.9, f is analytic in $\mathbb{C} \setminus \{1\}$ and furthermore

$$\lim_{z\to 1}(z-1)f(z) = \lim_{z\to 1}\left(1 - \frac{2}{2^z}\right)(z-1)\zeta(z) = \lim_{z\to 1}\left(1 - \frac{2}{2^z}\right)\lim_{z\to 1}(z-1)\zeta(z) = 0 \cdot 1 = 0,$$

so f has a removable singularity at $z = 1$. Therefore, f is entire which completes the proof of the problem. ∎

Problem 18.10

Bak and Newman Chapter 18 Exercise 10.

Proof. Let $\{p_k\}$ be the set of all primes. Since ζ has a singularity at $z = 1$, we have $\zeta(1+\epsilon) \to \infty$ as $\epsilon \to 0^+$ so that the identity [4, Eqn. (5), p. 269] ensures that

$$\prod_{k=1}^{\infty}\left(1 - \frac{1}{p_k}\right) \tag{18.12}$$

diverges to 0. Clearly, we observe that

$$\sum_{k=1}^{\infty}\frac{1}{p_k^2} \le \sum_{k=1}^{\infty}\frac{1}{k^2} < \infty.$$

Assume that $\displaystyle\sum_{k=1}^{\infty} -\frac{1}{p_k}$ was convergent. Then Problem 17.4 guarantees that the product (18.12) is convergent which is a contradiction. Hence the series

$$\sum_{k=1}^{\infty} -\frac{1}{p_k}$$

must diverge, completing the proof of the problem. ∎

Remark 18.1

For another proof of Problem 18.10, please refer to [26, Problem 8.10, p. 180].

---

[b]Read [4, p. 268].

CHAPTER **19**

Applications to Other Areas of Mathematics

Bak and Newman Chapter 19 Exercise 1.

Proof. Let $f(z) = z - \tan z$ and $N \in \mathbb{N}$. Consider the square S_N with vertices $N\pi(\pm 1 \pm i)$. Then f is meromorphic in S_N, so Theorem 10.8 implies that

$$\frac{1}{2\pi i} \int_{\partial S_N} \frac{\tan^2 z}{\tan z - z}\, \mathrm{d}z = \frac{1}{2\pi i} \int_{\partial S_N} \frac{\sec^2 z - 1}{\tan z - z}\, \mathrm{d}z = \mathbb{Z}_f - \mathbb{P}_f. \tag{19.1}$$

On the one hand, if we plot the graphs of $y = x$ and $y = \tan x$ on $[-N\pi, N\pi]$, then we see that the straight line $y = x$ has *one and only one* intersection with the curve $y = \tan x$ in each of the following intervals

$$[-(k+1/2)\pi, -k\pi] \quad \text{and} \quad [k\pi, (k+1/2)\pi],$$

where $k = 1, 2, \ldots, N - 1$, see Figure 19.1 for an example.

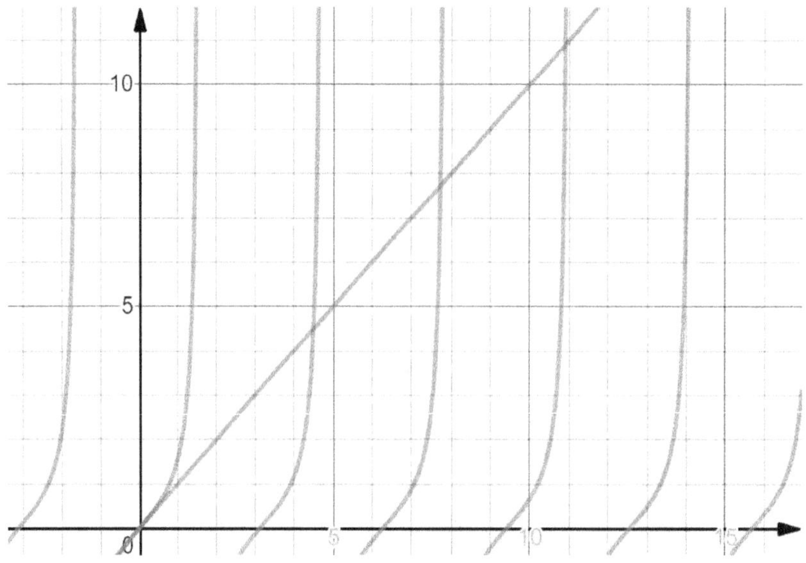

Figure 19.1: The intersections of $y = x$ and $y = \tan x$.

Furthermore, the power series of $\tan z$ (see [1, p. 184]) indicates that $f(0) = f'(0) = f''(0) = 0$ but $f^{(3)}(0) \neq 0$, so f has a zero of order 3 at $z = 0$. In conclusion, if we denote $\mathbb{Z}'_f$ to be the number of nonreal zeros of f, then we establish that

$$\mathbb{Z}_f = \mathbb{Z}'_f + 2(N-1) + 3 = \mathbb{Z}'_f + 2N + 1 \tag{19.2}$$

inside S_N. On the other hand, since $\tan z = \frac{\sin z}{\cos z}$, we have $\mathbb{P}_f = \mathbb{Z}_{\cos z}$ inside S_N. It is clear that $\cos(n + \frac{1}{2})\pi = 0$ for all $-N \leq n \leq N-1$, we know that

$$\mathbb{P}_f = 2N \tag{19.3}$$

inside S_N.

Putting the values (19.2) and (19.3) back into the expression (19.1), we get

$$\frac{1}{2\pi i} \int_{\partial S_N} \frac{\tan^2 z}{\tan z - z}\, dz = 1 + \mathbb{Z}'_f. \tag{19.4}$$

Now it remains to estimate the integral in the formula (19.4). On the vertical side $\operatorname{Re} z = N\pi$ and $y = \operatorname{Im} z \in [-N\pi, N\pi]$, we have

$$|\tan z| = \frac{|e^{iz} - e^{-iz}|}{|e^{iz} + e^{-iz}|} = \frac{|e^{-y} - e^{y}|}{|e^{-y} + e^{y}|} < 1. \tag{19.5}$$

Similarly, the bound (19.5) holds on the other vertical side. Next, if we take N large enough, then along the horizontal side $x = \operatorname{Re} z \in [-N\pi, N\pi]$ and $\operatorname{Im} z = N\pi$, we obtain

$$|\tan z| = \frac{|e^{-N\pi}e^{ix} - e^{N\pi}e^{-ix}|}{|e^{-N\pi}e^{ix} + e^{N\pi}e^{-ix}|} = \frac{|e^{2N\pi} - e^{2ix}|}{|e^{2N\pi} + e^{2ix}|} \leq \frac{1 + e^{-2N\pi}}{1 - e^{-2N\pi}} \leq \frac{10}{9}.$$

Consequently, we always have $|\tan z| \leq \frac{10}{9}$ on ∂S_N for large enough N. By the triangle inequality, $|\tan z - z| \geq |z| - |\tan z| \geq N - \frac{10}{9}$ on ∂S_N so that

$$\left| \frac{\tan^2 z}{\tan z - z} \right| \leq \frac{100}{81N - 90}$$

on ∂S_N. Since the length of ∂S_N is $8N$, it follows from Theorem 4.10 (The M-L Formula) that

$$\left| \frac{1}{2\pi i} \int_{\partial S_N} \frac{\tan^2 z}{\tan z - z}\, dz \right| \leq \frac{400N}{\pi(81N - 90)} < 2$$

for large enough N. Combining this estimate with the formula (19.4), we conclude at once that $\mathbb{Z}'_f = 0$, completing the proof of the problem. ∎

Problem 19.2

Bak and Newman Chapter 19 Exercise 2.

Proof. By the hypothesis, we know that

$$\int_{C_N} f_2(z)\, dz \to -\int_{C_N} \frac{dz}{z}$$

as $N \to \infty$. Applying Corollary 10.9 (The Argument Principle) to the function $f(z) = z$ to get

$$\int_{C_N} \frac{dz}{z} = 2\pi i.$$

Therefore, we conclude that

$$\int_{C_N} f_2(z)\,\mathrm{d}z \to -2\pi i$$

as $N \to \infty$.

By Problem 19.1, all the zeros of $\tan z - z$ are real. Denote the set of its zeros by $\{0, x_1, x_2, \ldots\}$. Thus it suffices to sum the values of $\frac{\sin^2 x}{x^2}$ at $\{0, x_1, x_2, \ldots\}$. Figure 19.2 illustrates that the curves $y = \frac{\sin^2 x}{x^2}$ and $y = \tan x - x$ intersect on the set $\{x_1, x_2, \ldots\}$.

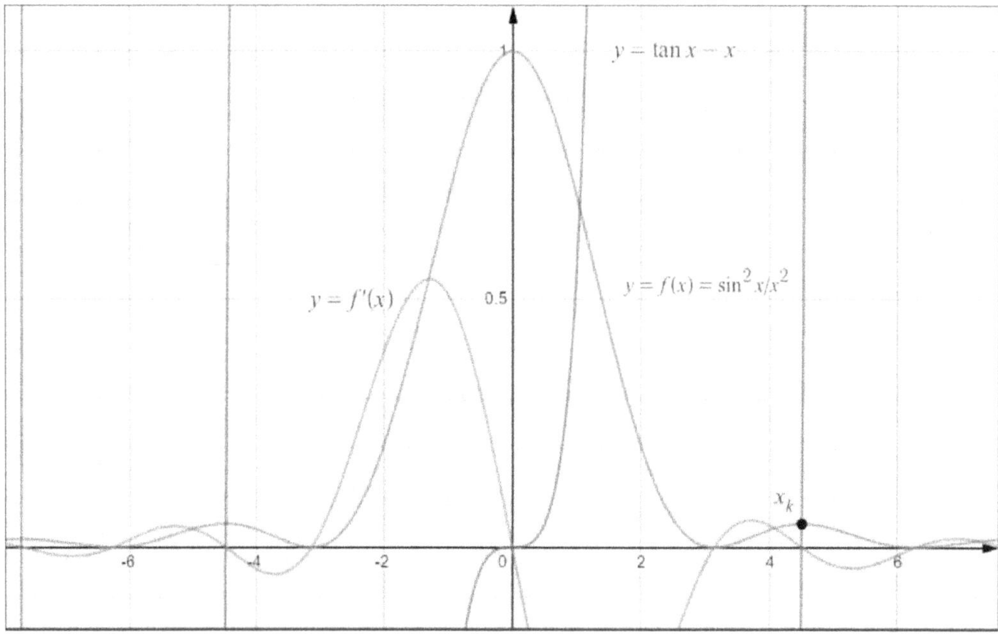

Figure 19.2: The intersections of $y = \frac{\sin^2 x}{x^2}$ and $y = \tan x - x$.

According to Theorem 10.5 (The Cauchy's Residue Theorem), we have

$$\int_{C_N} f_2(z)\,\mathrm{d}z = 2\pi i \left[\sum_{k=1}^{\infty} \operatorname{Res}\,(f_2; x_k) + \operatorname{Res}\,(f_2; i) + \operatorname{Res}\,(f_2; -i) + \operatorname{Res}\,(f_2; 0) \right]$$

$$= 2\pi i \left[\sum_{k=1}^{\infty} \frac{\sin^2 x_k}{x_k^2} + \operatorname{Res}\,(f_2; i) + \operatorname{Res}\,(f_2; -i) + \operatorname{Res}\,(f_2; 0) \right] \tag{19.6}$$

Since $\tan(\pm i) - (\pm i) \neq 0$, we deduce from [4, Eqn. (1), p. 129] that

$$\operatorname{Res}\,(f_2; i) = \frac{-1}{2i(\tan i - i)} = -\frac{1 + e^2}{4}. \tag{19.7}$$

Similarly, we have

$$\operatorname{Res}\,(f_2; -i) = -\frac{1 + e^2}{4}. \tag{19.8}$$

Next, recall from the proof of Problem 19.1 that $\tan z - z$ has a zero of order 3 at $z = 0$, so the function $f_2(z)$ has a simple pole at $z = 0$. Therefore, it follows from [1, p. 184] that $\tan z - z = \frac{z^3}{3} + \frac{2z^5}{15} + \cdots$, so we have

$$f_2(z) = \frac{1}{(1 + z^2)(\frac{z}{3} + \frac{2z^3}{15} + \cdots)}$$

and then [4, Eqn. (1), p. 129] implies that

$$\text{Res}\,(f_2;0) = 3. \tag{19.9}$$

Substituting all the residues (19.7), (19.8) and (19.9) into the formula (19.6), we establish

$$\sum_{k=1}^{\infty} \frac{\sin^2 x_k}{x_k^2} - \frac{1+\mathrm{e}^2}{2} + 3 = -1$$

$$\sum_{k=1}^{\infty} \frac{\sin^2 x_k}{x_k^2} = \frac{\mathrm{e}^2 - 7}{2}$$

so that

$$\text{Var}\left(\frac{\sin^2 x}{x^2}\right) = 2\left(\sum_{k=1}^{\infty} \frac{\sin^2 x_k}{x_k^2} + \lim_{x \to 0} \frac{\sin^2 x}{x^2}\right) = \mathrm{e}^2 - 5.$$

We complete the analysis of the problem.

> **Problem 19.3**
>
> *Bak and Newman Chapter 19 Exercise 3.*

Proof. To apply the idea of §19.1, we note that all the zeros of $\mathrm{e}^z - z$ are simple. Next, we recall that f'/f has residue at every simple zero of f, so we consider the function

$$f(z) = \frac{1}{z^2} \cdot \frac{\mathrm{e}^z - 1}{\mathrm{e}^z - z}.$$

Now it is clear from Theorem 10.5 (The Cauchy's Residue Theorem) that

$$\sum \frac{1}{z_k^2} = \frac{1}{2\pi i} \int_{C_N} f(z)\,\mathrm{d}z - \text{Res}\,(f;0), \tag{19.10}$$

where C_N is the square described in §19.1 and the summation counts all zeros of $\mathrm{e}^z - z$ inside C_N. Since f has a double pole at $z = 0$, we have

$$\text{Res}\,(f;0) = \frac{\mathrm{d}}{\mathrm{d}z}[z^2 f(z)]\Big|_{z=0} = 1$$

and then the expression (19.10) reduces to

$$\sum \frac{1}{z_k^2} = \frac{1}{2\pi i} \int_{C_N} f(z)\,\mathrm{d}z - 1. \tag{19.11}$$

It remains to find the estimate of the integral. In fact, it is obvious that $|z|^2 \geq N^2\pi^2$ on C_N. Let $z = x + iN\pi$ for $x \in [-N\pi, N\pi]$. The triangle inequality implies that $|\mathrm{e}^z - 1| \leq \mathrm{e}^x + 1$ on $[-N\pi, N\pi]$. Furthermore, if we take N to be even, then

$$|\mathrm{e}^z - z| = \sqrt{[(-1)^N \mathrm{e}^x - x]^2 + N^2\pi^2} \geq \mathrm{e}^x - x > 0.$$

so that

$$\left|\frac{\mathrm{e}^z - 1}{\mathrm{e}^z - z}\right| \leq \frac{\mathrm{e}^x + 1}{\mathrm{e}^x - x} \tag{19.12}$$

on $[-N\pi, N\pi]$. Let $F(x) = \frac{\mathrm{e}^x + 1}{\mathrm{e}^x - x}$. Then we have

$$F'(x) = \frac{1 - x\mathrm{e}^x}{(\mathrm{e}^x - x)^2}.$$

Thus $F'(x_0) = 0$ if and only if $x_0 e^{x_0} = 1$. Clearly, $x_0 \in (0,1)$, so we can establish from the First Derivative Test that F attains its absolute maximum at $x = x_0$ and the maximum value is given by

$$F(x_0) = \frac{e^{x_0} + 1}{e^{x_0} - x_0} = \frac{x_0(e^{x_0} + 1)}{x_0(e^{x_0} - x_0)} = \frac{1}{1 - x_0} > 0. \tag{19.13}$$

Combining the inequality (19.12) and the result (19.13), we obtain the bound

$$\left| \frac{e^z - 1}{e^z - z} \right| \leq \frac{1}{1 - x_0} < \infty \tag{19.14}$$

on the horizontal line segment $z = x + iN\pi$ for $x \in [-N\pi, N\pi]$. Actually, the same bound (19.14) also holds on the horizontal line segment $z = x - iN\pi$ for $x \in [-N\pi, N\pi]$.

On the vertical line segment $z = N\pi + iy$ for $y \in [-N\pi, N\pi]$, we see that

$$|e^z - 1| = \sqrt{e^{2N\pi} - 2e^{N\pi} \cos y + 1} \leq e^{N\pi} + 1 \tag{19.15}$$

and

$$\begin{aligned}
|e^z - z| &= \sqrt{(e^{N\pi} \cos y - N\pi)^2 + (e^{N\pi} \sin y - y)^2} \\
&= \sqrt{e^{2N\pi} - 2e^{N\pi}(N\pi \cos y + y \sin y) + N^2\pi^2 + y^2} \\
&\geq \sqrt{e^{2N\pi} - 4N\pi e^{N\pi}} \\
&\geq \frac{e^{N\pi}}{2}
\end{aligned} \tag{19.16}$$

for large even positive integer N. Therefore, it yields from the estimates (19.15) and (19.16) that

$$\left| \frac{e^z - 1}{e^z - z} \right| \leq 2\left(1 + \frac{1}{e^{N\pi}}\right) \leq 4$$

on $z = N\pi + iy$ for $y \in [-N\pi, N\pi]$, where N is a sufficiently large even positive integer. Similarly, if $z = -N\pi + iy$ for $y \in [-N\pi, N\pi]$, then

$$|e^z - 1| = \sqrt{e^{-2N\pi} - 2e^{-N\pi} \cos y + 1} \leq e^{-N\pi} + 1 \leq 2$$

and

$$\begin{aligned}
|e^z - z| &= \sqrt{(e^{-N\pi} \cos y + N\pi)^2 + (e^{-N\pi} \sin y - y)^2} \\
&= \sqrt{e^{-2N\pi} + 2e^{-N\pi}(N\pi \cos y - y \sin y) + N^2\pi^2 + y^2} \\
&\geq \frac{N\pi}{2}
\end{aligned}$$

for large even positive integer N. Consequently, we get

$$\left| \frac{e^z - 1}{e^z - z} \right| \leq \frac{4}{N\pi}$$

on $z = -N\pi + iy$ for $y \in [-N\pi, N\pi]$, where N is a sufficiently large even positive integer.

Hence the analysis in the previous two paragraphs guarantee that there exists a positive constant M such that

$$\left| \frac{e^z - 1}{e^z - z} \right| \leq M$$

on C_N for large enough even positive integer N. Now Theorem 4.10 (The M-L Formula) asserts that

$$\left| \int_{C_N} f(z)\,\mathrm{d}z \right| \le \frac{M}{N^2} \cdot 8N\pi \le \frac{8M}{N\pi}$$

for large enough even positive integer N so that

$$\lim_{N\to\infty} \int_{C_N} f(z)\,\mathrm{d}z = 0$$

and then the expression (19.11) implies that

$$\sum_{k=1}^{\infty} \frac{1}{z_k^2} = -1.$$

We have completed the proof of the problem. $\blacksquare$

Problem 19.4

Bak and Newman Chapter 19 Exercise 4.

Proof. The argument in §19.3 can be applied exactly the same here except the last two equations on [4, p. 278] are given by

$$1 + \alpha z + \frac{\alpha^2 z^2}{2!} + \cdots = e^{\alpha z} = 1 + a_1 z + \frac{a_2 z^2}{2!} + \cdots ,$$
$$1 + \gamma z + \frac{\gamma^2 z^2}{2!} + \cdots = e^{\gamma z} = 1 + b_1 z + \frac{b_2 z^2}{2!} + \cdots .$$

Thus, we have immediately that $a_1 = \alpha$ and $a_k = \alpha^k = a_1^k$ for all $k = 2, 3, \ldots$. Similarly, we obtain $b_1 = \gamma$ and $b_k = \gamma^k = b_1^k$ for all $k = 2, 3, \ldots$. Consequently, for every pair $a_1, b_1 \ge 0$ with $a_1 + b_1 = 2$, it gives a solution $a_k = a_1^k$ and $b_k = b_1^k$ to the system of equations in §19.3. This completes the proof of the problem. $\blacksquare$

Problem 19.5

Bak and Newman Chapter 19 Exercise 5.

Proof. Similar to the discussion in §19.4, we have the equation

$$\frac{z}{1-z} = \frac{z^{a_1}}{1-z^{d_1}} + \frac{z^{a_2}}{1-z^{d_2}} + \cdots + \frac{z^{a_k}}{1-z^{d_k}}. \tag{19.17}$$

Assume that d_1 was relatively prime to $d_2, d_3, \ldots, d_k$. Thus if $z \to \exp(\frac{2\pi i}{d_1}) \neq 1$, then we must have $z^{d_j} \to \exp(\frac{2\pi i d_j}{d_1}) \neq 1$ for all $j = 2, 3, \ldots, k$. In other words, as $z \to \exp(\frac{2\pi i}{d_1})$, the first term on the right-hand side of the equation (19.17) will approach infinity while all the remaining terms tend to a finite limit, a contradiction. Hence it is impossible to do so which ends the proof of the problem. $\blacksquare$

Problem 19.6

*Bak and Newman Chapter 19 Exercise 6.**

Proof.

(a) The cases for $n = 0$ and $n = 1$ are trivial. Assume that

$$c_k \leq 3^k \quad \text{and} \quad c_{k+1} \leq 3^{k+1} \tag{19.18}$$

for some $k \in \mathbb{N}$. If $n = k + 1$, then the assumption (19.18) certainly implies

$$c_{k+2} = c_{k+1} + 2c_k \leq 3^{k+1} + 2 \cdot 3^k = 3^k(3 + 2) = 5 \cdot 3^k \leq 3^{k+2}.$$

By induction, $c_n \leq 3^n$ holds for all $n \geq 0$. Since $\limsup_{n \to \infty} |c_n|^{\frac{1}{n}} \leq \limsup_{n \to \infty} 3 = 3$, it derives from Theorem 2.8 that the radius of convergence of $F(z) \geq \frac{1}{3}$.

(b) Direct computation gives

$$(1 - z - 2z^2)F(z) = (1 - z - 2z^2)\sum_{n=0}^{\infty} c_n z^n$$

$$= \sum_{n=0}^{\infty} c_n z^n - \sum_{n=0}^{\infty} c_n z^{n+1} - \sum_{n=0}^{\infty} 2c_n z^{n+2}$$

$$= c_0 + c_1 z + \sum_{n=2}^{\infty} c_n z^n - c_0 z - \sum_{n=1}^{\infty} c_n z^{n+1} - \sum_{n=0}^{\infty} 2c_n z^{n+2}$$

$$= 1 + \sum_{n=0}^{\infty} c_{n+2} z^{n+2} - \sum_{n=0}^{\infty} c_{n+1} z^{n+2} - \sum_{n=0}^{\infty} 2c_n z^{n+2}$$

$$= 1 + \sum_{n=0}^{\infty} (c_{n+2} - c_{n+1} - 2c_n) z^{n+2}$$

$$= 1$$

which implies that

$$F(z) = \frac{1}{1 - z - 2z^2} = -\frac{2}{3(2z - 1)} + \frac{1}{3(z + 1)}. \tag{19.19}$$

(c) By expressing the two rational functions in the expression (19.19) as power series, we follow that

$$F(z) = -\frac{1}{3} \cdot \frac{1}{z - \frac{1}{2}} + \frac{1}{3} \cdot \frac{1}{z - (-1)}$$

$$= \frac{2}{3}\sum_{n=0}^{\infty} 2^n z^n + \frac{1}{3}\sum_{n=0}^{\infty} (-1)^n z^n$$

$$= \sum_{n=0}^{\infty} \frac{2^{n+1} + (-1)^n}{3} z^n.$$

Hence we find that

$$c_n = \frac{2^{n+1} + (-1)^n}{3}$$

for all $n = 0, 1, 2, \ldots$.

This completes the analysis of the problem.

> **Problem 19.7**
>
> *Bak and Newman Chapter 19 Exercise 7.*[*]

Proof. We note that

$$(1 - z)F(z) = (1 - z)\sum_{n=1}^{\infty} c_n z^n$$

$$= \sum_{n=1}^{\infty} c_n z^n - \sum_{n=1}^{\infty} c_n z^{n+1}$$

$$= c_1 z + \sum_{n=2}^{\infty} c_n z^n - \sum_{n=1}^{\infty} c_n z^{n+1}$$

$$= z + \sum_{n=1}^{\infty} c_{n+1} z^{n+1} - \sum_{n=1}^{\infty} c_n z^{n+1}$$

$$= z + \sum_{n=2}^{\infty} (c_{n+1} - c_n) z^{n+1}$$

$$= \sum_{n=1}^{\infty} n^2 z^n. \tag{19.20}$$

By differentiating $\displaystyle\sum_{n=0}^{\infty} z^n = \frac{1}{1 - z}$, we have

$$\frac{z}{(1 - z)^2} = \sum_{n=0}^{\infty} n z^n.$$

Furthermore differentiation gives

$$\frac{z^2 + z}{(1 - z)^3} = \sum_{n=0}^{\infty} n^2 z^n. \tag{19.21}$$

Combining the expressions (19.20) and (19.21), we get

$$F(z) = \frac{z^2 + z}{(1 - z)^4} = \frac{1}{(1 - z)^2} - \frac{3}{(1 - z)^3} + \frac{2}{(1 - z)^4}.$$

Using the binomial series, we have

$$\frac{1}{(1 - z)^2} = \sum_{n=0}^{\infty} C_n^{n+1} z^n, \quad \frac{1}{(1 - z)^3} = \sum_{n=0}^{\infty} C_n^{n+2} z^n \quad \text{and} \quad \frac{1}{(1 - z)^4} = \sum_{n=0}^{\infty} C_n^{n+3} z^n.$$

Hence we obtain

$$c_n = C_n^{n+1} - 3C_n^{n+2} + 2C_n^{n+3}$$

$$= n + 1 - \frac{3(n + 1)(n + 2)}{2} + \frac{(n + 1)(n + 2)(n + 3)}{3}$$

$$= \frac{n(n + 1)(2n + 1)}{6},$$

completing the proof of the problem.

Proof. Following the idea in §19.4 II, we write

$$R(z) = \frac{1}{1-z^3} \cdot \frac{1}{1-z^4} = \prod_{n=1}^{3} \frac{1}{z-\alpha^n} \cdot \prod_{m=1}^{4} \frac{1}{z-\beta^m},$$

where $\alpha = \exp(\frac{2\pi i}{3})$ and $\beta = \exp(\frac{2\pi i}{4}) = i$. Since R has a double pole at $z = 1 (= \alpha^3 = \beta^4)$ and simple poles at its remaining singularities, R has a partial fraction decomposition of the form

$$R(z) = \frac{a_1}{z-1} + \frac{a_2}{(z-1)^2} + \sum_{j=1}^{5} \frac{b_j}{z-z_j},$$

where $z_1 = \alpha$, $z_2 = \alpha^2$, $z_3 = i$, $z_4 = -1$ and $z_5 = -i$.

Next, we know that

$$a_1 = \frac{\mathrm{d}}{\mathrm{d}z}[(z-1)^2 R(z)]\Big|_{z=1} = \frac{5}{24}, \quad a_2 = \lim_{z\to1}(z-1)^2 R(z) = \frac{1}{12}$$

and

$$b_j = \begin{cases} -\dfrac{1}{3z_j^2(1-z_j^4)}, & \text{if } j=1,2; \\[2ex] -\dfrac{1}{4z_j^3(1-z_j^3)}, & \text{if } j=3,4,5. \end{cases}$$

Simple calculation gives

$$b_1 = \frac{1}{6} - \frac{\sqrt{3}i}{18}, \quad b_2 = \frac{1}{6} + \frac{\sqrt{3}i}{18}, \quad b_3 = -\frac{1}{8} - \frac{i}{8}, \quad b_4 = \frac{1}{8} \quad \text{and} \quad b_5 = -\frac{1}{8} + \frac{i}{8}.$$

Consequently, we have

$$R(z) = \frac{5}{24(z-1)} + \frac{1}{12(z-1)^2} + \frac{\frac{1}{6} - \frac{\sqrt{3}i}{18}}{z-\alpha} + \frac{\frac{1}{6} + \frac{\sqrt{3}i}{18}}{z-\overline{\alpha}} + \frac{-\frac{1}{8} - \frac{i}{8}}{z-i}$$

$$+ \frac{1}{8(z+1)} + \frac{-\frac{1}{8} + \frac{i}{8}}{z+i}. \tag{19.22}$$

Finally, by comparing the coefficients of z^n on both sides of the expression (19.22), we achieve

$$C(n) = -\frac{5}{24} + \frac{n+1}{12} - \left(\frac{1}{6} - \frac{\sqrt{3}i}{18}\right) \cdot \overline{\alpha}^{n+1} - \left(\frac{1}{6} + \frac{\sqrt{3}i}{18}\right) \cdot \alpha^{n+1}$$

$$+ \left(\frac{1}{8} + \frac{i}{8}\right) \cdot (-i)^{n+1} + \frac{(-1)^n}{8} + \left(\frac{1}{8} - \frac{i}{8}\right) \cdot i^{n+1}.$$

This completes the proof of the problem. ∎

Proof. Let $z = x + iy$. Then we notice that

$$\sum_{n \geq 2} \left| \frac{1}{np^{nz}} \right| = \sum_{n \geq 2} \frac{1}{np^{nx}} \leq \frac{1}{2} \sum_{n \geq 2} \frac{1}{p^{nx}} = \frac{1}{2p^x(p^x - 1)} = \frac{1}{p^{2x}} \cdot \frac{p^x}{2(p^x - 1)}. \tag{19.23}$$

Since $p \geq 2$ and $x > \frac{1}{2}$, we have $p^x > \sqrt{2}$ which implies that $\frac{p^x}{2(p^x-1)} < 2$. Therefore, the inequality (19.23) reduces to

$$\sum_{n \geq 2} \left| \frac{1}{np^{nz}} \right| < \frac{2}{p^{2x}} \leq \frac{2}{\sqrt{p}} < \infty. \tag{19.24}$$

As a result, we have

$$\sum_{\substack{p \text{ prime} \\ n \geq 2}} \left| \frac{1}{np^{nz}} \right| < \sum_{p \text{ prime}} \frac{2}{p^{2x}}. \tag{19.25}$$

Suppose that K is a compact subset of $\operatorname{Re} z > \frac{1}{2}$ and $F = \{z \in \mathbb{C} \,|\, \operatorname{Re} z \leq \frac{1}{2}\}$. Now F is closed in $\mathbb{C}$ and $F \cap K = \varnothing$. By [22, Exercise 21, p. 101], there exists a $\delta > 0$ such that $d(K, F) > 2\delta$. Furthermore, we have $K \subseteq \{z \in \mathbb{C} \,|\, \operatorname{Re} z \geq \frac{1}{2} + \delta\}$. Given a prime $p \geq 2$, for every $n \in \mathbb{N}$, consider the function

$$f_{n,p}(z) = \frac{1}{np^{nz}}$$

which is clearly analytic in $\operatorname{Re} z > \frac{1}{2}$. Now it follows from Theorem 1.9 (The Weierstrass M-Test) and the inequality (19.24) that the sequence $\displaystyle\sum_{n=2}^{\infty} f_{n,p}(z)$ converges uniformly to a function $f_p(z)$ in $\operatorname{Re} z > \frac{1}{2}$. Hence we conclude from Theorem 7.6 that $f_p(z)$ is analytic in $\operatorname{Re} z > \frac{1}{2}$.

Next, if $\operatorname{Re} z \geq \frac{1}{2} + \delta$, then we get from the inequality (19.25) that

$$\sum_{p \text{ prime}} |f_p(z)| \leq \sum_{\substack{p \text{ prime} \\ n \geq 2}} \left| \frac{1}{np^{nz}} \right| < \sum_{p \text{ prime}} \frac{2}{p^{2x}} \leq \sum_{p \text{ prime}} \frac{2}{p^{1+2\delta}} < 2 \sum_{n=1}^{\infty} \frac{1}{n^{1+2\delta}} < \infty$$

which means that the series

$$\sum_{p \text{ prime}} f_p(z)$$

converges to a function $f(z)$ uniformly in $\operatorname{Re} z \geq \frac{1}{2} + \delta$, and then in K, by Theorem 1.9 (The Weierstrass M-Test). Finally, we apply Theorem 7.6 to establish the fact that $f(z)$ is analytic in $\operatorname{Re} z > \frac{1}{2}$, as desired. This completes the proof of the problem. ∎

Index

Bibliography

[1] L. Ahlfors, *Complex Analysis*, 3rd ed., Mc-Graw Hill Inc., 1979.

[2] T. M. Apostol, *Calculus Vol. 1: One-Variable Calculus, with an Introduction to Linear Algebra*, 2nd ed., John Wiley & Sons, Inc., 1967.

[3] N. H. Asmar and L. Grafakos, *Complex Analysis with Applications*, Cham, Springer International Publishing, 2018.

[4] J. Bak and D. J. Newman, *Complex Analysis*, 3rd ed., New York, N. Y.: Springer, 2010.

[5] R. G. Bartle and D. R. Sherbert, *Introduction to Real Analysis*, 4th ed., John Wiley & Sons, Inc., 2011.

[6] R. B. Burckel, *An Introduction to Classical Complex Analysis*, Vol. 1, Basel: Birkhäuser Basel, 1979.

[7] J. B. Conway, *Fucntions of One Complex Variable*, 2nd ed., New York: Springer-Verlag, 1978.

[8] D. S. Dummit and R. M. Foote, *Abstract Algebra*, 3rd ed., Hoboken, N. J.: Wiley, 2004.

[9] E. Freitag and B. Rolf, *Complex Analysis*, 2nd ed., London: Springer, 2009.

[10] T. W. Gamelin, *Complex Analysis*, New York, N. Y.: Springer, 2001.

[11] I. S. Gradshteyn and I. M. Ryzhik, *Table of Integrals, Series and Products*, 7th ed., San Diego: Academic Press, 2007.

[12] G. H. Hardy, *A Course of Pure Mathematics*, 10th ed., Cambridge University Press, 2002.

[13] P. K. Kythe, *Handbook of Conformal Mappings and Applications*, Boca Raton, Florida: CRC Press, 2019.

[14] S. Lang, *Complex Analysis*, 4th ed., New York, N. Y.: Springer, 1999.

[15] J. D. Lawrence, *A Catalog of Special Plane Curves*, New York: Dover, 1972.

[16] I. H. Lin, *Classical Complex Analysis: A Geometric Approach - Vol. 1*, Singapore: World Scientific Publishing Company Co. Pte. Ltd., 2010.

[17] M. Marden, *Geometry of Polynomials*, 2nd ed., Mathematical Surveys, No. 3, Providence, R. I.: Amer. Math. Soc., 1966.

[18] Q. G. Mohammad, On the zeros of polynomials, *Amer. Math. Monthly*, Vol. 72, No. 6, pp. 631 - 633, 1965.

[19] J. R. Munkres, *Topology*, 2nd ed., Upper Saddle River, N.J.: Prentice-Hall, 2000.

[20] T. Needham, *Visual Complex Analysis*, Oxford: Clarendon Press, 1997.

[21] G. Pólya and G. Szegö, *Problems and Theorems in Analysis II: Theory of Functions. Zeros. Polynomials. Determinants. Number Theory. Geometry*, Berlin, Heidelberg: Springer, 1998.

[22] W. Rudin, *Principles of Mathematical Analysis*, 3rd ed., Mc-Graw Hill Inc., 1976.

[23] W. Rudin, *Real and Complex Analysis*, 3rd ed., Mc-Graw Hill Inc., 1987.

[24] E. M. Stein and R. Shakarchi, *Complex Analysis*, Princeton, N. J.: Princeton University Press, 2003.

[25] T. Tao, *Analysis II*, 3rd ed., Singapore: Springer, 2016.

[26] K. W. Yu, *A Complete Solution Guide to Principles of Mathematical Analysis*, Amazon.com, 2018.

[27] K. W. Yu, *Problems and Solutions for Undergraduate Real Analysis I*, Amazon.com, 2018.

[28] K. W. Yu, *Problems and Solutions for Undergraduate Real Analysis II*, Amazon.com, 2019.